Celestial
BASIC
Astronomy
On Your Computer

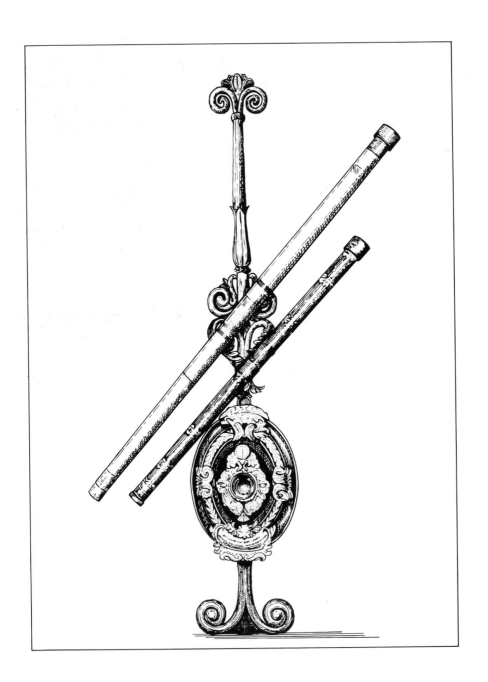

Celestial
BASIC
Astronomy
On Your Computer

Eric Burgess
Fellow of the Royal Astronomical Society

SYBEX

Berkeley • Paris • Düsseldorf

Cover photo courtesy of Jet Propulsion Laboratory.
Book design, layout, and illustrations by Jeanne E. Tennant.

Applesoft and Apple II are trademarks of Apple Computer, Inc.
Atari is a trademark of Atari, Inc.
IBM is a trademark of International Business Machines Corporation.
PET is a trademark of Commodore Business Machines.
Sorcerer is a trademark of Exidy Corporation.
TRS-80 is a trademark of Tandy Corporation.

Library of Congress Card Number: 82-60187
ISBN 0-89588-087-3
Printed in the United States of America
10 9 8 7 6 5 4 3 2 1

Contents

Acknowledgements   viii

Introduction   ix

PART 1   1
TIME

Program 1   **CALDR**
              Perpetual Calendar and Day of Week   3

Program 2   **EASTR**
              Date of Easter Sunday for Any Year   15

Program 3   **TIMES**
              Time Conversion: Local Mean to Sidereal, or
              Sidereal to Local Mean   19

Program 4   **JULDY**
              Calendar Date to Julian Day   29

Program 5   **CDATE**
              Julian Day to Calendar Date   33

Program 6   **EPOCH**
              Updating Star Coordinates   37

Program 7   **PSTAR**
              Transits and Elongations of Polaris   43

PART 2   49
THE MOON

Program 8   **RADEM**
              Right Ascension and Declination of Moon for
              Any Date   51

Program 9   **ECLIP**
             Umbral Lunar Eclipses for Any Year   59

Program 10   **PHASE**
             Approximate Phase of Moon for Any
             Date   65

PART **3**   73
THE PLANETS

Program 11   **RADEC**
             Right Ascension and Declination for All Planets
             for Any Date   75

Program 12   **MARSP**
             Angular Diameter and Distance of Mars for Any
             Date, and Next Opposition   83

Program 13   **MVENC and MERVE**
             Elongations, Phases, Angular Diameters, and
             Distances of Mercury and Venus, and Next
             Elongations   91

Program 14   **PRISE**
             Times of Rising and Setting of Mercury and
             Venus before or after the Sun for Any
             Date   109

Program 15   **RISES**
             Times of Rising, Transit, and Setting of Planets,
             Sun, and Moon for Any Date   117

Program 16   **SKYSET/SKYPLT**
             Horizon Plots of Visible Planets, Sun, Moon,
             and Stars for Any Date, Time,
             and Location   127

Program 17   **PLNTF**
             Finds and Plots Planets, Sun, and Moon in
             Constellations for Any Date and Time   143

Program 18   **JSATS**
             Positions of Galilean Satellites of Jupiter for Any
             Date and Time   163

PART 4   173
GENERAL AND TUTORIAL

Program 19   **ACONV**
Useful Astronomical Conversions   175

Program 20   **SSTAR**
Dates and Radiants of Annual Meteor
Showers   181

Program 21   **CONST and CONSH**
Recognizing the Constellations   191

Program 22   **PLNDT**
Solar System Data   215

Program 23   **PHOTO**
Photographing the Planets   235

APPENDIX   243

Program 16A   **SKYPLA**
Alternative Skyplot Program   245

Program 17A   **PLNTA**
Alternative Planet Finder Program   259

Program 20A   **SSTARA**
Alternative Meteor Shower Program   283

Observer's Guide to the Programs   **294**

Bibliography   **296**

Index   **297**

Acknowledgements

I gratefully acknowledge those members of the Manchester Astronomical Society who in the 1930s stimulated my interest in astronomy and the motions of the celestial bodies—Lawrence H. A. Carr, H. L. Dilks, Ronald Lister, William Porthouse, J. P. Rowlands, E. Denton Sherlock, and H. B. Tetlow.

I also express my thanks to Christopher Williams for arranging the SKYSET/SKYPLT programs to suit the Apple computer, and to Barbara Gordon for her untiring efforts in editing the manuscript.

Eric Burgess
Sebastopol, California
July 1982

Introduction

Microcomputers are becoming increasingly commonplace in homes and small offices. As the applications for which the computers were originally developed are mastered, new applications are springing up. One of the most recent and, indeed, most natural of the microcomputer's applications is to amateur astronomy. Now many amateur and armchair astronomers can have immediate access to useful astronomical information that before had to be obtained from almanacs, monthly astronomical magazines, or by laborious calculations with pocket calculators. Relatively simple programs, written in the BASIC language, can be used to relieve the chores of typical astronomical calculations.

Although these calculations can be made using a programmable hand-held calculator, the process is inherently more laborious and error prone, and the results are displayed less conveniently than with a computer. Alternatively, the amateur astronomer can subscribe to the astronomical almanacs or wait for the monthly issue of an astronomical magazine, but then he must extrapolate or interpolate by hand from the data provided in these publications. How much more convenient it is to load a program from a cassette file or a floppy disk into your home computer and, for example, have the configurations and names of the Galilean satellites of Jupiter identified before you go to your telescope, any time you choose to observe, anywhere in the world!

The programs in *Celestial BASIC* provide amateur and armchair astronomers and students with important aids to observing and understanding the configurations, motions, and apparitions of our Sun and Moon, the planets, and the stars. Planetary positions, as well as the rising and setting times of all the planets, Sun, and Moon, can be quickly found for any locality in the world, at any time, on any date. Astronomical (star) times can be converted to local times and vice versa. Circles can be set on equatorially mounted telescopes, and true north can be determined with the aid of Polaris. You can find out when the next lunar eclipse will occur and how much of our satellite will be covered by Earth's shadow. You can learn when Mars will be in opposition next and how large it will appear in a telescope, or when Venus or Mercury are favorably placed for observation in the morning or evening sky.

In the comfort of your own living room, these programs allow you to experience changes of hours and years in the configurations of the night sky previously impossible except in a planetarium. You can have the night (or day) sky displayed on the monitor screen with the planets, Moon, and meteor showers. You can see how the stars look from Antarctica, or see how the midnight sun appears on the oil fields of Alaska's North Slope. You can follow the diurnal and annual progression of the star sphere. And if you are not up to steam in recognizing the constellations, use a tutorial to learn how to identify the major constellations and their brightest stars.

Educators, who are increasingly using computers at all levels in teaching, will find the programs useful to illustrate the dynamics of observational astronomy. In addition to the tutorial on recognizing the constellations, other tutorials aid in converting astronomical measurements and in learning the latest details about the planets and their satellites.

This book originated as a practical tool. I began to share the programs with others, and the interest was such that I decided to incorporate them into a handbook. In addition to providing current information about the planets, these programs provide the convenience of checking observations in the past and predicting them into the future. I have compared the results of several of these programs with my observational notebooks that extend back to the 1930s, and I find them sufficiently accurate for everyday observations.

As mentioned in the introduction to the Bibliography, it is possible to add further steps within the structure of the program to increase the accuracy of the computer's readout. The references cited in the Bibliography provide the information necessary to achieve higher orders of accuracy. (Larger random access memories may be required to incorporate these refinements.)

The book is divided into four major sections and an Appendix. Part One covers times, dates, and conversions among them, updating of coordinates, and using Polaris to determine true north. Part Two concentrates on our Moon, and Part Three concentrates on the planets. Part Four provides general and tutorial programs. The Appendix consists of alternative programs not requiring use of Applesoft™ high resolution graphics for three major programs in the book. Written in the Exidy Sorcerer™ computer's BASIC format, these programs are provided as an aid to those readers who wish to modify them for use in computers other than the Apple II™.

There is also an Observer's Guide to the Programs, which can serve as a checklist of programs to run through to help you make the most of your observing time. A Bibliography is provided to refer you to further information on astronomical computations.

Originally written for an Exidy Sorcerer computer, the programs have been rewritten for an Apple II to make them available to a wider range of users. The programs are written in the popular BASIC language, and complex programming techniques have been avoided. This has been done to minimize complications of program structure so that the programs are readily understandable to the beginner and adaptable to other computers. There are also plenty of remark (REM) statements to help the reader understand what the program is asking the computer to do. All the programs are usable on computers such as the IBM™ Personal Computer, Radio Shack TRS-80™, Atari™, Commodore PET™, Exidy Sorcerer, and similar microprocessors, without major changes.

Some of the programs will require modifications, however, to suit specialized functions of different computers. These modifications will occur especially in connection with the graphic displays, as the various computers would handle such displays quite differently. For this reason the use of high resolution graphics peculiar to the Apple computer were not used more generally. Most of the graphic displays consist of characters from the ASCII character set; for example, periods, colons, X's, and asterisks represent the positions of the stars. For the program that uses Applesoft high resolution graphics to plot objects in the sky (SKYSET/SKYPLT), an alternative method of programming is given in the Appendix. Instruction lines that may need modification are identified in REM statements or in the preamble to each program. Most programs can be adapted relatively easily for use in the Southern Hemisphere, where constellations will need to be inverted. Several programs have the capability of offering Northern or Southern Hemisphere viewing without modification—SKYSET/SKYPLT, for example. All programs have been carefully edited, run, and checked against actual astronomical observations over many years.

The programs are designed so that large memories are not required. Amateur astronomers having computers with 48K of random access memory (RAM) can readily combine several of these programs. They may be conveniently placed in one package when needed concurrently, or they can be selected by menus.

I am grateful to the many people who have derived the equations forming the basic algorithms of the computer programs, many of which have been used in their fundamental forms for a long time. My intention is to provide a useful service to amateur and armchair astronomers, educators, and students by presenting the equations in a form that allows them to be manipulated on a small microprocessing system.

PART
1

TIME

THE SCIENCE OF ASTRONOMY is an outgrowth of man's need to measure time. Early studies of the sky resulted in finding a relationship between the motions of the heavenly bodies and the recurrence of seasonal and other natural phenomena. Time was measured, calendars were derived, and religious functions were set from observations of the motions of the Sun, Moon, and stars. As astronomy progressed and greater precision of observations became possible, more precise systems of time measurement were needed.

The following programs provide easy access to and conversions between the various systems for measuring time. Because the celestial bodies are not fixed in space relative to Earth-centered coordinates, two programs also provide a means to update the positions of the stars over the years.

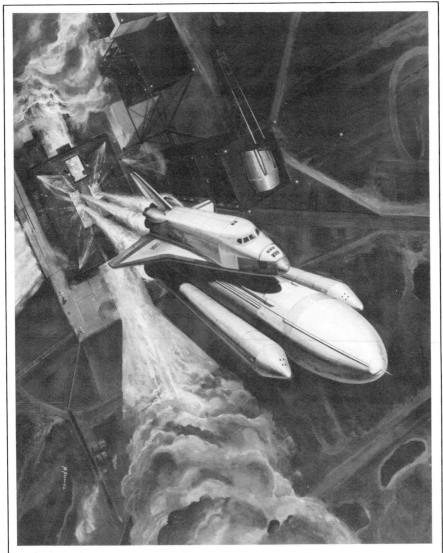

Mankind's need to measure time and establish calendars relied on careful observations of the heavens. The development of astronomy stimulated interest in other worlds and encouraged mankind to move out from Earth. In this painting a space shuttle lifts off the launch pad at the Kennedy Space Center in Florida, with main engines and solid rocket motors roaring. The space shuttle is the first reusable space vehicle, and it offers enormous potential for future manned operations in space.

Program 1: CALDR

Perpetual Calendar and Day of Week_____

All measurements of time have historically depended on astronomical observations—the day is measured from the rotation of Earth on its axis, the week approximates the changing phases of our Moon, the month is measured from the revolution of the Moon around Earth, and the year is measured from the revolution of Earth around the Sun.

Many ancient peoples, particularly the Babylonians, based their calendars on the cycles of the Moon, and the lunar measurement of years has been preserved in the modern Jewish, Chinese, and Moslem calendars. By contrast, the Egyptians based their calendars on the Sun, which also figured prominently in their religion. The Egyptian civilization depended upon the seasonal rising of the Nile, which was closely associated with the solar cycle. Ancient peoples determined the solar year by observing the helical rising of a bright star after it had been invisible because of the proximity of the Sun. Sirius was often used for this purpose. By averaging many such helical observations, the solar year was found to be very close to 365 days.

In ancient Rome, months were based on the lunar cycles. The pontifices watched for the first appearance of the thin crescent moon after the new moon so that they could declare the beginning of the new month. This first day, shouted from the steps of the Capitol, was termed *Kalendae*, which means *the calling*. Our word calendar is derived from this term.

Unfortunately for our measurement of time, the lunar cycle is not a whole number of days, nor is the time Earth takes to complete an orbit of the Sun relative to the stars. The Moon's cycle is 29.53059 days. Earth's orbit around the Sun takes 365.242196 . . . days. So 12 months are short of a year, and 13 months would give us a year that is too long. And our seven-day week (based on religion), although close to the lunar phases, is not a factor of the lunar period, the month, or the year.

When the Romans adopted the Egyptian solar year at the time of Julius Caesar, their own lunar-solar calendar was very much in error. Introduced to Rome by an astronomer, Sosigenes of Alexandria, the Egyptian calendar was ordered into official Roman use by Julius Caesar in 45 B.C. It was called the Julian calendar, and it was based on a solar year of 365.25 days. The year was divided into months, of which eleven contained 30 or 31 days and the twelfth had 28 days only. The first month was March and the last month was February. July is named after Julius Caesar and August after Augustus Caesar, both months being allocated the full 31 days, as befitted a caesar. The seventh month was named September, the eighth October, the ninth November, and the tenth December, after the Latin *septem, octo, novem,* and *decem,* for seven, eight, nine, and ten, respectively.

The Julian calendar lost approximately one-quarter day each year. This loss was corrected by adding an extra day to the twelfth month (February) every fourth year, which was the leap year. Nevertheless, this calendar gradually became out of step with the seasonal position of the Sun relative to the stars. The year of the Julian calendar was actually 11 minutes 4 seconds longer than the time it takes the apparent Sun to revolve to precisely the same position. By 1500 the error amounted to approximately 11 days. Christian religious festivities based on Easter assumed a fixed vernal equinox of March 21, and as a consequence they were becoming gradually out of step with the seasons. Accordingly, Pope Gregory XIII entrusted a reformation of the calendar to a German Jesuit, Christopher Schlussel, whose latinized name was Clavius. (Clavius is immortalized by a large crater that bears his name near the Moon's south pole.) Clavius used a scheme devised by a Neapolitan astronomer, Aloysius Lilius, in which centuries would not be leap years unless perfectly divisible by 400. To correct the calendar, Pope Gregory ordered that October 15, 1582, should follow October 4. Despite protests from angry mobs, who thought that ten days of their lives were being stolen, the correction was made and the new calendar was called the Gregorian calendar. The new calendar also moved the beginning of the year from March 25 to January 1.

The Gregorian calendar was adopted by most of the Roman Catholic countries and by Denmark and the Netherlands in 1582. But it was nearly two centuries before it was generally accepted. During that time a traveler could leave England in February 1679, for example, and find that it was February 1680 in some parts of Europe and in Scotland. The day of the month was also different between England and some parts of Europe.

Finally, other countries began to accept the new calendar. The Protestants in Germany and Switzerland adopted it in 1700. Britain and the American colonies adopted the Gregorian calendar in 1752, omitting eleven days between September 2 and 14. Prussia began to use the new calendar in 1778. Other countries followed—Ireland in 1782, Russia in 1902. Following the French Revolution, a new calendar was adopted in France, the first day of the year being September 22, 1792. This calendar was used until December 31, 1805, when France accepted the Gregorian calendar again.

Other calendars are still in use, however, particularly in regard to religious events. The Jewish calendar uses a lunar and a solar cycle. The months are lunar months, but they are about 11 days short of a solar year. A thirteenth month periodically has to be intercalated to maintain some synchronism with the solar cycle. The Moslem calendar ignores the solar cycle entirely and is tied to lunar cycles with alternate months of 30 and 29 days. The year begins at different seasons over a 32.5-year cycle. Prior to World War II there was an attempt among some businesses in Europe to introduce a 13-month calendar in which all months would have four weeks. This business calendar would have allowed more meaningful financial comparisons, but it did not receive wide acceptance.

The CALDR program offers several alternatives. You can choose to ask the computer to tell you the day of the week for any date, provide you with a calendar for any month of any year, or provide you with a calendar for the holidays of any year.

The program automatically adjusts for leap years. It adjusts all months of the calendar, the number of days from the beginning of the year to the beginning of each month, and the number of days remaining in the year at the end of any month, and it displays them as well.

The program starts by asking you to select one of the routines. If you select the calendar routine, it then asks you to select the month of the year that you wish to have displayed. It jumps to a subroutine that finds the day of the week for the first day of the year and then runs through the months to the month requested, which it displays on the screen. While the program could be simplified by jumping to the single month and finding the day of the week for the start of that month, the present form allows you to modify the program if you wish so that it would display a calendar for a whole year, month by month.

The routine to print the day of the week for any date is a standard routine.

The third routine displays the holidays in any year and is somewhat more complicated. It uses the day-of-the-week routine and a subroutine based on EASTR (Program 2). Since many holidays are based on actual dates, such as Christmas, the program easily adds the day of the week to the date. Other holidays, such as Thanksgiving, are related to a particular day of the week in a given month. The program ascertains the first Thursday (in the case of Thanksgiving) in the required month and then adds the appropriate number of weeks to find such holidays. Other holidays, such as

Ash Wednesday, are related to Easter. For these the program uses the EASTR subroutine and then adds or subtracts the appropriate number of days and corrects for the cases when the date moves into another month. Then it adds the day of the week to the date found. A typical screen display of the holidays is shown in Figure 1.1.

The instruction HOME in this and other programs is to clear the screen and put the cursor in the top left-hand corner. This statement may have to be changed for your computer. For example, on a TRS-80 you would use the instruction CLS; on other computers it may be PRINT CHR$(XX). Also, the references to the RETURN key may have to be changed to the ENTER key.

The listing for the CALDR program follows.

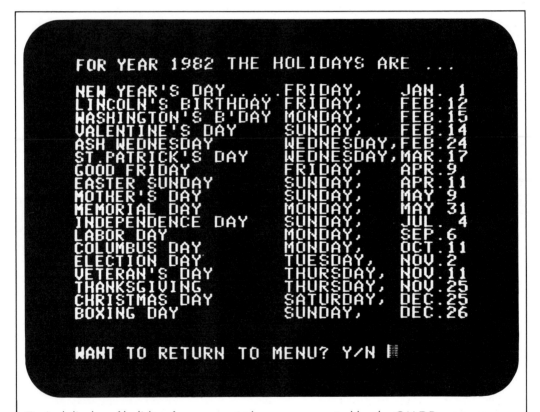

Typical display of holidays for a requested year as generated by the CALDR program

Figure 1.1

```
10   HOME : PRINT
20   PRINT : PRINT
30   PRINT  TAB( 10)"-----------------"
40   PRINT  TAB( 10)"I    CALENDAR    I"
50   PRINT  TAB( 10)"-----------------"
60   PRINT : PRINT : PRINT
70   PRINT  TAB( 2)"A PERPETUAL CALENDAR, DAY-OF-WEEK,"
80   PRINT  TAB( 7)"AND HOLIDAY PROGRAM"
90   PRINT : PRINT
100   PRINT  TAB( 5)"BY ERIC BURGESS F.R.A.S."
110   FOR K = 1500 TO 0 STEP  - 1: NEXT K
120   HOME
130   PRINT : PRINT : PRINT
140   DIM M(12)
150   PRINT "YOU CAN SELECT A CALENDAR FOR ANY MONTH"
160   PRINT "OR THE DAY OF THE WEEK FOR ANY DATE"
170   PRINT "OR THE HOLIDAYS FOR ANY YEAR."
180   PRINT : PRINT : PRINT
190   PRINT : PRINT : PRINT
200   PRINT "SELECT CALENDAR     (C)"
210   PRINT "        DAY OF WEEK (D)"
220   PRINT "        OR HOLIDAYS (H)"
230   PRINT : INPUT "TYPE C, D, OR H ";B$
240   IF B$ = "D" GOTO 1840
250   IF B$ = "H" GOTO 2250
260   IF B$ <  > "C" THEN  PRINT "INVALID ENTRY": PRINT : GOTO 230
270   HOME : PRINT : PRINT
280   GOTO 460
290   REM  SUB TO SELECT MONTH FOR PRINTING
300   ON N GOTO 310,320,350,360,370,380,390,400,410,420,430,440
310   PRINT "JANUARY";: GOTO 450
320   IF LY = 1 THEN X = 1
330   IF H = 1 AND W <  > N THEN  GOTO 1180
340   PRINT "FEBRUARY";: GOTO 450
350   PRINT "MARCH";: GOTO 450
360   PRINT "APRIL";: GOTO 450
370   PRINT "MAY";: GOTO 450
380   PRINT "JUNE";: GOTO 450
390   PRINT "JULY";: GOTO 450
400   PRINT "AUGUST";: GOTO 450
410   PRINT "SEPTEMBER";: GOTO 450
420   PRINT "OCTOBER";: GOTO 450
430   PRINT "NOVEMBER";: GOTO 450
440   PRINT "DECEMBER";: GOTO 450
450   RETURN
460   PRINT : PRINT : PRINT
470   RESTORE
480   INPUT "YEAR REQUIRED ";Y
490   HOME : PRINT : PRINT : PRINT
500   IF Y > 1752 THEN 520
510   PRINT : PRINT "YEAR MUST BE AFTER 1752": GOTO 470
520   IF Y / 4 -  INT (Y / 4) = 0 AND Y / 100 -  INT (Y / 100) <  > 0 THEN
      LY = 1
530   PRINT : INPUT "MONTH REQUIRED ";W
```

CALDR *(continued)*

```
540   IF W < 1 OR W > 12 THEN  PRINT "INVALID RESPONSE": PRINT : GOTO 530
550  H = 1
560   GOTO 730
570  Q = 1:E = 1
580   GOSUB 600
590   GOTO 710
600  K =   INT (0.6 + (1 / Q))
610  L = Y - K
620  O = Q + 12 * K
630  P = L / 100
640  Z1 =   INT (P / 4):Z2 =   INT (P):Z3 =   INT ((5 * L) / 4)
650  Z4 =   INT (13 * (O + 1) / 5)
660  Z = Z4 + Z3 - Z2 + Z1 + E - 1
670  Z = Z - (7 *  INT (Z / 7)) + 1
680   IF B$ = "H" THEN  RETURN
690  D =   - (Z - 1)
700   RETURN
710  S = 0
720   DATA  0,31,28,31,30,31,30,31,31,30,31,30,31
730   HOME : PRINT : PRINT : PRINT : PRINT
740  Q = 1:E = 1: GOSUB 600
750   PRINT "CALENDAR FOR ";Y
760   FOR F = 1500 TO 0 STEP  - 1: NEXT F
770   FOR N = 0 TO 12: READ M(N): NEXT N
780   FOR N = 1 TO 12
790   PRINT : PRINT : PRINT : PRINT : IF (N - 1) < > W GOTO 800
800   HOME : PRINT : PRINT "PLEASE WAIT"
810  S = S + M(N - 1)
820   IF M(N) < > 28 THEN 840
830   IF LY = 1 THEN 940
840   IF H = 1 AND W < > N THEN  GOTO 320
850   IF LY = 1 AND H = 1 AND W > 2 THEN  GOTO 870
860   GOTO 910
870   HOME : PRINT : PRINT
880   PRINT "- ";S + 1; TAB( 7);
890   IF H = 1 AND W < > N THEN  GOTO 320
900   GOTO 980
910   HOME : PRINT : PRINT
920   PRINT "- ";S; TAB( 7);
930   GOTO 980
940   IF H = 1 AND W < > N THEN  GOTO 320
950   HOME : PRINT : PRINT
960   PRINT "- ";31; TAB( 7):S = S + 1
970  V = 0
980   FOR I = 1 TO 10: PRINT "-";: NEXT I
990   IF H = 1 GOTO 1700
1000   GOSUB 300
1010   GOTO 1160
1020   PRINT "JANUARY";: GOTO 1160
1030   IF LY = 1 THEN X = 1
1040   IF H = 1 AND W < > N GOTO 1180
1050   PRINT "FEBRUARY";: GOTO 1160
1060   PRINT "MARCH";: GOTO 1160
1070   PRINT "APRIL";: GOTO 1160
1080   PRINT "MAY";: GOTO 1160
```

CALDR *(continued)*

```
1090  PRINT "JUNE";: GOTO 1160
1100  PRINT "JULY";: GOTO 1160
1110  PRINT "AUGUST";: GOTO 1160
1120  PRINT "SEPTEMBER";: GOTO 1160
1130  PRINT "OCTOBER";: GOTO 1160
1140  PRINT "NOVEMBER";: GOTO 1160
1150  PRINT "DECEMBER";: GOTO 1160
1160  FOR I = 1 TO 10: PRINT "-";: NEXT I
1170  PRINT " ";
1180  IF H = 1 AND W < > N GOTO 1290
1190  IF LY = 1 THEN  GOTO 1220
1200  PRINT 365 - S;" -";
1210  GOTO 1290
1220  IF LY = 1 AND H = 1 AND W > 2 THEN 1280
1230  IF M(N) < > 28 THEN  GOTO 1260
1240  PRINT 335;" -";
1250  GOTO 1290
1260  PRINT 366 - S;" -";
1270  GOTO 1290
1280  PRINT 366 - S - 1;" -";
1290  REM  PRINT HEADING OF DAYS OF WEEK
1300  IF H = 1 AND W < > N THEN GOTO 1350
1310  PRINT  CHR$ (10)
1320  PRINT : PRINT " S      M      T      W      T      F      S"
1330  PRINT
1340  FOR I = 1 TO 40: PRINT "-";: NEXT I
1350  FOR C = 1 TO 6
1360  IF N < > W THEN 1380
1370  PRINT
1380  IF H = 1 AND W < > N THEN GOTO 1400
1390  PRINT  TAB( 2)
1400  FOR G = 1 TO 7
1410  D = D + 1
1420  IF N > 2 AND LY = 1 AND H = 1 THEN  GOTO 1440
1430  IF N > 1 AND LY = 1 THEN  GOTO 1460
1440  D2 = D - S
1450  GOTO 1470
1460  D2 = D - (S - 1)
1470  IF D2 > M(N) + X THEN  GOTO 1550
1480  IF H = 1 AND W < > N THEN GOTO 1520
1490  IF LY = .1 AND N > 2 THEN D2 = D2 - 1
1500  IF D2 > 0 THEN  PRINT D2;
1510  PRINT  TAB( 2 + 6 * G);
1520  NEXT G
1530  IF D2 = M(N) + X THEN  GOTO 1560
1540  NEXT C
1550  D = D - G
1560  IF H = 1 AND W < > N THEN GOTO 1600
1570  PRINT : PRINT
1580  INPUT "PRESS RETURN TO CONTINUE ";A$
1590  GOTO 1630
1600  NEXT N
1610  IF H = 1 AND W < > N THEN GOTO 1630
1620  FOR I = 1 TO 6: PRINT ;: NEXT I
1630  PRINT : PRINT
```

CALDR *(continued)*

```
1640   HOME : PRINT : PRINT : PRINT
1650   INPUT "WANT TO RETURN TO MENU? Y/N ";A$
1660   PRINT
1670   IF A$ = "Y" THEN   CLEAR : GOTO 120
1680   IF A$ < > "N" THEN 1650
1690   GOTO 1820
1700   IF W = 1 THEN 1020
1710   IF W = 2 THEN 1030
1720   IF W = 3 THEN 1060
1730   IF W = 4 THEN 1070
1740   IF W = 5 THEN 1080
1750   IF W = 6 THEN 1090
1760   IF W = 7 THEN 1100
1770   IF W = 8 THEN 1110
1780   IF W = 9 THEN 1120
1790   IF W = 10 THEN 1130
1800   IF W = 11 THEN 1140
1810   IF W = 12 THEN 1150
1820   HOME
1830   GOTO 2230
1840   PRINT
1850   HOME
1860   PRINT : PRINT : PRINT
1870   J$(1) = "SUNDAY"
1880   J$(2) = "MONDAY"
1890   J$(3) = "TUESDAY"
1900   J$(4) = "WEDNESDAY"
1910   J$(5) = "THURSDAY"
1920   J$(6) = "FRIDAY"
1930   J$(7) = "SATURDAY"
1940   IF HY = 1 THEN   GOTO 2020
1950   HOME : PRINT : PRINT : PRINT
1960   PRINT : PRINT : PRINT
1970   PRINT "ENTER DATE"
1980   PRINT : PRINT
1990   INPUT "    THE YEAR   ";Y
2000   INPUT "    THE MONTH ";M
2010   INPUT "    THE DAY    ";D
2020   Q = M:E = D
2030   IF HY = 1 THEN   GOTO 2080
2040   IF Y > 1752 THEN   GOTO 2080
2050   HOME : PRINT : PRINT : PRINT
2060   PRINT "YEAR MUST BE AFTER 1752"
2070   GOTO 1950
2080   GOSUB 600
2090   IF HY = 1 THEN   RETURN
2100   HOME : PRINT : PRINT : PRINT
2110   VTAB 10
2120   PRINT Y;", ";M;", ";E
2130   PRINT
2140   PRINT "......IS A ";J$(Z)
2150   PRINT : PRINT : PRINT
2160   INPUT "WANT ANOTHER DATE? Y/N ";D$
2170   IF D$ = "Y" THEN   GOTO 1950
2180   IF D$ = "N" THEN   GOTO 2200
```

CALDR *(continued)*

```
2190   PRINT "INVALID RESPONSE ": PRINT : GOTO 2160
2200   PRINT : INPUT "WANT TO RETURN TO MENU? Y/N ";C$
2210   IF C$ = "Y" THEN  CLEAR : GOTO 120
2220   IF C$ < > "N" THEN  PRINT "INVALID RESPONSE": PRINT : GOTO 2200
2230   HOME
2240   END
2250   REM  HOLIDAY ROUTINES
2260   HOME : PRINT : PRINT
2270   INPUT "YEAR REQUIRED ";Y$:Y = VAL (Y$)
2280   IF Y / 4 - INT (Y / 4) = 0 AND Y / 100 - INT (Y / 100) < > 0 THEN
       LY = 1
2290   REM  GET DAY OF FIRST DAY OF YEAR
2300   D = 1:M = 1:HY = 1: GOSUB 1870
2310   HOME
2320   PRINT
2330   PRINT "FOR YEAR ";Y;" THE HOLIDAYS ARE ..."
2340   PRINT
2350   PRINT "NEW YEAR'S DAY.....";J$(Z);",";
2360   PRINT  TAB( 30)"JAN. 1"
2370   REM  LINCOLN'S BIRTHDAY
2380   M = 2:D = 12
2390   GOSUB 2020
2400   PRINT "LINCOLN'S BIRTHDAY ";J$(Z);",";
2410   PRINT  TAB( 30)"FEB.12"
2420   REM  WASHINGTON'S BIRTHDAY
2430   M = 2:ZW = 2:D = 1
2440   GOSUB 2020
2450   IF Z < > ZW THEN D = D + 1: GOTO 2440
2460   D = D + 14
2470   PRINT "WASHINGTON'S B'DAY MONDAY,";
2480   PRINT  TAB( 30)"FEB.";D
2490   REM  VALENTINE'S DAY
2500   D = 14:M = 2: GOSUB 2020
2510   PRINT "VALENTINE'S DAY    ";J$(Z);",";
2520   PRINT  TAB( 30)"FEB.14"
2530   REM  ASH WEDNESDAY
2540   GOSUB 3220
2550   D = D - 46
2560   IF M = 4 THEN D = D + 31:M = 3: GOTO 2590
2570   IF LY = 1 AND M = 3 THEN D = D + 29:M = 2: GOTO 2590
2580   IF M = 3 THEN D = D + 28:M = 2
2590   IF LY = 1 AND D < 1 THEN D = D + 29:M = 3: GOTO 2610
2600   IF D < 1 THEN D = D + 28:M = 2
2610   GOSUB 3480
2620   PRINT "ASH WEDNESDAY      WEDNESDAY,"; TAB( 30)M$;D
2630   D = 17:M = 3: GOSUB 2020
2640   PRINT "ST.PATRICK'S DAY    ";J$(Z);","; TAB( 30)"MAR.17"
2650   REM  EASTER
2660   GOSUB 3220
2670   MX = M
2680   DF = D - 2
2690   IF DF < 1 THEN M = M - 1:DF = DF - 31
2700   GOSUB 3480
2710   PRINT "GOOD FRIDAY       FRIDAY, "; TAB( 30)M$;DF
2720   M = MX
```

CALDR *(continued)*

```
2730  PRINT "EASTER SUNDAY      SUNDAY, "; TAB( 30)M$;D
2740  REM  MOTHER'S DAY
2750  M = 5:ZW = 1:D = 1
2760  GOSUB 2020
2770  IF Z < > ZW THEN D = D + 1: GOTO 2760
2780  D = D + 7
2790  PRINT "MOTHER'S DAY       SUNDAY, "; TAB( 30)"MAY ";D
2800  D = 1:M = 5:ZW = 2
2810  GOSUB 2020
2820  IF Z < > ZW THEN D = D + 1: GOTO 2810
2830  D = D + 28
2840  IF D > 31 THEN D = D - 7
2850  PRINT "MEMORIAL DAY       MONDAY, " TAB( 30)"MAY ";D
2860  D = 4:M = 7: GOSUB 2020
2870  PRINT "INDEPENDENCE DAY   ";J$(Z);", "; TAB( 30)"JUL. 4"
2880  REM  LABOR DAY
2890  M = 9:ZW = 2:D = 1
2900  GOSUB 2020
2910  IF Z < > ZW THEN D = D + 1: GOTO 2900
2920  PRINT "LABOR DAY          MONDAY, "; TAB( 30)"SEP.";D
2930  REM  COLUMBUS DAY
2940  M = 10:ZW = 2:D = 1
2950  GOSUB 2020
2960  IF Z < > ZW THEN D = D + 1: GOTO 2950
2970  D = D + 7
2980  PRINT "COLUMBUS DAY       MONDAY, "; TAB( 30)"OCT.";D
2990  D = D - 9: IF D < 2 THEN D = D + 7
3000  PRINT "ELECTION DAY       TUESDAY, "; TAB( 30)"NOV.";D
3010  D = 11:M = 11: GOSUB 2020
3020  PRINT "VETERAN'S DAY      ";J$(Z);", "; TAB( 30)"NOV.11"
3030  REM  THANKSGIVING
3040  M = 11:ZW = 5:D = 1
3050  GOSUB 2020
3060  IF Z < > ZW THEN D = D + 1: GOTO 3050
3070  D = D + 21
3080  PRINT "THANKSGIVING       THURSDAY, "; TAB( 30)"NOV.";D
3090  REM  CHRISTMAS
3100  D = 25:M = 12
3110  GOSUB 2020
3120  PRINT "CHRISTMAS DAY      ";J$(Z);", "; TAB( 30)"DEC.25"
3130  D = 26:M = 12
3140  GOSUB 2020
3150  PRINT "BOXING DAY         ";J$(Z);", "; TAB( 30)"DEC.26"
3160  PRINT
3170  PRINT
3180  INPUT "WANT TO RETURN TO MENU? Y/N ";A$
3190  IF A$ = "Y" THEN  CLEAR : GOTO 120
3200  IF A$ < > "N" THEN  PRINT "INVALID RESPONSE": PRINT : GOTO 3180
3210  GOTO 2230
3220  REM  SUB TO CALC D AND M FOR EASTER SUNDAY
3230  N = Y - 1900
3240  A = N / 19
3250  A = 19 * (A -  INT (A))
3260  B =  INT ((7 * A + 1) / 19)
3270  M = 0
```

Astronauts leaving Earth see the launch site on the Florida coastline and the distant curvature of Earth's horizon. Views like this put a new perspective on our planet and our attitude toward it.

Program 2: EASTR

Date of Easter Sunday for Any Year_____

The date of Easter is very important in the Christian ecclesiastical calendar. It governs events over almost a third of each year, from Septuagesima Sunday (nine weeks before Easter Sunday) to Trinity Sunday (eight weeks after).

The date of Easter Sunday is derived astronomically. At first Easter was synchronized with the Jewish Passover, but this, although accepted by the Eastern Church, was rejected by the Church in Rome. In the year 325 it was decreed at the Council of Nicaea that Easter should be celebrated on the same date by all Christians. That date was decided to be the first Sunday following the first full moon on or after the vernal (spring) equinox. The vernal equinox occurs when the Sun passes the point at which the ecliptic crosses Earth's equatorial plane from south to north. At the time of the Council of Nicaea, the vernal equinox was assumed to be fixed at March 21.

As mentioned in connection with the CALDR program, the old Julian calendar had a year that was too long. By the sixteenth century the vernal equinox was actually occurring on March 11, not March 21. Thus the celebration of Easter would inevitably move toward the summer season. To stop this, Pope Gregory XIII introduced his revised calendar to maintain the Easter celebration in the spring season and to maintain a better approximation of the solar year.

Today Easter Sunday is calculated as the Sunday following the first full moon after the vernal equinox, which occurs on March 21. It can thus fall as early as March 22 or as late as April 25. Passover is also governed by the vernal equinox full moon, but while Easter intentionally falls *after* the full moon, Passover coincides with the full moon. Consequently, Passover cannot begin on Easter Sunday.

In 1966 Thomas H. O'Beirne published an algorithm to calculate the date of Easter for any year between 1900 and 2099. His algorithm is much simpler than the complex algorithms developed in the early 1800s by Gauss. Gauss's theories for the congruence of numbers allowed calendrical problems to be solved, but some of his theories were quite complex and required special rules to allow for exceptions. The O'Beirne algorithm, which works for two centuries only in the form used in this book, divides the year by 19 and a multiple of the remainder by 19. Then it uses a multiple of that remainder for another division and further manipulates the remainders. This short program uses the O'Beirne algorithm (instructions 170 through 320). It quickly provides the date of Easter Sunday for any year between 1900 and 2099. However, even this algorithm needs to be corrected for certain years (1974, 1984, and 1994). The date given by the program has been adjusted by adding 7 days to the date for Easter 1974 and 1984. For Easter 1994 the date has been adjusted by adding 7 days, subtracting 31 days, and changing the month to April.

The listing for the EASTR program follows.

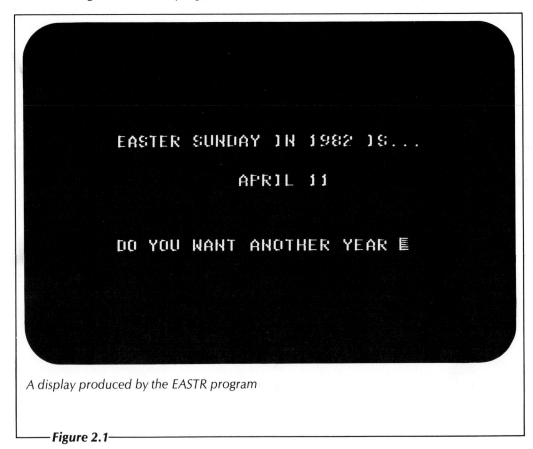

A display produced by the EASTR program

Figure 2.1

```
                                                            ┌─ EASTR ─┐

10    HOME
20    PRINT : PRINT : PRINT : PRINT
30    PRINT "THIS PROGRAM CALCULATES THE DATE"
40    PRINT "OF EASTER SUNDAY FOR ANY YEAR"
50    PRINT "      BETWEEN 1900 AND 2099"
60    PRINT : PRINT : PRINT
70    PRINT "      BY ERIC BURGESS F.R.A.S."
80    PRINT : PRINT : PRINT
90    PRINT  TAB( 6)"ALL RIGHTS RESERVED BY"
100   PRINT  TAB( 6)"S & T SOFTWARE SERVICES"
110   FOR J = 3000 TO 1 STEP  - 1: NEXT
120   HOME : PRINT : PRINT : PRINT
130   INPUT "YEAR REQUIRED ";Y$
140   Y =  VAL (Y$)
150   IF Y < 1900 OR Y > 2099 THEN PRINT "INVALID RESPONSE": PRINT : GOTO 130
160   IF Y = 0 THEN  PRINT "INVALID RESPONSE" PRINT : GOTO 130
170   N = Y - 1900
180   A = N / 19
190   A = 19 * (A -  INT (A))
200   B =  INT ((7 * A + 1) / 19)
210   M = 0
220   M = (11 * A + 4.00001 - B) / 29
230   X = M -  INT (M)
240   IF X = 1 THEN  GOTO 270
250   IF X < > 1 THEN M = 29 * X
260   GOTO 280
270   M = 0
280   Q =  INT (N / 4)
290   W = (N + Q + 31 - M) / 7
300   W = 7 * (W -  INT (W))
310   W =  INT (W)
320   DE =  INT (25 - M - W)
330   HOME
340   PRINT : PRINT : PRINT : PRINT : PRINT
350   PRINT  TAB( 5)"EASTER SUNDAY IN ";Y;" IS..."
360   PRINT : PRINT : PRINT  TAB( 15);
370   IF DE > 0 THEN M$ = "APRIL"
380   IF DE < 0 THEN M$ = "MARCH"
390   IF DE = 0 THEN  PRINT "MARCH 31": GOTO 460
400   IF DE <  - 9 THEN DE = DE + 9: GOTO 400
410   IF DE < 0 THEN D = 31 -  ABS (DE)
420   IF DE > 0 THEN D = DE
430   IF Y = 1974 OR Y = 1984 THEN D = D + 7
440   IF Y = 1994 THEN D = D + 7 - 31:M$ = "APRIL"
450   PRINT " ";M$;" ";D
460   FOR J = 2000 TO 1 STEP  - 1: NEXT
470   PRINT : PRINT : PRINT : PRINT
480   PRINT  TAB( 5)"";
490   INPUT "DO YOU WANT ANOTHER YEAR ";A$
500   IF A$ = "Y" THEN  GOTO 120
510   IF A$ < > "N" THEN  PRINT "INVALID REPLY": GOTO 490
520   HOME
530   END
```

Photo Credit: NASA/Johnson

Our perspective changes when we see Earth from space. Our blue-green planet becomes an astronomical body like the other worlds of space. Watching our planet spin on its axis gave a new dimension to our measurement of time. This photograph was obtained by the astronauts in Apollo 13 on their journey home from the Moon. It shows the southwestern United States through a great clearing of the cloud deck surrounding our planet. Part of Mexico, the peninsula of Baja California, and the Gulf of California are also clearly visible.

Program 3: TIMES

Time Conversion: Local Mean to Sidereal, or Sidereal to Local Mean

The star sphere as viewed from Earth is identified by a system of grid lines known as circles of right ascension and declination (see Figure 3.1a—b). The poles of the celestial sphere, directly over Earth's poles, are 90 degrees north and south declination. The celestial equator, the plane of Earth's equator projected on the celestial sphere, is 0 degrees declination. The zero of right ascension is the point at which the plane of the ecliptic (the plane of Earth's orbit) crosses the plane of Earth's equator from south to north. This point is called the vernal equinox or the first point of Aries. Because Earth wobbles on its axis, none of these points is fixed in space. Over a period of some 26,000 years the axis of spin moves to point around a circle in the heavens. Likewise, the vernal equinox moves through the zodiacal constellations. Although it is known as the first point of Aries, it is now located in the constellation Pisces.

At any instant, a given locality's sidereal time (star time) is the time that has elapsed since the most recent meridian passage of the vernal equinox. The meridian is the circle projected on the celestial sphere that passes through the celestial pole, the observer's zenith, and the south point on the horizon. In the Southern Hemisphere it passes through the north point on the horizon.

Like local time, sidereal time is measured in hours, minutes, and seconds, which are, however, shorter than those of solar time. While Earth spins on its axis relative to the Sun in 24 hours (the mean solar day), Earth also goes around the Sun in very nearly 365.25 days. So relative to the stars, Earth rotates on its axis in less than 24 hours. The sidereal day is accordingly 3 minutes 55.91 seconds shorter than the mean solar day. In the orbital period of 365.25 days this difference adds up to 24 hours.

Sidereal time provides the hour of right ascension on the meridian at any local time. Thus if the sidereal time is 23 hours, stars or planets having a right ascension of 23 hours will be on the meridian at that time. The hour angle of a star or a planet is the angular distance measured westward along the celestial equator from the meridian to the hour circle of the star (see Figure 3.1b). It is found by subtracting the star's right ascension from the local sidereal time. If the sidereal time is known at a particular instant, together with the right ascension, the position of a star in the sky can also be found. In this way, using graduated circles (known as setting circles) on

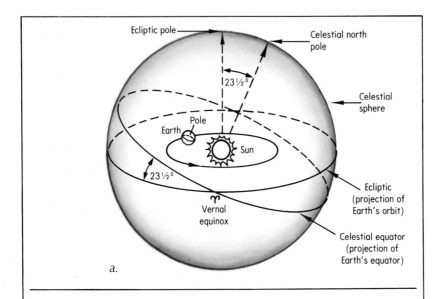

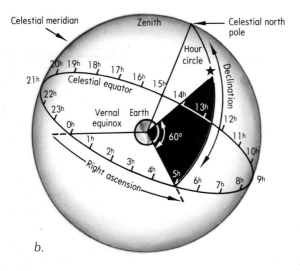

The celestial sphere as viewed from Earth. Part (a) illustrates the projection of Earth's poles, equator, and orbit on the celestial sphere. The vernal equinox is the point at which the plane of Earth's orbit crosses the plane of Earth's equator from south to north. Part (b) illustrates how these projections determine the grid lines of right ascension and declination.

Figure 3.1

the axes of an equatorially mounted telescope, astronomers can direct a telescope to point toward faint objects whose right ascension and declination are given in star atlases.

This program computes local sidereal time given local clock time, or vice versa. It applies worldwide; you can enter location parameters (time zone, town name, longitude) each time you run the program, or you can set up your local parameters when you key in the program (instruction 190) and change them when asked during a run, still leaving your own locality permanently in the program. This would be useful if you wish to compute time for some other locality, such as a vacation observing site. A typical display produced by this program is shown in Figure 3.2.

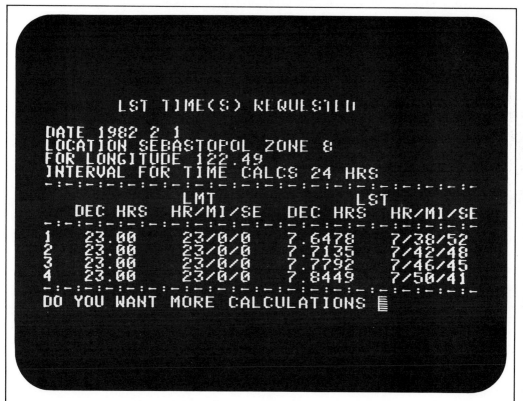

This figure shows a display produced by the TIMES program, giving conversions for a series of three local mean times separated by 24 hours. Times are expressed in decimal hours and hours/minutes/seconds irrespective of how you input the first time in the series.

Figure 3.2

The program asks for your time zone and longitude and applies a correction to clock time to take into account your location within your time zone. You can convert between universal time (U.T.) and sidereal time at 0 longitude by entering time zone 0 and longitude 0 when asked. The program can be modified easily if you wish to accept longitude only, ignoring the time zone. Alternatively, you can have it ignore longitude and calculate times more approximately from the time zone input only. This also applies to other programs in this book where both time zone and longitude are shown as inputs.

Times are computed within a few seconds' accuracy, suitable for setting circles on equatorial mounts. You can use decimal hours or hours, minutes, and seconds for inputs. The program does not correct for daylight saving time; you must adjust inputs and outputs when this is in force at your locality or at the locality for which you are computing data.

The program is written for and tested on an Apple II. If you are using a TRS-80, substitute CLS for HOME to clear the screen (in instructions 20, 140, 270, etc.). For other computers you must use an ASCII instruction to clear the screen, such as PRINT CHR$(12) or CHR$(24). TRS-80 users also should substitute the ENTER key for the RETURN key in instruction 510.

The listing of the TIMES program follows.

— TIMES —

```
10 FLG = 0
20  HOME : PRINT : PRINT : PRINT
30  PRINT : PRINT : PRINT
40  PRINT  TAB( 12)"TIME CONVERSIONS"
50  PRINT
60  PRINT  TAB( 10)"A COMPUTER PROGRAM"
70  PRINT  TAB( 12)"FOR ASTRONOMERS"
80  PRINT
90  PRINT  TAB( 8)"BY ERIC BURGESS F.R.A.S."
100  PRINT
110  PRINT  TAB( 8)"ALL RIGHTS RESERVED BY"
120  PRINT  TAB( 8)"S & T SOFTWARE SERVICE"
130  FOR J = 3000 TO 1 STEP  - 1: NEXT
140  HOME : PRINT : PRINT : PRINT
150  REM  SET INITIAL CONDITIONS ZN TIME ZONE
160  REM  - IF EAST OF GREENWICH L$ NAME OF
170  REM  YOUR LOCALITY LO IS LONGITUDE - IF EAST
180  REM  EXAMPLE IS FOR WEST COAST
190 ZN = 8:L$ = "SEBASTOPOL":LO = 122.49
200  PRINT "INITIAL CONDITIONS SET FOR "L$
210  : PRINT : PRINT  TAB( 10)"TIME ZONE ";ZN
220  PRINT  TAB( 10)"LONGITUDE ";LO
230  PRINT
240  INPUT "DO YOU WANT TO CHANGE CONDITIONS (Y/N) ";A$
250  IF A$ = "N" THEN 330
260  IF A$ < > "Y" THEN  PRINT "INVALID REPLY": PRINT : GOTO 240
270  HOME : PRINT : PRINT
280  INPUT "LOCATION NAME ";L$
290  INPUT "TIME ZONE ";ZN$
300 ZN =  VAL (ZN$)
310  INPUT "LONGITUDE ";LO$
320 LO =  VAL (LO$)
330  HOME : PRINT : PRINT : PRINT
340 Z = ZN
350  IF FLG = 1 THEN 820
360 A$ = "":A1$ = "":A2$ = "":A3$ = ""
370  PRINT "PROGRAM COMPUTES LOCAL MEAN TIME (LMT)"
380  PRINT
390  PRINT "  WHEN GIVEN LOCAL SIDEREAL TIME (LST)"
400  PRINT
410  PRINT "OR LOCAL SIDEREAL TIME WHEN GIVEN"
420  PRINT "        LOCAL MEAN TIME"
430  PRINT
440  PRINT "AND YOU CAN REQUEST CALCULATIONS"
450  PRINT "FOR A NUMBER OF TIME INTERVALS": PRINT
460  PRINT "TIME MAY BE ENTERED AND DISPLAYED IN"
470  PRINT "DECIMAL HOURS OR HRS. MIN. SEC"
480  PRINT "     USING 24 HOUR CLOCK": PRINT
490  PRINT "TIME INTERVALS MUST BE IN DECIMAL HOURS"
500  PRINT : PRINT ":::::::::::::::::::::::"
510  INPUT "TO CONTINUE PRESS RETURN KEY";A$
520  HOME : PRINT : PRINT : PRINT
530  INPUT "DO YOU WANT TO CALCULATE LMT OR LST ";T$
```

┌─ **TIMES** *(continued)* ─────────────────────────────

```
540   PRINT : IF T$ = "LMT" GOTO 570
550   IF T$ = "LST" GOTO 570
560   PRINT "INVALID REPLY": PRINT : GOTO 530
570   PRINT "ENTER THE DATE ": PRINT
580   INPUT "THE YEAR ";YD$:Y =  VAL (YD$)
590   IF Y = 0 THEN  PRINT "INVALID REPLY": PRINT : GOTO 580
600   IF Y > 1800 GOTO 650
610   PRINT "IS ";Y;" THE CORRECT YEAR"
620   INPUT Y$
630   IF Y$ = "Y" THEN 650
640   IF Y$ = "N" THEN  PRINT : GOTO 570
650   PRINT : INPUT "THE MONTH ";MD$:M =  VAL (MD$)
660   IF M = 0 OR M > 12 THEN  PRINT "INVALID REPLY": PRINT : GOTO 650
670   PRINT : INPUT "THE DAY ";DD$:D =  VAL (DD$)
680   IF D = 0 OR D > 31 THEN  PRINT "INVALID REPLY": PRINT : GOTO 670
690   IF M = 2 AND D > 29 THEN  PRINT "INVALID REPLY": PRINT : GOTO 670
700   PRINT : PRINT
710   INPUT "WANT TO SEE LOCATION CONDITIONS? ";A$
720   IF A$ = "Y" THEN 750
730   IF A$ = "N" THEN 790
740   PRINT "INVALID REPLY": PRINT : GOTO 710
750   PRINT "LOCATION CONDITIONS ARE..."
760   PRINT "    LOCATION NAME... ";L$
770   PRINT "    TIME ZONE....... ";ZN
780   PRINT "    LONGITUDE....... ";LO
790   PRINT
800   INPUT "DO YOU WANT TO CHANGE THEM ";A5$
810   IF A5$ = "Y" THEN FLG = 1: GOTO 270
820   FLG = 0
830   GOSUB 1650
840   IF T$ = "LST" THEN U$ = "LMT": GOTO 860
850   U$ = "LST"
860   GOSUB 1900
870   INPUT "SPECIFY NUMBER OF CALCULATIONS ";NO
880   PRINT : INPUT "AND TIME INTERVALS ";IN
890   HOME
900   REM   CALCS FOR DAYS FROM EPOCH
910   EP = 722895: GOSUB 1950:DE = DG:ND = DE - EP
920   REM   CALC AND PRINT TIME TABLE
930   PRINT  TAB( 8)T$;" TIME(S) REQUESTED": PRINT
940   TC = .065753
950   PRINT "DATE "Y;" ";M;" ";D
960   PRINT "LOCATION ";L$;" ZONE ";ZN - LGC / 15
970   PRINT "FOR LONGITUDE ";LO
980   PRINT "INTERVAL FOR TIME CALCS ";IN;" HRS"
990   PRINT "-:-:-:-:-:-:-:-:-:-:-:-:-:-:-:-:-:-:-"
1000   PRINT "             ";U$;"             ";T$
1010   PRINT "  DEC HRS  HR/MI/SE   DEC HRS  HR/MI/SE"
1020   PRINT "-:-:-:-:-:-:-:-:-:-:-:-:-:-:-:-:-:-:-"
1030   FOR K = 1 TO NO
1040   H1 =  INT (T1):M1 =  INT ((T1 - H1) * 100)
1050   S1 =  INT ((((T1 - H1) * 100) - M1) * 100)
1060   GOSUB 1230
1070   IF  INT (T2) > 24 THEN T2 = T2 - 24: GOTO 1080
1080   H2 =  INT (T2):M2 =  INT ((T2 - H2) * 100)
```

└───

```
1090 S2 =   INT (((((T2 - H2) * 100) - M2) * 100)
1100 HR2 =   INT (T2):MI2 = (T2 - INT (T2)) * 60
1110 SE2 = (MI2 -   INT (MI2)) * 60:SE2 =   INT (SE2)
1120 MI2 =   INT (MI2): GOSUB 2050
1130 H1$ =   STR$ (H1):M1$ =   STR$ (M1):S1$ =   STR$ (S1):P$ = "."
1140 HA$ = H1$ + P$ + M1$ + S1$
1150 H2$ =   STR$ (H2):M2$ =   STR$ (M2):S2$ =   STR$ (S2)
1160 HB$ = H2$ + P$ + M2$ + S2$
1170  PRINT K; TAB( 5);HA$; TAB( 14)RR;"/";IM;"/";ES;
1180  PRINT  TAB( 23);HB$; TAB( 32)HR2;"/";MI2;"/";SE2
1190 T1 = T1 + IN
1200  IF T1 > 24 THEN T1 = T1 - 24:ND = ND + 1: GOTO 1200
1210  NEXT K
1220  GOTO 1450
1230  REM   PARAMETERS FOR TIME CALCS
1240  REM   GMST AT EPOCH 1979 MARCH 22
1250 GC = 11.927485
1260  REM   DAILY RATE OF CHANGE OF GMST
1270 TC = .065711
1280  IF T$ = "LMT" THEN 1370
1290  REM   CONVERSION TO LST
1300 T2 = TC * ND + GC + (((ZN + T1) / 24) * TC) + T1
1310  REM   CONVERSION FOR NON MID TIME ZONE
1320 T2 = T2 + (.0656667 * LGC)
1330  IF T2 > 24 THEN T2 = T2 - 24: GOTO 1330
1340  IF T2 <  - 24 THEN T2 = T2 + 24: GOTO 1340
1350  IF T2 < 0 THEN T2 = T2 + 24
1360  RETURN
1370  REM   CONVERSION TO LMT
1380 T2 = T1 - (TC * ND + GC) - (ZN / 24 * TC)
1390  REM   CONVERSION TO NON MID TIME ZONE
1400 T2 = T2 - (.0656667 * LGC)
1410  IF T2 < 0 THEN T2 = T2 + 24: GOTO 1410
1420  IF T2 > 24 THEN T2 = T2 - 24: GOTO 1420
1430 T2 = T2 - (T2 / 24 * TC)
1440  RETURN
1450  PRINT "-:-:-:-:-:-:-:-:-:-:-:-:-:-:-:-:-:-:-:-:-"
1460  INPUT "DO YOU WANT MORE CALCULATIONS ";A$
1470  HOME : PRINT : PRINT : PRINT
1480  IF A$ = "N" THEN  GOTO 1630
1490  IF A$ <  > "Y" THEN  PRINT "INVALID REPLY": PRINT : GOTO 1460
1500 H1$ = "":M1$ = "":S1$ = "":HA$ = "":H2$ = "":M2$ = "":S2$ = "":HB$ = ""
1510 T1 = 0:T2 = 0
1520  HOME : PRINT : PRINT : PRINT
1530  PRINT "DO YOU STILL WANT TO CALCULATE ";T$
1540  INPUT A$
1550  IF A$ = "N" AND T$ = "LST" THEN T$ = "LMT": GOTO 1580
1560  IF A$ = "N" AND T$ = "LMT" THEN T$ = "LST": GOTO 1580
1570  IF A$ <  > "Y" THEN  PRINT "INVALID REPLY": PRINT : GOTO 1530
1580  HOME : PRINT : PRINT : PRINT
1590  INPUT "DO YOU WANT TO CHANGE DATE ";A$
1600  IF A$ = "N" THEN ZN = Z: GOTO 700
1610  IF A$ <  > "Y" THEN  PRINT "INVALID REPLY": PRINT : GOTO 1590
1620  PRINT :ZN = Z: GOTO 570
1630  HOME
```

TIMES *(continued)*

```
1640   END
1650   HOME : PRINT : PRINT : PRINT
1660   PRINT "DO YOU WANT INPUT IN DEC. HRS (1)"
1670   PRINT "          OR IN HRS MIN SEC (2)"
1680   PRINT : PRINT
1690   PRINT  TAB( 10)" ";: INPUT PT$
1700   PT =  VAL (PT$): PRINT : PRINT
1710   IF PT = 1 THEN  GOTO 1750
1720   IF PT = 2 THEN  GOTO 1820
1730   PRINT "INVALID REPLY": PRINT : GOTO 1650
1740   IF T$ = "LMT" THEN T$ = "LST"
1750   PRINT "    WHAT IS THE INPUT TIME"
1760   PRINT "    HR.XXXX (24-HR CLOCK)"
1770   PRINT : PRINT
1780   PRINT "      ";: INPUT T1: PRINT
1790   IF T1 > 23.9999 THEN  PRINT : PRINT "INVALID ENTRY": GOTO 1780
1800   PRINT : PRINT : PRINT
1810   GOTO 1890
1820   PRINT "    WHAT IS THE INPUT TIME"
1830   PRINT "    HR,MIN,SEC (24-HR CLOCK)"
1840   PRINT : PRINT
1850   PRINT "        ";: INPUT HR,MI,SE
1860   PRINT : PRINT : PRINT
1870   IF HR > 23 OR MI > 59 OR SE > 59.99 THEN  PRINT "INVALID ENTRY":
       PRINT :GOTO 1850
1880   T1 = HR + MI / 60 + SE / 3600
1890   RETURN
1900   LGC = (ZN * 15) - LO
1910   IF LGC < 0 THEN  GOTO 1930
1920   ZN = ZN +  ABS (LGC / 15): GOTO 1940
1930   ZN = ZN + LGC / 15
1940   RETURN
1950   REM  CALC GREGORIAN DAYS TO DATE REQUESTED
1960   IF M >  = 3 GOTO 2020
1970   REM  CALCS FOR JAN AND FEB
1980   DG = 365 * Y + D
1990   DG = DG + ((M - 1) * 31) +  INT ((Y - 1) / 4)
       -  INT ((.75) * INT ((Y - 1) / 100 + 1))
2000   RETURN
2010   REM  CALCS FOR MAR THRU DEC
2020   DG = 365 * Y + D + ((M - 1) * 31) -  INT (M * .4 + 2.3)
2030   DG = DG +  INT (Y / 4) -  INT ((.75) *  INT ((Y / 100) + 1))
2040   RETURN
2050   REM  SUB FOR PRINTING HR,MI,SE
2060   RR =  INT (T1)
2070   IM = (T1 -  INT (T1)) * 60:ES = (IM -  INT (IM)) * 60
2080   ES =  INT (ES):IM =  INT (IM)
2090   RETURN
```

Photo Credit: NASA/Ames

Until recently, astronomers had to be content with looking at the other worlds in our universe through Earth's atmosphere. With the advent of the space shuttle, a new breed of telescopes and other sophisticated instruments of observation will be placed into space and enable us to look much farther afield and back into time. This drawing shows the new infrared space telescope that will be put into operation by the space shuttle. It will penetrate the dust of space to enormous distances, almost to the edge of the knowable universe.

Program 4: JULDY

Calendar Date to Julian Day_____

Many astronomical events are referred to by their Julian day, which measures an arbitrary time period starting from 1 January 4713 B.C., the first day being Julian day 0. Julian day does not refer to the Julian calendar. It is a method of counting days that was introduced in 1582 by a mathematician named Scaliger. The name was derived from Scaliger's father, Julius. Use of this day number by astronomers avoids confusion that might arise from the use of different calendars at different times and places. The Julian day begins at noon, 12 hours later than our calendar day.

This program provides the Julian day for any Gregorian calendar day between 1100 and 2200. The program takes the Julian day for the beginning of the year 1900 and calculates Gregorian days from that date to the date selected. It then either adds or subtracts, as appropriate, the number of Gregorian days to or from the Julian day at the epoch 1900. The program can be modified easily to apply to other centuries also. This program can be merged with CDATE, its complement. See Program 5 for instructions.

The listing of the JULDY program follows.

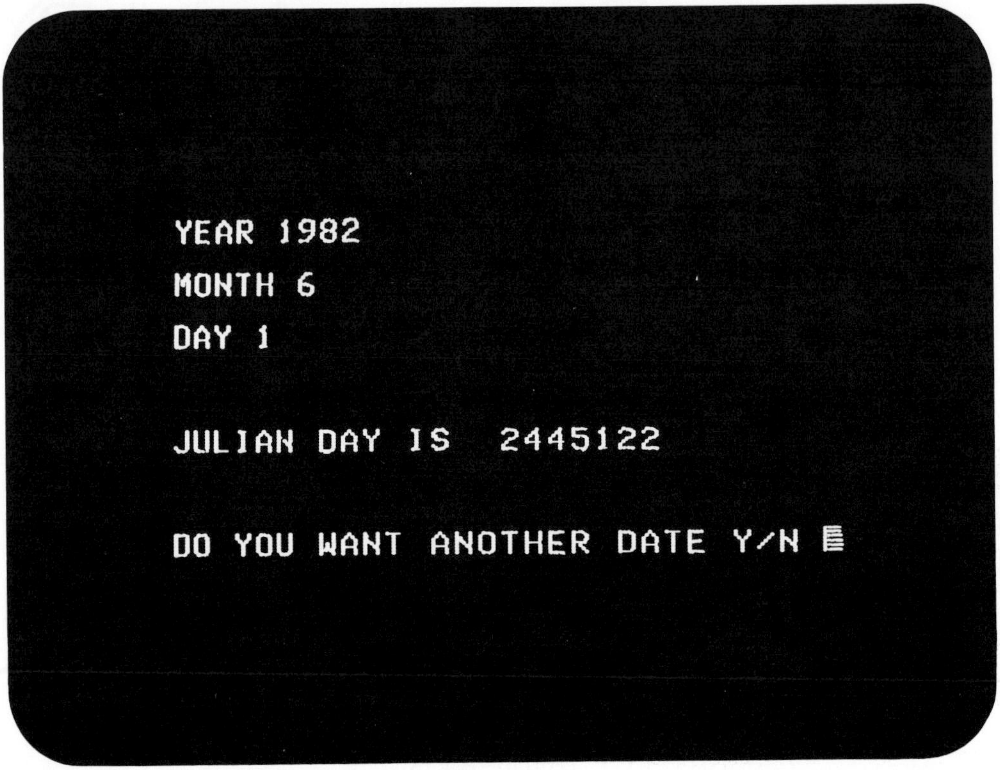

YEAR 1982
MONTH 6
DAY 1

JULIAN DAY IS   2445122

DO YOU WANT ANOTHER DATE Y/N ▤

A typical display generated by the JULDY program

Figure 4.1

─── **JULDY** ───

```
20   HOME
30   PRINT : PRINT : PRINT
40   PRINT  TAB( 5)"--------------------------"
50   PRINT  TAB( 5)"I      JULIAN DAY        I"
60   PRINT  TAB( 5)"--------------------------"
70   PRINT : PRINT : PRINT
80   PRINT  TAB( 10)"AN ASTRONOMY PROGRAM"
90   PRINT : PRINT  TAB( 9)"BY ERIC BURGESS F.R.A.S."
100  PRINT : PRINT : PRINT
110  PRINT  TAB( 9)"ALL RIGHTS RESERVED BY"
120  PRINT  TAB( 9)"S & T SOFTWARE SERVICE"
130  PRINT : PRINT : PRINT
140  PRINT  TAB( 3)"THIS PROGRAM PROVIDES THE JULIAN DAY"
150  PRINT  TAB( 5)"FOR A CALENDAR DATE 1100 TO 2200"
160  FOR K = 1 TO 3000: NEXT K
170  HOME : PRINT : PRINT : PRINT
180  INPUT "YEAR ";Y1
190  PRINT
200  IF Y1 < 0 THEN Y1 = Y1 - 100
210  Y = Y1 - 1900
220  INPUT "MONTH ";M1
230  PRINT
240  INPUT "DAY ";D1
250  PRINT : PRINT : PRINT
260  M2 = M1 - 1
270  IF M2 = 1 THEN DY = 31
280  IF M2 = 2 THEN DY = 59
290  IF M2 = 3 THEN DY = 90
300  IF M2 = 4 THEN DY = 120
310  IF M2 = 5 THEN DY = 151
320  IF M2 = 6 THEN DY = 181
330  IF M2 = 7 THEN DY = 212
340  IF M2 = 8 THEN DY = 243
350  IF M2 = 9 THEN DY = 273
360  IF M2 = 10 THEN DY = 304
370  IF M2 = 11 THEN DY = 334
380  D3 = D1 + DY
390  D4 = 365 * Y +  INT (Y / 4)
400  IF Y1 > 1999 THEN D4 = D4 - 2
410  D5 = 15020 +  INT (D4) + D3
420  JD = D5
430  IF Y / 4 -  INT (Y / 4) = 0 AND M1 < 3 THEN JD = JD - 1
440  JD = JD + 2400000
450  JD = JD -  INT ((Y1 - 1900) / 100)
460  IF Y1 < 1583 THEN JD = JD + (10 -  INT ((1583 - Y1) / 100))
470  IF Y1 > 1999 THEN JD = JD + 3
480  PRINT "JULIAN DAY IS  ";JD
490  PRINT : PRINT : PRINT
500  INPUT "DO YOU WANT ANOTHER DATE Y/N ";A$
510  IF A$ = "Y" THEN 170
520  IF A$ < > "N" THEN  PRINT "INVALID RESPONSE": PRINT : GOTO 500
530  HOME
540  END
```

Photo Credit: NASA/Johnson

As man moves out into space it becomes imperative that he forget some of the differences that have divided people on Earth. Common measurements of time and space have to be adapted and used. This was first demonstrated in the Apollo-Soyuz program, in which United States' and Russian spacecraft had to be capable of docking and working with one another. This picture shows the Soyuz spacecraft of the USSR, which docked with an Apollo command module in the first international manned space venture.

Program 5: CDATE

Julian Day to Calendar Date_____

This program complements the JULDY program. It provides a Gregorian calendar date for any Julian day between 1100 and 2000. Again, it is easily modified for application in other centuries. These two programs, JULDY and CDATE, can be merged into one if you wish, with a selection of one or the other at the beginning of the merged program. If you intend to do this, you should increase all the line numbers in CDATE by 1000 when you key them in so that you can merge the two programs later. Then you must add an appropriate selection and branch routine at the beginning of JULDY.

The listing of the CDATE program follows.

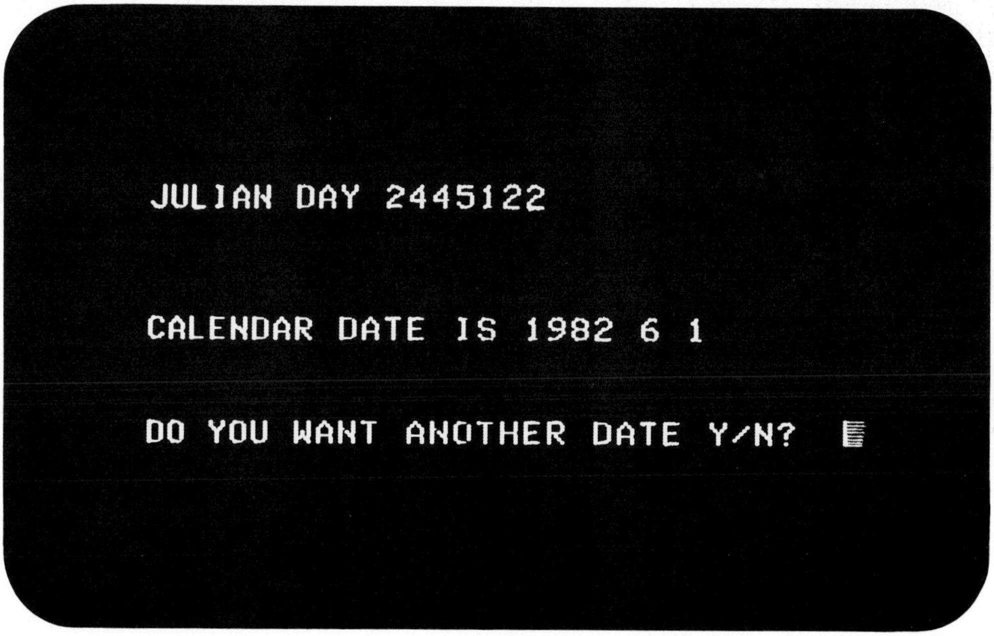

```
JULIAN DAY 2445122


CALENDAR DATE IS 1982 6 1


DO YOU WANT ANOTHER DATE Y/N?
```

This display generated by the CDATE program is the converse of the display in Figure 4.1.

Figure 5.1

```
┌─ CDATE ─────────────────────────────────────────────────────────────┐

 20   HOME
 30   PRINT : PRINT : PRINT
 40   PRINT  TAB( 5)"------------------------------"
 50   PRINT  TAB( 5)"I JULIAN DAY TO CALENDAR DATE I"
 60   PRINT  TAB( 5)"------------------------------"
 70   PRINT : PRINT : PRINT
 80   PRINT  TAB( 10)"AN ASTRONOMY PROGRAM"
 90   PRINT : PRINT  TAB( 9)"BY ERIC BURGESS F.R.A.S."
100   PRINT : PRINT : PRINT
110   PRINT  TAB( 9)"ALL RIGHTS RESERVED BY"
120   PRINT  TAB( 9)"S&T SOFTWARE SERVICE"
130   PRINT : PRINT : PRINT
140   PRINT "THIS PROGRAM PROVIDES THE CALENDAR DATE"
150   PRINT  TAB( 2)"FOR A JULIAN DAY BETWEEN 1100-2000"
160   FOR K = 1 TO 2000: NEXT K
170   HOME : PRINT : PRINT : PRINT : PRINT
180   INPUT "JULIAN DAY ";JD
190   JD = JD - 2400000
200   PRINT
210   ND = JD - 15018
220   Y1 = ND / 365.25
230   Y = 1900 +  INT (Y1)
240   IF Y / 4 -  INT (Y / 4) = 0 AND Y / 100 -  INT (Y / 100) < > 0 THEN
      LY = 1
250   D = 365.25 * (Y1 -  INT (Y1))
260   IF D -  INT (D) > .5 THEN D = D + 1
270   D =  INT (D)
280   IF LY = 0 AND M < 3 THEN D = D - 1
290   D = D +  INT ((Y - 2000) / 100)
300   IF Y < 1583 THEN D = D - (10 +  INT ((Y - 1500) / 100))
310   IF D - 31 < 0 THEN M = 1:D = D: GOTO 430
320   IF D - 59 < 0 THEN M = 2:D = D - 31: GOTO 430
330   IF D - 90 < 0 THEN M = 3:D = D - 59: GOTO 430
340   IF D - 120 < 0 THEN M = 4:D = D - 90: GOTO 430
350   IF D - 151 < 0 THEN M = 5:D = D - 120: GOTO 430
360   IF D - 181 < 0 THEN M = 6:D = D - 151: GOTO 430
370   IF D - 212 < 0 THEN M = 7:D = D - 181: GOTO 430
380   IF D - 243 < 0 THEN M = 8:D = D - 212: GOTO 430
390   IF D - 273 < 0 THEN M = 9:D = D - 243: GOTO 430
400   IF D - 304 < 0 THEN M = 10:D = D - 273: GOTO 430
410   IF D - 334 < 0 THEN M = 11:D = D - 304: GOTO 430
420   IF D - 365 < 0 THEN M = 12:D = D - 334
430   IF LY = 1 AND M > 2 THEN D = D - 1
440   IF LY = 1 AND M > 2 THEN D = D - 1
450   IF D = > 1 THEN  GOTO 510
460   IF D < 1 AND LY = 1 AND M = 3 THEN D = 29:M = 2: GOTO 510
470   IF D < 1 AND LY = 0 AND M = 3 THEN D = 28:M = 2: GOTO 510
480   IF D < 1 AND M = 1 THEN D = 31:M = 12:Y = Y - 1: GOTO 510
490   IF D < 1 AND M = 2 OR M = 4 OR M = 6 OR M = 9 OR M = 11 THEN D = 31:
      M = M - 1: GOTO 510
500   IF D < 1 THEN D = 30:M = M - 1
510   IF D = 365 THEN D = 31:M = 12
520   IF D = 366 THEN D = 1:M = 1:Y = Y + 1

└──────────────────────────────────────────────────────────────────────┘
```

CDATE *(continued)*

```
530    PRINT : PRINT : PRINT
540    PRINT "CALENDAR DATE IS ";Y;" ";M;" ";D
550    IF Y < > 1582 THEN  GOTO 600
560    PRINT
570    PRINT "(NOTE: IN 1582 DATES BEFORE OCT 15"
580    PRINT "MUST BE DECREASED BY 10 DAYS"
590    PRINT "TO MATCH THE JULIAN CALENDAR)"
600    PRINT : PRINT : PRINT
610    INPUT "DO YOU WANT ANOTHER DATE Y/N? ";A$
620    IF A$ = "Y" THEN LY = 0:Y = 0:D = 0:M = 0: GOTO 170
630    IF A$ < > "N" THEN  PRINT "          INVALID RESPONSE": PRINT :
       GOTO 610
640    HOME
650    END
```

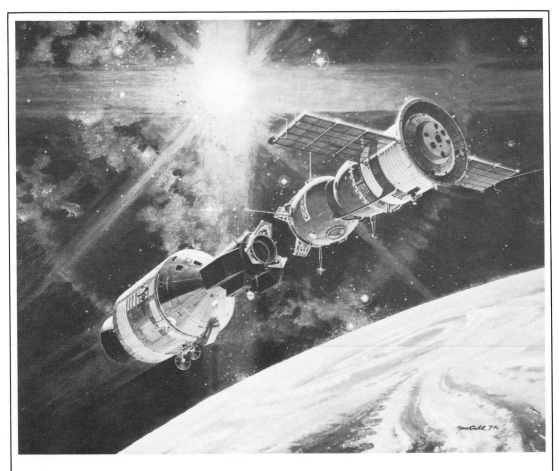

Photo Credit: NASA/Johnson

This picture portrays the American and Russian spacecraft in orbit around Earth. During their voyage the crews passed freely from one spacecraft to the other, unhindered by artificial national boundaries, boundaries that are invisible when one looks at Earth from space. There were high hopes that this cooperation in space would be the beginning of a new epoch in the history of mankind.

Program 6: EPOCH

Updating Star Coordinates_____

Earth acts like a wobbling gyroscope or a spinning top, and over a period of some 26,000 years the axis of spin moves to point around a circle in the heavens. At present the north axis points nearly toward the star Polaris in Ursa Minor, which is our pole star. But 12,000 years ago it pointed toward the bright star Vega in the constellation Lyra. At the time of ancient Egypt, the pole star was a star in the constellation of Draco.

Because Earth wobbles on its axis, there are changes from year to year in the positions of all stars relative to the right ascension and declination grid (see Program 3 for a discussion of right ascension and declination). Right ascensions and declinations are listed in nebula and star tables for a given epoch, say 1950. If you want to find a faint stellar object by setting to the circles of an equatorially mounted telescope, you will need to update positions to the current epoch of observation. This program does this for you with sufficient accuracy to position the stellar object within the field of view of a typical finder telescope (see Figure 6.1).

You must input the epoch of the star table you are using, your present epoch, and then each right ascension and declination you need updated. The program provides the updated values by computing the following equations.

The change in right ascension is given by:

$$W = .0042 \times T \times [X + (Z \times \sin R_1 \times \tan D_1)]$$

and the change in declination is given by:

$$D = .00028 \times T \times Y \times \cos R_1$$

where R_1 is right ascension at epoch, D_1 is declination at epoch, and T is the time in years from the first epoch. X, Y, and Z are determined from the time between the epochs as follows:

$$X = 3.07234 + (.00186 \times T_2)$$
$$Y = 20.0468 - (.0085 \times T_2)$$
$$Z = Y/15$$

where

$$T_2 = \frac{[(Y_1 + Y_2)/2] - 1900}{100}$$

where Y_1 is the first epoch year and Y_2 is the second epoch year.

The listing of the EPOCH program follows.

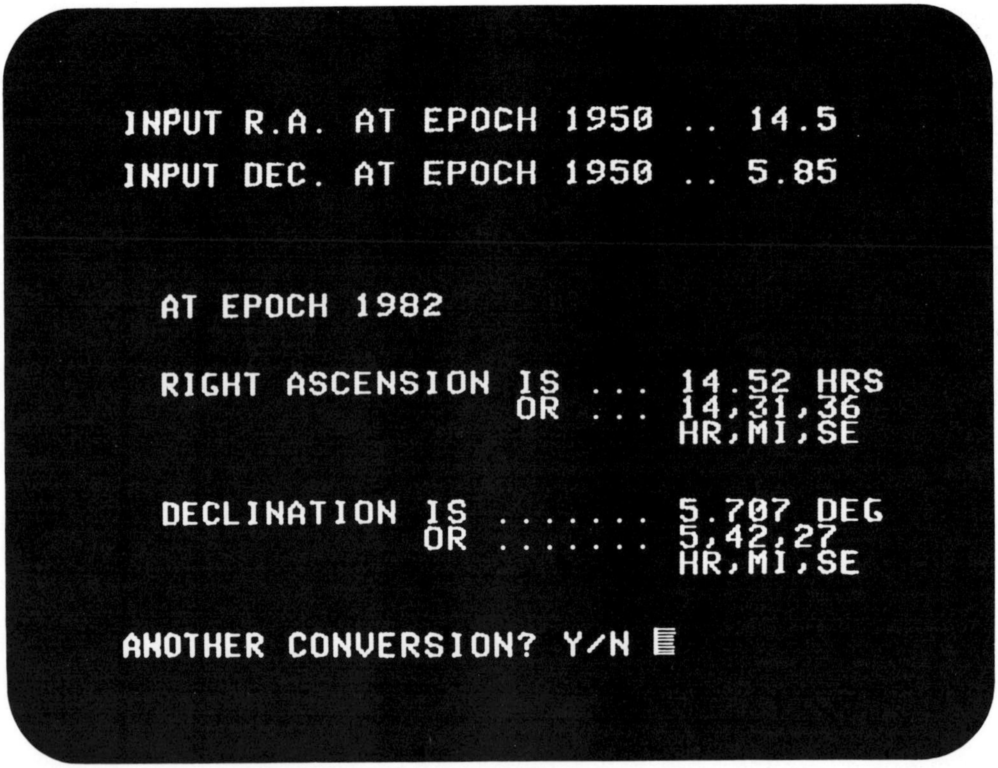

```
INPUT R.A. AT EPOCH 1950 .. 14.5
INPUT DEC. AT EPOCH 1950 .. 5.85


AT EPOCH 1982

RIGHT ASCENSION IS ... 14.52 HRS
              OR ... 14,31,36
                      HR,MI,SE


DECLINATION IS ...... 5.707 DEG
           OR ...... 5,42,27
                      HR,MI,SE


ANOTHER CONVERSION? Y/N ▤
```

The program EPOCH updates star coordinates to allow for the precession of the equinox. This is a typical display generated by the program.

Figure 6.1

```
                                                                 EPOCH

10    CLEAR
20    HOME : PRINT : PRINT
30    DEF  FN RAD(X) = X / 57.29878
40    PRINT : PRINT : PRINT
50    PRINT  TAB( 10)"------------------"
60    PRINT  TAB( 10)"I     EPOCH      I"
70    PRINT  TAB( 10)"------------------"
80    PRINT : PRINT
90    PRINT  TAB( 9)"AN ASTRONOMY PROGRAM"
100   PRINT
110   PRINT  TAB( 7)"BY ERIC BURGESS F.R.A.S."
120   PRINT : PRINT
130   PRINT  TAB( 8)"ALL RIGHTS RESERVED BY"
140   PRINT  TAB( 8)"S & T SOFTWARE SERVICE"
150   PRINT
160   FOR J = 1 TO 2000: NEXT J
170   HOME : PRINT : PRINT : PRINT
180   PRINT : PRINT : PRINT : PRINT
190   PRINT  TAB( 9)"THIS PROGRAM COMPUTES"
200   PRINT  TAB( 5)"RIGHT ASCENSION AND DECLINATION"
210   PRINT  TAB( 4)"FOR AN EPOCH WHEN GIVEN RA AND DEC"
220   PRINT  TAB( 5)"FOR ANOTHER EPOCH, REDUCING FOR"
230   PRINT  TAB( 14)"PRECESSION"
240   FOR J = 1 TO 3000: NEXT J
250   HOME
260   PRINT : PRINT : PRINT
270   INPUT "WHAT IS FIRST EPOCH (YEAR) ";Y1
280   PRINT
290   INPUT "WHAT IS SECOND EPOCH ";Y2
300   T2 = ((Y2 + Y1) / 2 - 1900) /100
310   X = 3.07234 + (.00186 * T2)
320   Y = 20.0468 - (.0085 * T2)
330   Z = Y / 15
340   PRINT : PRINT : PRINT
350   REM  CALC DAYS BETWEEN EPOCHS
360   T = Y2 - Y1
370   PRINT "PICK MODE OF INPUT:"
380   PRINT : PRINT
390   PRINT " R.A. IN DECIMAL HRS   (1)"
400   INPUT "        OR IN HR,MI,SE  (2) ";RA$
410   RA = VAL (RA$)
420   IF RA = 0 OR RA > 2 THEN  PRINT "INVALID RESPONSE": PRINT : GOTO 390
430   PRINT : PRINT
440   PRINT " DEC. IN DECIMAL DEG. (3)"
450   INPUT "        OR IN DE,MI,SE  (4) ";DE$
460   DE = VAL (DE$)
470   IF DE < 3 OR DE > 4 THEN  PRINT "INVALID RESPONSE": PRINT : GOTO 440
480   HOME : PRINT : PRINT : PRINT
490   IF RA = 2 THEN 510
500   PRINT "INPUT R.A. AT EPOCH ";Y1;" ";: INPUT ".. ";R1: GOTO 530
510   PRINT "INPUT R.A. AT EPOCH ";Y1;: INPUT " .. ";RI,MI,SI
520   IF DE = 4 THEN 540
530   PRINT : PRINT "INPUT DEC. AT EPOCH ";Y1;: INPUT " .. ";D1: GOTO 550
```

┌──── **EPOCH** (continued) ──┐

```
540   PRINT "INPUT DEC. AT EPOCH ";Y1;: INPUT " .. ";DC,MC,SC
550   IF RA = 2 THEN R1 = RI + MI / 60 + SI / 3600
560   IF DE = 4 THEN D1 = DC + MC / 60 + SC / 3600
570  R1 = R1 * 15
580   PRINT : PRINT
590  W = .0042 * T * (X + (Z *  SIN ( FN RAD(R1)) *  TAN ( FN RAD(D1))))
600  R2 = R1 + W
610  D2 = D1 + .00028 * T * Y *  COS ( FN RAD(R1))
620   PRINT : PRINT
630   PRINT "  AT EPOCH ";Y2
640   PRINT : PRINT
650  R2 = R2 / 15
660   IF R2 > 24 THEN R2 = R2 - 24
670   IF R2 < 0 THEN R2 = R2 + 24
680  RC =  VAL ( LEFT$ ( STR$ (R2),5))
690   PRINT "  RIGHT ASCENSION IS ... ";RC;" HRS"
700  R3$ =  STR$ ( INT (R2))
710  M2 = 60 * (R2 -  INT (R2))
720  M2$ =  STR$ ( INT (M2))
730  S2 = 60 * (M2 -  INT (M2))
740  S2 =  INT (S2)
750  S2$ =  STR$ (S2)
760  R4$ = R3$ + "," + M2$ + "," + S2$
770   PRINT "                     OR ... ";R4$
780   PRINT  TAB( 26)"HR,MI,SE"
790   PRINT : PRINT
800   IF D2 <  - 90 THEN D2 =  - 90 +  ABS (D2) -  INT ( ABS (D2))
810   IF D2 < 0 THEN DD =  VAL ( LEFT$ ( STR$ (D2),6)): GOTO 840
820   IF D2 > 90 THEN DD = 90 - (D2 -  INT (D2)): GOTO 840
830  DD =  VAL ( LEFT$ ( STR$ (D2),5))
840   PRINT "  DECLINATION IS ....... ";DD;" DEG"
850   IF D2 < 0 THEN D4 =  INT (D2) + 1: GOTO 870
860  D4 =  INT (D2)
870   IF D2 < 0 THEN ME = 1 + 60 * (D2 -  INT (D2)): GOTO 890
880  ME = 60 * (D2 -  INT (D2)): GOTO 900
890  ME = 60 - ME
900  ME$ =  STR$ ( INT (ME))
910  SE = 60 * (ME -  INT (ME))
920  SE =  INT (SE)
930  SE$ =  STR$ (SE)
940  ED$ =  STR$ (D4) + "," + ME$ + "," + SE$
950   PRINT "                     OR ....... ";ED$
960   PRINT  TAB( 26)"HR,MI,SE"
970   PRINT : PRINT
980   INPUT "ANOTHER CONVERSION? Y/N ";A$
990   IF A$ = "Y" THEN  HOME : PRINT : PRINT : GOTO 480
1000   PRINT : PRINT
1010   INPUT "ANOTHER EPOCH? Y/N ";A$
1020   IF A$ = "Y" THEN  GOTO 250
1030   HOME
1040   END
```

└──┘

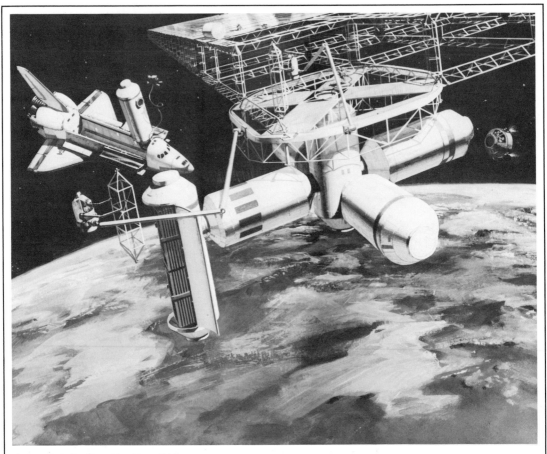

Photo Credit: Rockwell International Space Division

As more nations achieve the capability of developing space, large structures will be built in orbit around Earth. Huge telescopes, space hotels, solar power satellites, space settlements, and space manufacturing facilities will become common sights. A new era will dawn, and people will begin to move out into the unlimited economic commons of the solar system and start to reach for the stars.

Program 7: PSTAR

Transits and Elongations of Polaris

Our pole star, Polaris, is not exactly at the north pole of the celestial sphere, i.e., at 90 degrees north declination. So it oscillates in the course of each day and year to the east and west of true north.

When an equatorial mount of a telescope is set up, the polar axis should point to the celestial pole. The exact elevation of the pole at any observing site can be determined from the elevation of Polaris when it is at east or west elongation, and the azimuth of true north can be determined by observing Polaris when the star is at upper or lower transit. The elevation of the pole is also the latitude of the observer.

This program determines these times (see Figure 7.1) within ten minutes for any date, which is sufficiently accurate to set a polar axis, except for a very large telescope.

Starting from epoch 1980, the program uses the right ascension of Polaris, the hour angle, and the sidereal time for a given date to determine the local time at which the transits and elongations take place.

The listing of the PSTAR program follows.

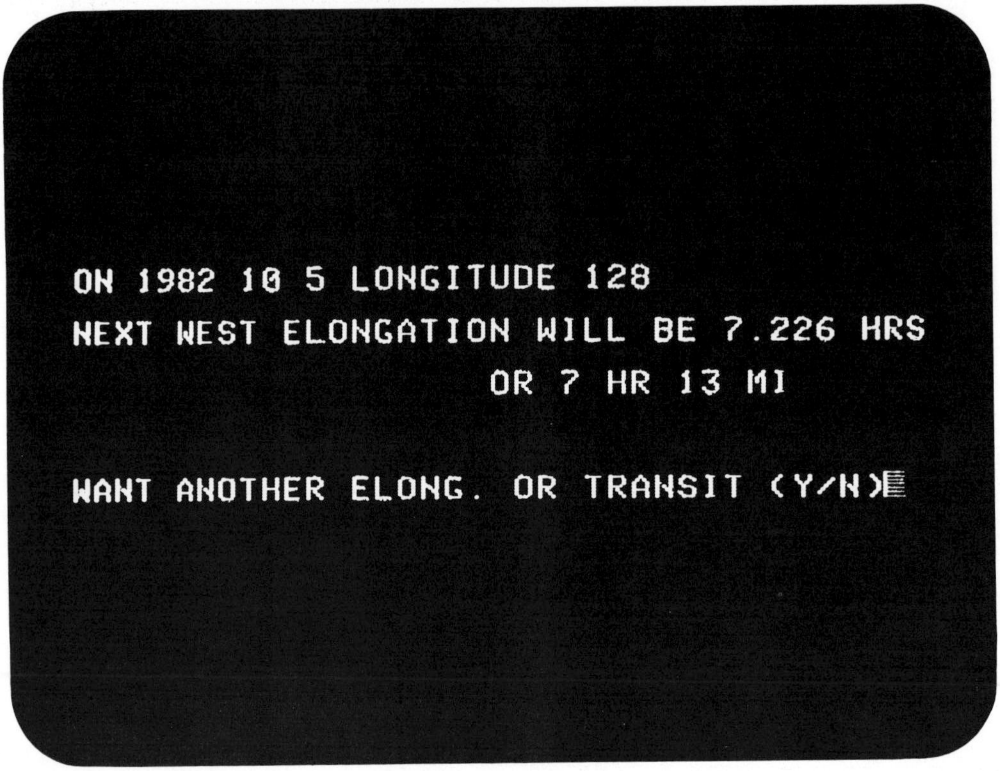

ON 1982 10 5 LONGITUDE 128
NEXT WEST ELONGATION WILL BE 7.226 HRS
                    OR 7 HR 13 MI

WANT ANOTHER ELONG. OR TRANSIT (Y/N)

By observing the transits and elongations of Polaris you can ascertain true north and your latitude. This is a typical display generated by the PSTAR program.

Figure 7.1

```
                                                              ─PSTAR─




 10   REM   PSTAR POSITIONS
 20   HOME : PRINT : PRINT : PRINT : PRINT
 30   PRINT   TAB( 8)"--------------------"
 40   PRINT   TAB( 8)"I       POLARIS       I"
 50   PRINT   TAB( 8)"--------------------"
 60   PRINT : PRINT
 70   PRINT   TAB( 8)"AN ASTRONOMY PROGRAM"
 80   PRINT   TAB( 7)"BY ERIC BURGESS F.R.A.S."
 90   PRINT : PRINT
 100  PRINT   TAB( 8)"ALL RIGHTS RESERVED BY"
 110  PRINT   TAB( 8)"S & T SOFTWARE SERVICE"
 120  PRINT : PRINT
 130  FOR K = 3000 TO 1 STEP  - 1: NEXT
 140  HOME : PRINT : PRINT : PRINT
 150  PRINT   TAB( 5)"THIS PROGRAM PROVIDES THE TIMES OF"
 160  PRINT
 170  PRINT   TAB( 10)"ELONGATIONS AND TRANSITS"
 180  PRINT
 190  PRINT   TAB( 10)"OF POLARIS FOR ANY DATE"
 200  PRINT
 210  PRINT   TAB( 10)"WITH SUFFICIENT ACCURACY"
 220  PRINT
 230  PRINT   TAB( 13)"(WITHIN 10 MINUTES)"
 240  PRINT
 250  PRINT   TAB( 10)"FOR SETTING OR CHECKING": PRINT
 260  PRINT   TAB( 12)"THE ALIGNMENT OF THE": PRINT
 270  PRINT   TAB( 6)"EQUATORIAL MOUNT OF A TELESCOPE"
 280  PRINT : PRINT : PRINT
 290  INPUT "PRESS RETURN WHEN READY";A$
 300  HOME : PRINT : PRINT : PRINT
 310  INPUT "YEAR ";Y
 320  IF Y / 4 -  INT (Y / 4) = 0 AND Y / 100 -  INT (Y / 100) <  > 0 THEN
      LY = 1
 330  PRINT
 340  INPUT "MONTH ";M
 350  D2 = 0
 360  IF M = 2 THEN D2 = 31
 370  IF M = 3 THEN D2 = 59
 380  IF M = 4 THEN D2 = 90
 390  IF M = 5 THEN D2 = 120
 400  IF M = 6 THEN D2 = 151
 410  IF M = 7 THEN D2 = 181
 420  IF M = 8 THEN D2 = 212
 430  IF M = 9 THEN D2 = 243
 440  IF M = 10 THEN D2 = 273
 450  IF M = 11 THEN D2 = 304
 460  IF M = 12 THEN D2 = 334
 470  PRINT : INPUT "DAY ";D1
 480  D = D1 + D2
 490  IF LY = 1 AND M > 2 THEN D = D + 1
 500  REM   GST AT 0 HR OF EPOCH 1980
 510  REM   ADJUST FOR YEAR
 520  G = Y - 1980
```

┌─ **PSTAR** *(continued)* ─────────────────────────────────

```
530 YC = - .01638889 * G
540 GST = 6.65422 + YC
550  REM  ADJUST FOR DAILY GAIN OF SID TIME
560 DG = D * .0657096
570 GT = GST + DG
580  IF GT > 24 THEN GT = GT - 24
590  IF GT < 0 THEN GT = GT + 24
600  PRINT : INPUT "LONGITUDE ";LO
610  REM  ADJUST ST FOR LONGITUDE
620 CF = .065556 * LO / 360
630  REM  GET HOUR ANGLE
640 GT = GT + GT * CF
650 HA = 2.183333 - GT
660  IF HA < 0 THEN HA = HA + 24
670  REM  ADJUST FOR DAILY LOSS OF MEAN TIME
680 MT = HA
690 QD = 5.98362:QE = 5.933333
700 QW = 5.933333:QL = 11.967222
710  HOME : PRINT : PRINT
720  PRINT : PRINT "WHICH ELONGATION OR TRANSIT"
730  PRINT : PRINT
740  PRINT  TAB( 4)"WEST ELONGATION NEXT (1)"
750  PRINT  TAB( 4)"               PREVIOUS (2)"
760  PRINT  TAB( 4)"EAST ELONGATION NEXT (3)"
770  PRINT  TAB( 4)"               PREVIOUS (4)"
780  PRINT  TAB( 4)"LOWER TRANSIT   NEXT (5)"
790  PRINT  TAB( 4)"UPPER TRANSIT   NEXT (6)"
800  PRINT : PRINT
810  INPUT "SELECT BY NUMBER ";S$
820 S =  VAL (S$)
830  IF S = 0 OR S > 6 THEN  PRINT "INVALID RESPONSE": PRINT : GOTO 710
840  HOME : PRINT : PRINT : PRINT : PRINT
850  PRINT : "ON ";Y;" ";M;" ";D1;" LONGITUDE ";LO
860  PRINT
870  ON S GOTO 880,920,960,1000,1040,1080
880 WN = MT + QW:IN = WN: GOSUB 1280: GOSUB 1190:WN = IN
890 WN =  VAL ( LEFT$ ( STR$ (WN),5))
900  PRINT "NEXT WEST ELONGATION WILL BE ";WN;" HRS"
910  GOSUB 1250: GOTO 1110
920 WP = MT - QL - QE:IN = WP: GOSUB 1280: GOSUB 1190:WP = IN
930 WP =  VAL ( LEFT$ ( STR$ (WP),5))
940  PRINT "PREVIOUS WEST ELONGATION WAS ";WP;" HRS"
950  GOSUB 1250: GOTO 1110
960 EN = MT + QL + QW:IN = EN: GOSUB 1280: GOSUB 1190:EN = IN
970 EN =  VAL ( LEFT$ ( STR$ (EN),5))
980  PRINT "NEXT EAST ELONGATION WILL BE ";EN;" HRS"
990  GOSUB 1250: GOTO 1110
1000 EP = MT - QW:IN = EP: GOSUB 1280: GOSUB 1190:EP = IN
1010 EP =  VAL ( LEFT$ ( STR$ (EP),5))
1020  PRINT "PREVIOUS EAST ELONGATION WAS ";EP;" HRS"
1030  GOSUB 1250: GOTO 1110
1040 NLT = MT + QL:IN = NLT: GOSUB 1280: GOSUB 1190:NLT = IN
1050 NLT =  VAL ( LEFT$ ( STR$ (NLT),5))
1060  PRINT "NEXT LOWER TRANSIT WILL BE ";NLT;" HRS"
1070  GOSUB 1250: GOTO 1110
```

PSTAR (continued)

```
1080 NUT = HA:IN = NUT: GOSUB 1280: GOSUB 1190:NUT = IN
1090 NUT =   VAL ( LEFT$ ( STR$ (NUT),5))
1100  PRINT "NEXT UPPER TRANSIT WILL BE ";NUT;" HRS": GOSUB 1250
1110  PRINT : PRINT : PRINT
1120  INPUT "WANT ANOTHER ELONG. OR TRANSIT (Y/N)";A$
1130  IF A$ = "Y" THEN  HOME : PRINT : PRINT : GOTO 720
1140  PRINT
1150  INPUT "WANT ANOTHER DATE (Y/N)";A$
1160  IF A$ = "Y" THEN 300
1170  HOME
1180  GOTO 1310
1190  REM  CONVERT TO HR,MIN,SEC
1200 IN = IN + IN * .00273043
1210 HM =   INT (IN)
1220 MI = 60 * (IN -   INT (IN))
1230 M2 =   INT (MI)
1240  RETURN
1250  PRINT
1260  PRINT  TAB( 20)"OR ";HM;" HR ";M2;" MI"
1270  RETURN
1280  IF IN > 24 THEN IN = IN - 24
1290  IF IN < 0 THEN IN = IN + 24
1300  RETURN
1310  END
```

PART
2

THE MOON

THE MOON HAS ALWAYS beckoned and intrigued mankind; its phases and cycles are recorded on bones recovered from the sites of prehistoric settlements. The Moon has provided us with light, it has served as a timekeeper, and it has been a basis for romantic ideas and flights of fanciful fiction.

With the advent of the computer age and space exploration, the Moon has become much less of a mystery to us. The programs in this section allow you to determine the phases of the Moon, the dates of lunar eclipses, and the position of the Moon relative to the stars for any date.

The closest heavenly body to Earth, through even a small telescope the Moon is revealed as a fascinating world of great craters, mountains, and plains.

Program 8: RADEM

Right Ascension and Declination of Moon for Any Date

This program computes approximate right ascension and declination of the Moon for any requested date. This data enables you to fix the Moon's position relative to the stars and also to determine its rising and setting (described in later programs).

Astronomers in ancient times paid great attention to observing the motions of the Moon. Calendars and many religious festivals were based on these motions. The Moon's orbit, however, is difficult to accurately predict; its orbit is not circular, and the Moon is actually more strongly bound to the Sun than to Earth. Earth and Moon behave somewhat as two planets gyrating together around the Sun. The Moon is about 226,000 miles from Earth at perigee and 252,000 miles away at apogee. Moreover, the positions of apogee and perigee gradually move around Earth. The orbit is also inclined to the orbit of Earth around the Sun, so the Moon moves above and below the apparent path of the Sun through the stars (called the ecliptic). The lunar orbit also wobbles, so that the points where the Moon crosses the ecliptic plane (the nodes) also move around Earth. In addition, there are perturbations due to the bulge of the Earth, smaller perturbations from the planets, and perturbations from Earth and Moon following an elliptical orbit around the Sun. Obviously, the motion of the Moon is complex. Nevertheless, it can be approximated to a level of accuracy suitable for most observations or for finding the position of the Moon among the stars.

Right ascension is expressed in sidereal hours and is measured eastward along the celestial equator from the vernal equinox (first point of Aries). One sidereal hour is equal to 15 degrees on the celestial sphere. Celestial objects that are on the meridian (in other words, they culminate) at the same time as the vernal equinox have a right ascension of 0 hours. Those that culminate 1 hour later have a right ascension of 1 hour, and so on up to 24 hours when the right ascension becomes zero again.

Declination is measured in degrees north (+) or south (−) of the celestial equator. (For a more complete explanation of right ascension and declination, see TIMES—Program 3.)

The program determines the position of Moon in its orbit by counting from its position at epoch 1960 (instructions 1010 through 1030) to the current date. It eliminates complete revolutions (instructions 1120 through 1160). It also adjusts for the motion of the nodes and the perigee position during the period since or before the epoch (instructions 1080 through 1280). Then it adjusts for the eccentricity of the Moon's orbit and for the inclination of the orbit to the plane of the ecliptic (instructions 1290 and 1300). The program prints out the right ascension and declination and repeats for the requested number of intervals (see Figure 8.1).

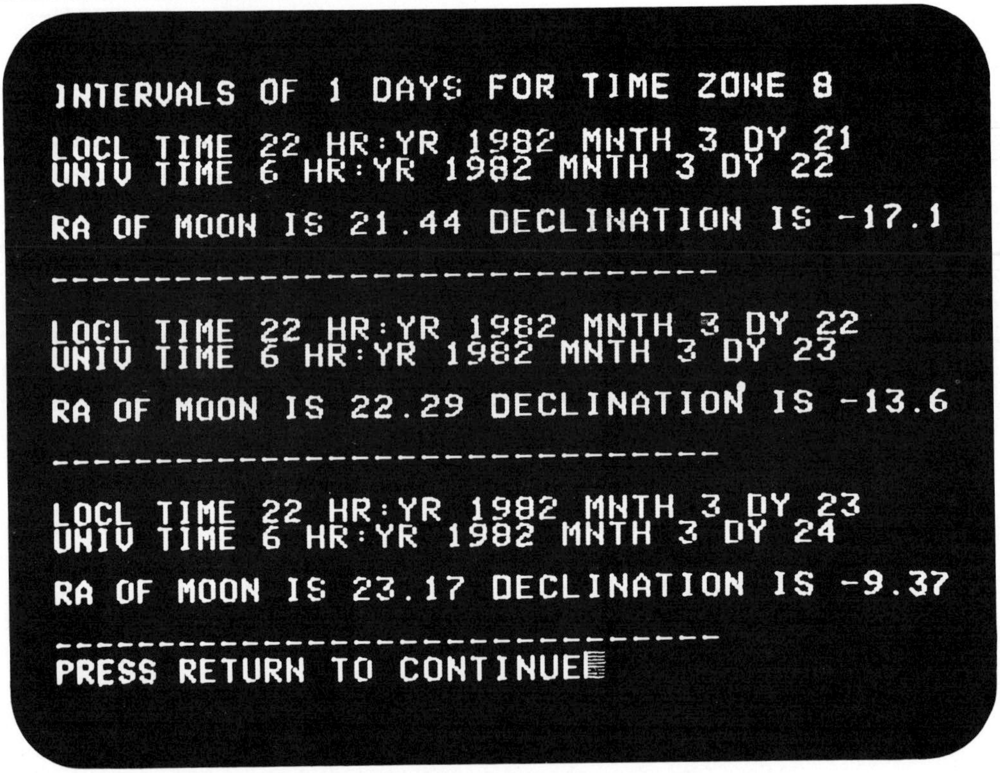

```
INTERVALS OF 1 DAYS FOR TIME ZONE 8

LOCL TIME 22 HR:YR 1982 MNTH 3 DY 21
UNIV TIME 6 HR:YR 1982 MNTH 3 DY 22

RA OF MOON IS 21.44 DECLINATION IS -17.1

------------------------------------

LOCL TIME 22 HR:YR 1982 MNTH 3 DY 22
UNIV TIME 6 HR:YR 1982 MNTH 3 DY 23

RA OF MOON IS 22.29 DECLINATION IS -13.6

------------------------------------

LOCL TIME 22 HR:YR 1982 MNTH 3 DY 23
UNIV TIME 6 HR:YR 1982 MNTH 3 DY 24

RA OF MOON IS 23.17 DECLINATION IS -9.37

------------------------------------

PRESS RETURN TO CONTINUE
```

The RADEM program calculates the right ascension and declination of the Moon for any date and time. It generates the type of display shown here.

Figure 8.1

The Moon's orbit is perturbed in many ways and other corrections can be added to achieve higher orders of accuracy. But for most purposes this program provides sufficient accuracy in locating the approximate position of the Moon among the stars for any date and time.

As with earlier programs, for computers other than Apple you may need to modify HOME statements to CLS or PRINT CHR$(12) statements and use the ENTER key instead of the RETURN key.

If you wish, this program can be merged with RADEC, the program for finding right ascensions and declinations of the planets. If you intend to do this, you should key in this program using line numbers starting at 2000. Then you can use it as a subroutine for RADEC.

The listing of the RADEM program follows.

┌─RADEM─

```
20   REM   RA AND DEC OF MOON FOR ANY DATE
30   DEF   FN RAD(X) = .01745328 * (X)
40   HOME : PRINT : PRINT : PRINT : PRINT
50   PRINT  TAB( 9)"ASTRONOMY PROGRAM": PRINT
60   PRINT  TAB( 9)"----------------"
70   PRINT  TAB( 9)"I     RADEM     I"
80   PRINT  TAB( 9)"----------------"
90   PRINT
100  PRINT  TAB( 7)"BY ERIC BURGESS F.R.A.S."
110  PRINT : PRINT
120  FL = 0
130  PRINT  TAB( 7)"ALL RIGHTS RESERVED BY"
140  PRINT  TAB( 7)"S & T SOFTWARE SERVICE"
150  PRINT
160  PRINT  TAB( 10)"VERSION 4/82"
170  PRINT : PRINT : PRINT
180  INPUT "DO YOU WANT INSTRUCTIONS Y/N ";A$
190  IF A$ = "N" THEN  HOME : PRINT : PRINT : GOTO 390
200  IF A$ < > "Y" THEN  PRINT : PRINT "INVALID RESPONSE: GOTO 180
210  HOME : PRINT : PRINT : PRINT
220  PRINT "THIS PROGRAM DISPLAYS THE APPROXIMATE"
230  PRINT "RIGHT ASCENSION AND DECLINATION"
240  PRINT "OF THE MOON FOR A REQUESTED DATE"
250  PRINT  TAB( 12)"AND TIME": PRINT
260  PRINT "FOR A SERIES OF INTERVALS"
270  PRINT : PRINT
280  PRINT "ENTER YEAR, MONTH, DAY, TIME ZONE,"
290  PRINT "      AND LOCAL TIME"
300  PRINT "ALSO TIME INTERVALS IN DAYS AND THE"
310  PRINT "NUMBERS OF INTERVALS REQUIRED"
320  PRINT : PRINT
330  PRINT "THE PROGRAM WILL DISPLAY UNIVERSAL TIME"
340  PRINT "AND THE APPROXIMATE RA AND DECLINATION"
350  PRINT "FOR EACH DAY AT THE SELECTED INTERVALS"
360  PRINT : PRINT : PRINT
370  INPUT "WHEN READY PRESS RETURN";A$
380  HOME : PRINT : PRINT : PRINT
390  PRINT "ENTER THE DATE": PRINT
400  INPUT "THE YEAR ";YD$:Y =  VAL (YD$)
410  IF Y = 0 THEN  PRINT "INVALID RESPONSE"; PRINT : GOTO 400
420  IF Y > 1800 GOTO 480
430  PRINT "IS ";Y;" THE CORRECT YEAR?"
440  INPUT Y$
450  IF Y$ = "Y" THEN 480
460  IF Y$ < > "N" THEN  PRINT "INVALID RESPONSE": PRINT : GOTO 430
470  IF Y$ = "N" THEN  PRINT : GOTO 400
480  LY = 0
490  IF Y / 4 -  INT (Y / 4) = 0 AND Y / 100 -  INT (Y / 100) < > 0 THEN
     LY = 1
500  PRINT : INPUT "THE MONTH ";MD$:M =  VAL (MD$)
510  IF M = 0 OR M > 12 THEN  PRINT "INVALID RESPONSE": PRINT : GOTO 500
520  PRINT : INPUT "THE DAY ";DD$:D =  VAL (DD$)
530  IF D = 0 OR D > 31 THEN  PRINT "INVALID RESPONSE": PRINT : GOTO 520
```

RADEM *(continued)*

```
540   IF M = 2 AND D > 29 THEN  PRINT "INVALID RESPONSE": PRINT : GOTO 520
550   IF LY < > 1 AND M = 2 AND D > 28 THEN  PRINT "INVALID RESPONSE": PRINT
      : GOTO 520
560   PRINT : INPUT "TIME ZONE ";TZ
570   PRINT : INPUT "TIME HRS ";T1
580   REM  STORE INITIAL DATE AND TIME
590   YP = Y:MP = M:DP = D
600   DU = D:TU = TZ + T1: IF TU > 24 THEN TU = TU - 24:DU = DU + 1
610   DU = DU + TU / 24
620   IF (LY = 1 AND M = 2 AND  INT (DU) > 29) THEN M = 3:DU = DU - 29:
      GOTO 670
630   IF (LY = 0 AND M = 2 AND  INT (DU) > 28) THEN M = 3:DU = DU - 28:
      GOTO 670
640   IF DU < 31 GOTO 670
650   IF (M = 4 OR M = 6 OR M = 9 OR M = 11) THEN M = M + 1:DU = DU - 30:
      GOTO 670
660   IF  INT (DU) > 31 THEN M = M + 1:DU = DU - 31
670   IF M = 13 THEN M = 1:Y = Y +1
680   IF F9 = 1 THEN 710
690   T2 = TU:D2 = DU:Y2 = Y:M2 = M
700   IF FL = 5 THEN  GOTO 860
710   PRINT : PRINT : PRINT
720   PRINT "SELECT INTERVALS AND HOW MANY"
730   PRINT "ENTER 1'S IF YOU NEED ONE PLOT ONLY"
740   PRINT "UP TO THREE INTERVALS ARE DISPLAYED"
750   PRINT "ON SCREEN BEFORE SCROLLING"
760   PRINT : PRINT
770   INPUT "WHAT IS THE TIME INTERVAL (DAYS) ";TI$
780   PRINT
790   TI =  VAL (TI$)
800   IF TI = 0 THEN  PRINT "INVALID RESPONSE": PRINT : GOTO 770
810   INPUT "HOW MANY INTERVALS ";IN$: PRINT
820   IN =  VAL (IN$)
830   IF IN = 0 THEN  PRINT "INVALID RESPONSE": PRINT : GOTO 810
840   REM  SETS INTERVAL COUNT AT 1
850   NC = 1
860   REM  CALC DAYS FROM EPOCH 1960,1,1
870   IF NC > 1 GOTO 900
880   HOME : PRINT
890   PRINT "INTERVALS OF ";TI;" DAYS FOR TIME ZONE ";TZ
900   DG = 365 * Y + DU + ((M - 1) * 31)
910   IF M > = 3 GOTO 950
920   REM  CALC IF JAN OR FEB
930   DG = DG +  INT ((Y - 1) / 4) -  INT ((.75) *  INT ((Y - 1) / 100 + 1))
940   GOTO 970
950   REM  CALC FOR MAR THRU DEC
960   DG = DG -  INT (M * .4 + 2.3) +  INT (Y / 4)
      -  INT ((.75) *  INT ((Y / 100) + 1))
970   NM = DG - 715875
980   NM = NM - .5
990   REM  FIND RA AND DEC OF MOON
1000  REM  LONG OF MOON
1010  LZ = 311.1687
1020  LE = 178.699
1030  LP = 255.7433
```

RADEM *(continued)*

```
1040 PG = .111404 * NM + LP
1050  IF PG <  - 360 THEN PG = PG + 360: GOTO 1050
1060  IF PG < 0 THEN PG = PG + 360
1070  IF PG > 360 THEN PG = PG - 360: GOTO 1070
1080 LMD = LZ + 360 * NM / 27.321582
1090 PG = LMD - PG
1100 DR = 6.2886 *  SIN (.01745328 * PG)
1110 LMD = LMD + DR
1120  IF LMD <  - 3600 THEN LMD = LMD + 3600: GOTO 1120
1130  IF LMD <  - 360 THEN LMD = LMD + 360: GOTO 1130
1140  IF LMD < 0 THEN LMD = LMD + 360: GOTO 1140
1150  IF LMD > 3600 THEN LMD = LMD - 3600: GOTO 1150
1160  IF LMD > 360 THEN LMD = LMD - 360: GOTO 1160
1170 RM = LMD / 15
1180  IF RM > 24 THEN RM = RM - 24: GOTO 1180
1190  IF RM < 0 THEN RM = RM + 24
1200 AL = LE - NM * .052954
1210  IF AL <  - 3600 THEN AL = AL + 3600: GOTO 1210
1220  IF AL <  - 360 THEN AL = AL + 360: GOTO 1220
1230  IF AL < 0 THEN AL = AL + 360: GOTO 1230
1240  IF AL > 3600 THEN AL = AL -3600: GOTO 1240
1250  IF AL > 360 THEN AL = AL - 360: GOTO 1250
1260 AL = LMD - AL
1270  IF AL < 0 THEN AL = AL + 360
1280  IF AL > 360 THEN AL = AL - 360
1290 HE = 5.1454 *  SIN (AL * 3.14159 / 180)
1300 DM = HE + 23.1444 *  SIN (LMD * 3.14159 / 180)
1310  PRINT
1320 RA$ =  STR$ (RM):DE$ =  STR$ (DM)
1330 RA$ =  LEFT$ (RA$,5):DE$ =  LEFT$ (DE$,5)
1340  IF  VAL (RA$) < 10 THEN RA$ =  LEFT$ (RA$,4)
1350  IF  VAL (DE$) <  - 9 THEN DE$ =  LEFT$ (DE$,5): GOTO 1370
1360  IF  VAL (DE$) > 10 AND  VAL (DE$) >  - 10 THEN DE$ =  LEFT$ (DE$,4)
1370  PRINT "LOCL TIME ";T1;" HR:YR ";YP;" MNTH ";MP;" DY ";DP
1380  PRINT "UNIV TIME ";TU;" HR:YR ";Y2;" MNTH ";M2;" DY "; INT (D2)
1390  PRINT
1400  PRINT "RA OF MOON IS ";RA$;" DECLINATION IS ";DE$
1410  PRINT "-----------------------------"
1420  IF NC / 3 -  INT (NC / 3) =0 THEN  INPUT "PRESS RETURN TO CONTINUE";Z$
1430  IF NC < IN GOTO 1450
1440  GOTO 1550
1450 NC = NC + 1:D = DP + TI
1460  Y = YP:M = MP
1470  IF (LY = 1 AND M = 2 AND D > 29) THEN M = 3:D = D - 29: GOTO 1530
1480  IF (LY = 0 AND M = 2 AND D > 28) THEN M = 3:D = D - 28: GOTO 1530
1490  IF D < 31 GOTO 1530
1500  IF (M = 4 OR M = 6 OR M = 9 OR M = 11) THEN M = M + 1:D = D - 30:
     GOTO 1530
1510  IF D > 31 THEN M = M + 1:D = D - 31
1520  IF M = 2 THEN  GOTO 1470
1530  IF M = 13 THEN M = 1:Y = Y + 1
1540 FL = 5: GOTO 590
1550  PRINT
```

Seen from orbit, the surface of the Moon is awe-inspiring—a series of vast circular mountain rings, great clefts running for hundreds of miles across the smooth plains, and innumerable crater pits. Its surface has been bombarded from space by countless natural projectiles and repeatedly churned into rubble—the lunar regolith. This picture shows the crater Triesnecker and its associated system of rilles.

Program 9: ECLIP

Umbral Lunar Eclipses for Any Year_____

Umbral eclipses occur when the Moon passes through the umbra of Earth's shadow. The umbra is the central conical portion of Earth's shadow, geometrically excluding all light from the Sun. It stretches some 857,000 miles from Earth, always extending beyond the distance of the Moon. However, the inclination of the Moon's orbit relative to Earth's orbit usually causes the Moon at full (when such eclipses occur) to be above or below Earth's shadow. Under favorable circumstances, when the Moon is at perigee, Earth is at aphelion, and the path of the Moon goes through the center of Earth's shadow, a total lunar eclipse can last for two hours. The partial phases can extend two more hours on either side of the period of totality. (Perigee is the point nearest Earth in the lunar orbit; aphelion is the point farthest from the Sun in Earth's orbit.)

Even when the Moon passes through the umbra or central portion of Earth's shadow, its surface does not become completely dark. This is because some sunlight is refracted through Earth's atmosphere and into the cone of the umbra. As this refracted light passes through the atmosphere it is reddened—a phenomenon similar to our seeing a red Sun at sunrise and sunset—and thereby causes the eclipsed Moon to appear as a dark red globe, an awe-inspiring sight in the night sky.

Lunar eclipses occur in concurrent sets. Each set begins with a partial lunar eclipse repeated with increasing magnitude each 18-year 11.3-day eclipse period. After 13 or 14 renewals the eclipse becomes total and repeats as total for 22 or 23 times. Then there is another series of partial eclipses until the end of the 865.5-year eclipse cycle, which consists of very nearly 48 saros. A saros (the 18.04-year eclipse period known to the Chaldeans) includes on the average 29 lunar eclipses and covers 223 lunations (the period of time between two successive new moons, 29.530588 days).

This program computes the date of the first umbral eclipse of the Moon in any year requested and shows the magnitude of the eclipse, that is, the fraction of the Moon's disk covered by the umbral shadow of Earth. When this fraction reaches or exceeds 1, the eclipse is total. Since the obscuring effects of Earth's penumbra are barely perceptible, penumbral eclipses are not identified. If there is no umbral eclipse in the year requested, the program continues until it finds the first eclipse in subsequent years. You can also ask the computer to display subsequent umbral eclipses. If there are no more eclipses in the year requested, the computer will search for and display the first umbral eclipse in subsequent years. A screen display generated by this program is shown in Figure 9.1.

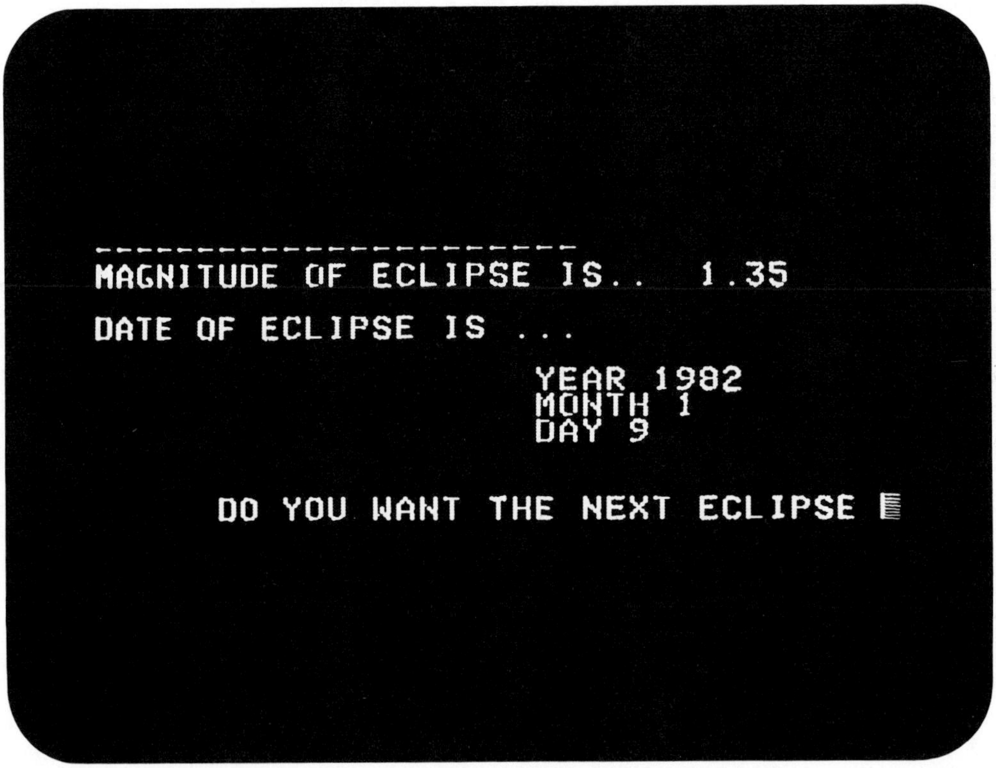

```
MAGNITUDE OF ECLIPSE IS..   1.35
DATE OF ECLIPSE IS ...
                        YEAR 1982
                        MONTH 1
                        DAY 9


        DO YOU WANT THE NEXT ECLIPSE ▤
```

By checking if a full moon occurs close to a node of the Moon's orbit, the program ECLIP tells you when eclipses, if any, take place in any year. This is a typical display generated by the program.

Figure 9.1

The program checks each full moon from the date requested to find if it occurs within the required distance from a node, and cycles until it finds the Julian day of the first full moon occurring close to the node. It calculates the amount of the Moon's disk covered by Earth's shadow, and determines the date on which the eclipse occurs by converting Julian day to calendar date.

The only modifications to the program required by computers other than the Apple II are to lines 10, 70, 190, 420, 850, and 870. In these lines HOME to clear the screen will need to be changed to your own computer's instruction to clear the screen.

The listing for the ECLIP program follows.

ECLIP

```
10   HOME : PRINT : PRINT : PRINT : PRINT
20   PRINT  TAB( 10)"LUNAR UMBRAL ECLIPSES"
30   PRINT  TAB( 10)"BY ERIC BURGESS F.R.A.S."
40   PRINT : PRINT  TAB( 10)"ALL RIGHTS RESERVED BY"
50   PRINT  TAB( 10)"S & T SOFTWARE SERVICES"
60   FOR KZ = 2000 TO 1 STEP  - 1: NEXT KZ
70   HOME
80   PRINT : PRINT : PRINT : PRINT
90   REM   ECLIPSE
100   PRINT  TAB( 5)"THIS PROGRAM GIVES MAGNITUDE"
110   PRINT  TAB( 5)"AND DATE OF LUNAR UMBRAL ECLIPSES"
120   PRINT  TAB( 5)"STARTING AT ANY YEAR REQUESTED"
130   FOR J5 = 2000 TO 1 STEP  - 1: NEXT J5
140   PRINT : PRINT
150   PRINT "PLEASE STATE THE"
160   PRINT
170   INPUT "YEAR TO START   ";Y
180  FL = 0
190   HOME : PRINT : PRINT : PRINT : PRINT
200   PRINT "RUNNING....PLEASE WAIT"
210 Z = Y - 1900
220 ZD = (Z * 12.368267) - 2
230 A =  INT (ZD)
240  DEF  FN RAD(X) = X * 3.141592 / 180
250 A = A + 1
260 B = 29.1053561 * A
270 C = B + 13.7774
280 D = (25.81691806 * A) + 138.94
290 E = (30.670565 * A) + 216.6378
300 F = E - ( SIN ( FN RAD(D))) * .412
310 G = F + ( SIN ( FN RAD(2 * D))) / 8.8
320 H = G + ( SIN ( FN RAD(C))) * 2.2265
330 I = H + ( SIN ( FN RAD(2 * E))) * .13
340 I =  SIN ( FN RAD(I))
350 J = 0.7128 - ( COS ( FN RAD(D))) / 36
360 W = I * 10 ^ J
370  IF W < 0 THEN W = 1.84769 + W * 1.8216: GOTO 390
380  IF W > 0 THEN W = 1.84769 - W * 1.8216
390 K = W + ( COS ( FN RAD(D))) / 30
400  IF K < 0 THEN  GOTO 250
410  IF FL = 1 GOTO 430
420  HOME : PRINT : PRINT : PRINT : PRINT : PRINT
430  PRINT : PRINT "--------------------"
440  PRINT "MAGNITUDE OF ECLIPSE IS..  ";
450 K =  VAL ( LEFT$ ( STR$ (K),4))
460  PRINT K
470  PRINT
480 L = 2415036.025 + (A * 29.53058868)
490 L = L - (.406 *  SIN ( FN RAD(D))) + (.174 *  SIN ( FN RAD(C)))
500 L = L + ( SIN ( FN RAD(2 * D))) / 62
510 L =  INT (L - ( SIN ( FN RAD(2 * E))) / 97)
520  IF L < 2299161 THEN  GOTO 580
530 L2 =  INT ((2299161 - 1867216.25) / 36525.25)
```

──────── **ECLIP** *(continued)* ────────

```
540  L = L2 + L
550  M =   INT (L2 / 4)
560  L = L - M
570  L = L + 1
580  N = L - 1720995
590  O =   INT ((N - 122.1) / 365.25)
600  P =   INT (O * 365.25)
610  Q =   INT ((N - P) / 30.6001)
620  R = ((N - P) -   INT (Q * 30.6001)) / 10000
630   IF Q < 13.7774 THEN   GOTO 650
640  S = Q - 12 - 1: GOTO 660
650  S = Q - 1
660  T = S
670  T2 = T + R
680   IF   SQR (5) < S GOTO 700
690  O = O + 1
700  U = O + T2
710  R = 3 + R * 10000
720   IF T = 2 AND R > 28 THEN T = T + 1:R = R - 28
730   IF T = 3 AND R > 31 THEN T = T + 1:R = R - 31
740   PRINT "DATE OF ECLIPSE IS ..."
750   PRINT
760   PRINT "                    YEAR ";O
770   PRINT "                    MONTH ";T
780   PRINT "                    DAY ";R
790   PRINT : PRINT
800   INPUT "      DO YOU WANT THE NEXT ECLIPSE ";A$
810   PRINT : PRINT
820   IF A$ = "Y" THEN   PRINT :FL = 1: GOTO 250
830   IF A$ <  > "N" THEN   PRINT "INVALID REPLY": PRINT : GOTO 800
840   INPUT "      DO YOU WANT ANOTHER YEAR ";A$
850   IF A$ = "Y" THEN   HOME : PRINT : PRINT : GOTO 170
860   IF A$ <  > "N" THEN   PRINT "INVALID REPLY": PRINT : GOTO 840
870   HOME
880   END
```

By all terrestrial standards, the Moon is an in-hospitable world. Black skies are dominated on the Earth-facing hemisphere by the huge Earth. The terrain is relatively smooth, but broken by scattered rocks. Hills are rounded by their bombardment from space. This photo shows one of the Apollo astronauts making scientific measurements near a huge boulder in the Taurus-Littrow region of the Moon.

Program 10: PHASE

Approximate Phase of Moon for Any Date_____

In many religious and calendrical systems the time of the new moon is most important, since from it the beginning of each lunar cycle can be established. Actually, the new moon cannot be observed—it occurs when the Moon is between Earth and the Sun. The time of the new moon has to be computed backward from the first observation of the extremely thin crescent (often referred to as the new moon) following the astronomical new moon.

The period from one new moon to the next is known as the synodic month. It is 29.530589 days in length and is divided into quarters—new moon, first quarter, full moon, and last quarter. The sidereal month is the passage of the Moon around its orbit to the right ascension of a certain fixed star. This is 27.321662 days. There are two other important lunar cycles. One is the draconic month, which is the period from node to node of the lunar orbit. Each 27.212221 days the Moon crosses Earth's orbital plane from south to north, and that crossing defines the draconic month. The period from perigee (closest approach to Earth) to perigee is the anomalistic month (27.554550 days), and it is used to determine the distance of the Moon from Earth. These various numbers are used in several of the programs in this book, adjusted where necessary to agree with epochs relating to the calculations.

Phases of the Moon were very important to people before the advent of street lighting. Without the light of the Moon, activities outdoors at night were greatly curtailed. The waxing moon is that period between new moon and full moon when more and more of the hemisphere of the Moon facing Earth is illuminated by the Sun. From a thin crescent, the Moon day by day expands to a half moon, then to a gibbous moon, and finally to a full moon rising opposite to the setting sun. The half moon illuminates the evening sky, the full moon brightens most of the night, and the waning half moon lights up the predawn hours.

Many artists hopelessly confuse the Moon's phases in their illustrations, indicating a late night scene by a crescent moon. If the crescent-shaped moon is regarded as a bow, an arrow placed in that bow ready to be shot always points toward the Sun. A thin crescent moon can appear only for a short period after sunset or before sunrise.

Because of the inclination of the ecliptic to Earth's equator, different phases of the Moon are more readily observed at certain seasons. The new crescent is best observed in the spring months because it sets well after the Sun compared with in the fall. The waning crescent is best placed for observation in fall, when it appears high in the sky before sunrise. In summer the full moon is low in the southern sky, while in winter it rides high. In fall, during the period around the full moon, the Moon moves along the ecliptic higher into the sky each evening, so moonrise occurs only a few minutes later each night. For nearly a week, an almost full moon rises shortly after sunset each day. People referred to this as the period of the harvest moon, because light is provided for harvesting after the Sun has set. You can visualize these effects by running SKYSET (Program 16).

The four lunar phases are separated by about 7.4 days on the average. Because the Moon's orbit is elliptical, the Moon does not travel at a constant speed around its orbit, so the interval between the phases varies at different lunations.

This program calculates the date of each new moon and then interpolates for the other phases. The date of the new moon is accurate within one day; the others are approximate only. Starting at instruction 380, the program changes the calendar date to a day number and then proceeds to calculate the next new moon from the date requested, making use of the synodic month (period from one new moon to the next). It derives the calendar date for this new moon and loops to calculate the calendar date for the next new moon. The program then interpolates for the other phases. It repeats this sequence for the number of months requested. Subroutines take care of month ends and year ends and the effects of leap years on the derived dates.

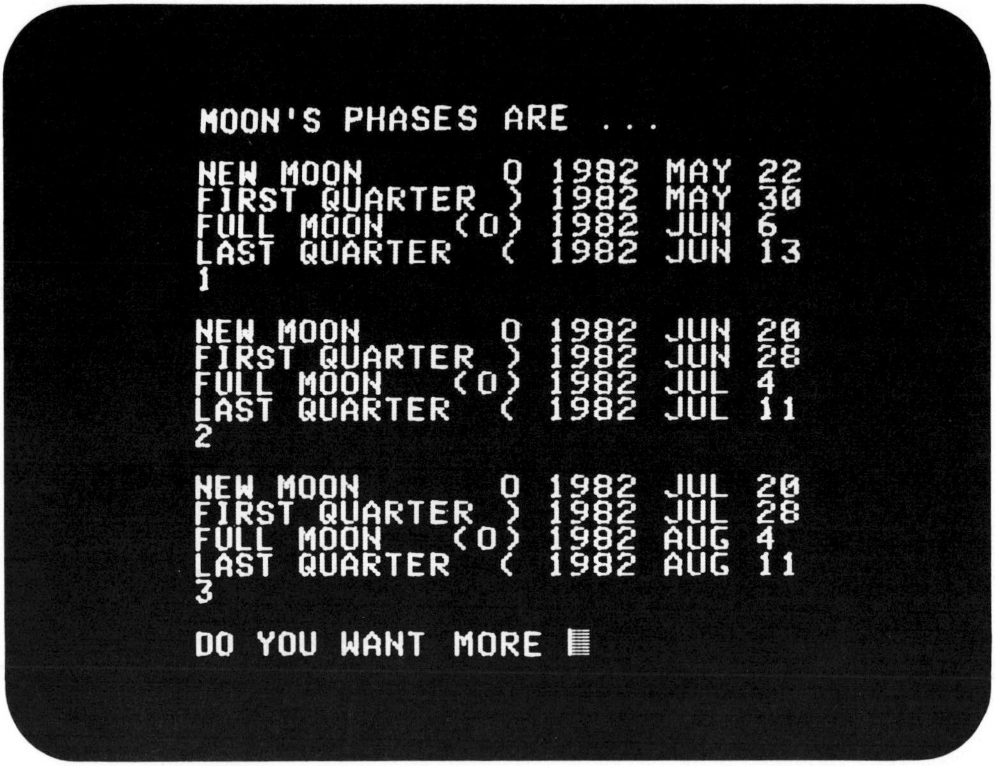

```
MOON'S PHASES ARE ...
NEW MOON        O 1982 MAY 22
FIRST QUARTER  ) 1982 MAY 30
FULL MOON    (O) 1982 JUN 6
LAST QUARTER  ( 1982 JUN 13
1

NEW MOON        O 1982 JUN 20
FIRST QUARTER  ) 1982 JUN 28
FULL MOON    (O) 1982 JUL 4
LAST QUARTER  ( 1982 JUL 11
2

NEW MOON        O 1982 JUL 20
FIRST QUARTER  ) 1982 JUL 28
FULL MOON    (O) 1982 AUG 4
LAST QUARTER  ( 1982 AUG 11
3

DO YOU WANT MORE ▓
```

The PHASE program provides a table showing the approximate dates of the phases of the Moon for any period of up to three months.

Figure 10.1

PHASE

```
10   REM   LUNAR PHASES
20   DEF   FN RAD(X) = .01745328 * (X)
30   HOME : PRINT : PRINT : PRINT : PRINT
40   PRINT : PRINT : PRINT
50   PRINT   TAB( 10)"---------------------"
60   PRINT   TAB( 10)"I   PHASES OF MOON   I"
70   PRINT   TAB( 10)"---------------------"
80   PRINT : PRINT : PRINT
90   PRINT   TAB( 9)"BY ERIC BURGESS F.R.A.S."
100  PRINT : PRINT
110  PRINT   TAB( 10)"ALL RIGHTS RESERVED BY"
120  PRINT   TAB( 10)"S & T SOFTWARE SERVICE"
130  FOR J = 3000 TO 1 STEP  - 1: NEXT
140  HOME : PRINT : PRINT : PRINT
150  PRINT   TAB( 5)"THIS PROGRAM PROVIDES DATES FOR"
160  PRINT   TAB( 5)"PHASES OF THE MOON STARTING AT"
170  PRINT   TAB( 5)"ANY MONTH OR ANY DATE AND FOR"
180  PRINT   TAB( 5)"A PERIOD OF UP TO THREE MONTHS"
190  PRINT   TAB( 8)"   (WITHIN ONE DAY)"
200  PRINT : PRINT : PRINT
210  PRINT "START DATE REQUIRED"
220  PRINT : INPUT "YEAR ";Y$:Y = VAL (Y$)
230  IF Y = 0 THEN  PRINT "INVALID RESPONSE": PRINT : GOTO 220
240  PRINT : INPUT "MONTH ";M$:M = VAL (M$)
250  IF M = 0 THEN  PRINT "INVALID RESPONSE": PRINT : GOTO 240
260  PRINT : INPUT "NUMBER OF MONTHS REQUIRED (1 TO 3) ";N$
270  N =  VAL (N$)
280  IF N > 3 THEN N = 3
290  IF N = 0 THEN  PRINT "INVALID RESPONSE"; PRINT : GOTO 260
300  HOME :NS = 1: PRINT
310  PRINT "MOON'S PHASES ARE ... "
320  MS = M
330  IN = 0
340  IF XP = 1 THEN M = M + 1
350  YP = Y
360  TY = Y + M / 100
370  IF 1582.10 > TY THEN FL = 2
380  P2 = M
390  P1 = Y
400  P3 = D
410  P4 = M / 100 + D / 10000
420  IF  SQR (5) = < P2 THEN  GOTO 440
430  P1 = P1 - 1:P2 = P2 + 12
440  P5 =  INT (365.25 * P1) +  INT ((P2 + 1) * 30.6001) + P3 + 1720995
450  IF FL = 2 GOTO 480
460  P0 =  INT (P1 / 100)
470  P5 = P5 - P0 + 2 +  INT (P0 / 4)
480  P6 = .0338631922
490  Q = (P6 * P5) + .67094
500  Q1 = 1 - (Q -  INT (Q))
510  P7 = Q1 / P6
520  P5 = (P5 + P7) * .985600267
530  P7 = P7 + (.1743 * ( SIN ( FN RAD(P5 + 73.63))))
```

┌─── **PHASE** *(continued)* ──────────────────────────────────

```
540  P8 = (13.06499245 * P5) + 271.5
550   P7 = P7 - (.4089 * ( SIN ( FN RAD(P8))))
560   P7 = (( SIN ( FN RAD(2 * P8))) * .0161) + P7
570   P8 = ((P7 - .5) / 10000) + P4
580    IF IN = 0 THEN P9 = P8
590    IF IN < > 0 THEN M = M - 1: GOTO 610
600   IN = IN + 1:M = M + 1:P1 = 0:P2 = 0:P3 = 0:P4 = 0: GOTO 360
610   Z = P9 - P8
620   Z = Z * 10000
630   Z =  ABS (30.6001 * Z)
640   PH = Z / 400
650   PM =   INT (P9 * 100)
660   PD =   INT ((P9 * 100 -  INT (P9 * 100)) * 100)
670    IF NS = 4 OR NS = 8 THEN PD = PD + 1
680    IF PD = 0 THEN PM = PM - 1:PD = 31
690    IF PD > 31 THEN PD = PD - 31: GOTO 690:PD = PD - 6:PM = PM + 1:Q = 1
700    IF PM > 12 THEN PM = PM - 12:Y = Y + 1
710    IF XP = 1 THEN PM = PM - 1
720    GOSUB 1050
730    PRINT
740    PRINT "NEW MOON       0 ";Y;" ";PM$;" ";PD
750    IF Q = 1 THEN PM = PM - 1:Q = 0
760   PD = PD + 1
770    IF PM = 2 AND PD + PH > 28 THEN PD = PD + 3
780    IF PD + PH > 31 THEN PD = PD - 31:PM = PM + 1
790    IF PM > 12 THEN PM = PM - 12:Y = Y + 1
800    IF PH -  INT (PH) < .5 THEN PH =  INT (PH) + 1
810   PH =  INT (PH)
820    GOSUB 1050
830    PRINT "FIRST QUARTER ) ";Y;" ";PM$;" ";PD + PH
840    IF PM = 2 AND PD + 2 * PH > 28 THEN PD = PD + 3
850    IF PD + 2 * PH > 31 THEN PD = PD - 31:PM = PM + 1
860    IF PM > 12 THEN PM = PM - 12:Y = Y + 1
870    GOSUB 1050
880    PRINT "FULL MOON    (O) ";Y;" ";PM$;" ";PD + 2 * PH
890    IF PM = 2 AND PD + 3 * PH > 28 THEN PD = PD + 3
900    IF PD + 3 * PH > 31 THEN PD = PD - 31:PM = PM + 1
910    IF PM > 12 THEN PM = PM - 12:Y = Y + 1
920    GOSUB 1050
930    IF XP = 1 THEN M = M - 1
940    PRINT "LAST QUARTER  ( ";Y;" ";PM$;" ";PD + 3 * PH
950    PRINT NS
960    IF PD + 3 * PH + 7 = 30 OR PD + 3 * PH + 7 = 31 THEN  GOTO 1170
970   Y = YP
980   M = M + 1
990    IF NS < N THEN NS = NS + 1:IN = 0:P9 = 0: GOTO 330
1000   PRINT : INPUT "DO YOU WANT MORE ";A$
1010   IF A$ = "Y" THEN  GOTO 140
1020   IF A$ < > "N" THEN  PRINT "ANSWER Y OR N ": GOTO 1000
1030   HOME
1040   END
1050   IF PM = 1 THEN PM$ = "JAN": RETURN
1060   IF PM = 2 THEN PM$ = "FEB": RETURN
1070   IF PM = 3 THEN PM$ = "MAR": RETURN
1080   IF PM = 4 THEN PM$ = "APR": RETURN
```

└───

PHASE *(continued)*

```
1090   IF PM = 5 THEN PM$ = "MAY": RETURN
1100   IF PM = 6 THEN PM$ = "JUN": RETURN
1110   IF PM = 7 THEN PM$ = "JUL": RETURN
1120   IF PM = 8 THEN PM$ = "AUG": RETURN
1130   IF PM = 9 THEN PM$ = "SEP": RETURN
1140   IF PM = 10 THEN PM$ = "OCT": RETURN
1150   IF PM = 11 THEN PM$ = "NOV": RETURN
1160   IF PM = 12 THEN PM$ = "DEC": RETURN
1170   PRINT :PZ = 30
1180   PRINT "NEW MOON       O ";Y;" ";PM$;" ";PZ
1190   IF NS = N THEN  GOTO 1270
1200 PM = PM + 1: GOSUB 1050
1210   PRINT "FIRST QUARTER ) ";Y;" ";PM$;" ";PZ - 23
1220   PRINT "FULL MOON     (O)";Y;" ";PM$;" ";PZ - 15
1230   PRINT "LAST QUARTER  ( ";Y;" ";PM$;" ";PZ - 8
1240   IF NF = 1 THEN NF = 0:NS = NS + 1: GOTO 1270
1250   NF = 1:PZ = 30 - 1: PRINT :NS = NS + 1: GOTO 1180
1260   PZ = 0:PD = 0
1270   XP = 1:M = M + 1
1280   PRINT NS: GOTO 970
```

PART 3

THE PLANETS

PLANETS WERE KNOWN as wandering stars to the ancient Greeks, because they appear to move relatively quickly with respect to the fixed background of stars. Observations of the motions of the planets led to new explanations. These new theories of planetary motion eventually caused the demise of the dogma of an Earth-centered universe and spurred the Copernican revolution of human thought. This was part of the transformation that ultimately resulted in the Industrial Revolution. Continued developments in physics and astronomy played an important part in accelerating the pace of the Industrial Revolution.

The programs contained in this section all deal with the motions of the planets. There are programs to find the planets among the stars of the zodiac, programs that determine when planets are in certain positions, such as when Mars is closest to Earth, and programs to aid in planetary observations.

Photo Credit: NASA/Jet Propulsion Laboratory

Spacecraft have now visited all the planets of the solar system known to the ancients; the most recent flyby was that of Voyager past Saturn. This huge planet possesses a bizarre collection of satellites and an extremely complex ring system. This photograph shows the shadow of the planet on the rings and some of the many gaps between the individual ringlets making up the system. Voyager showed that far from being empty gaps, they are actually filled with even fainter and narrower ringlets. One of Saturn's small satellites, Dione, is the bright spot close to the south pole of the planet.

Program 11: RADEC

Right Ascension and Declination for All Planets for Any Date_____

The location of any star or other celestial object can be placed on the celestial sphere in terms of its right ascension and declination, just as the location of any city on Earth can be described in terms of latitude and longitude. The right ascension of a planet is analogous to longitude on Earth; it is measured eastward along the celestial equator from the vernal equinox (see TIMES, Program 3). One sidereal hour is equal to 15 degrees on the celestial sphere. Celestial objects that are on the meridian (that is, they culminate) at the same time as the vernal equinox have a right ascension of 0 hours. Those that culminate 1 hour later have a right ascension of 1 hour, and so on.

The declination of a planet is analogous to latitude on Earth. It is measured in degrees north (+) or south (−) of the celestial equator.

The right ascension and declination of a planet allows you to place the planet in relation to the constellations and thus find where it can be seen in the night sky. You can plot a planet's position by using the right ascension/declination grid of a star map. By asking the computer to provide right ascensions and declinations for different days or months, you can plot the apparent motion of the planets relative to the stars and to each other. Other programs in this book do some of these plots for you (for example, Programs 16 and 17).

This program computes right ascensions and declinations of the planets Mercury, Venus, Mars, Jupiter, Saturn, Uranus, Neptune, and Pluto for any requested date. An example of the display provided by the program is shown in Figure 11.1.

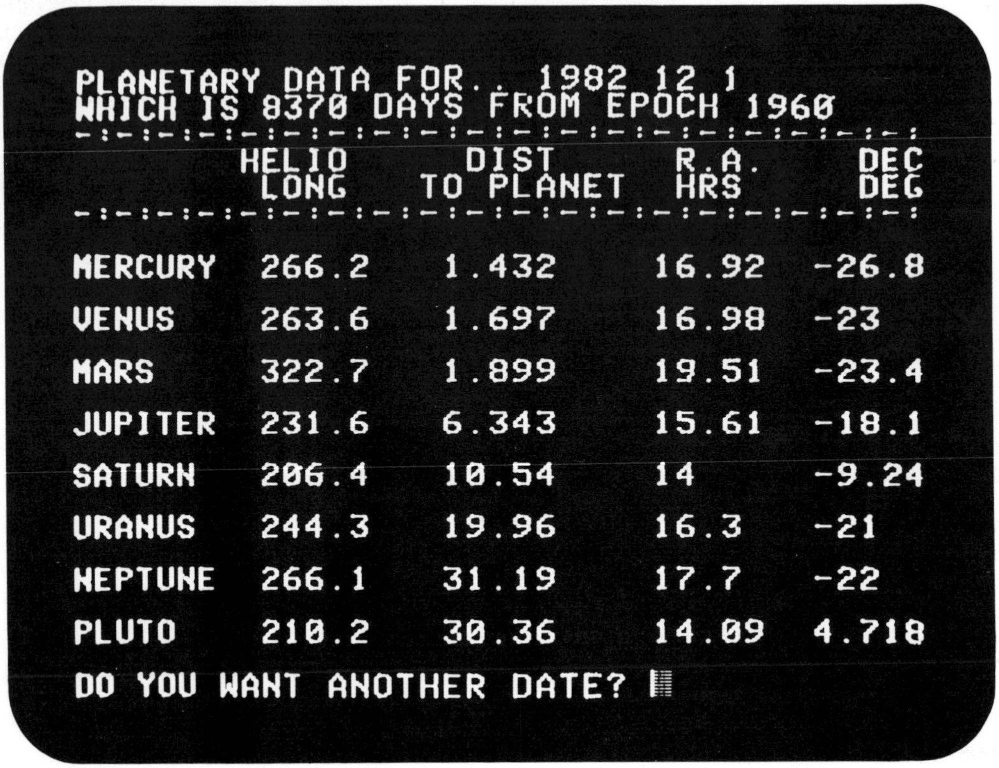

```
PLANETARY DATA FOR.  1982 12 1
WHICH IS 8370 DAYS FROM EPOCH 1960
-:-:-:-:-:-:-:-:-:-:-:-:-:-:-:-:-:-:-:
          HELIO      DIST      R.A.       DEC
          LONG   TO PLANET    HRS.        DEG
-:-:-:-:-:-:-:-:-:-:-:-:-:-:-:-:-:-:-:

MERCURY  266.2     1.432      16.92     -26.8

VENUS    263.6     1.697      16.98     -23

MARS     322.7     1.899      19.51     -23.4

JUPITER  231.6     6.343      15.61     -18.1

SATURN   206.4    10.54       14        -9.24

URANUS   244.3    19.96       16.3      -21

NEPTUNE  266.1    31.19       17.7      -22

PLUTO    210.2    30.36       14.09      4.718

DO YOU WANT ANOTHER DATE?
```

The program RADEC provides a table of right ascension and declination of all the planets for any requested date.

Figure 11.1

Calculations are based on the 1960 epoch (about midcentury to allow forward and backward calculations without perturbations becoming too large). The program calculates the number of days from the epoch to the requested date. It uses planetary orbital data from the data statements, in which angles are expressed in radians. These statements provide information about each planet at the epoch, such as average motion per day, position at the epoch, eccentricity of the orbit, longitude of the perihelion, length in astronomical units of the semi-major axis, inclination of the orbit, and longitude of the ascending node. As indicated in the remark (REM) statements, the program uses standard trigonometrical formulas to derive for each planet the current heliocentric longitude, the distance from the

Sun in astronomical units, the angular distance above or below the ecliptic plane, and the distance in astronomical units from Earth. The angular distance from the Sun is then derived from the three sides of the triangle of distances of Earth, planet, and Sun. From this are derived the geocentric coordinates of each planet. These are then converted to right ascension and declination by changing degrees (measured from the first point of Aries) to hours of right ascension, and using the inclination of Earth's axis to derive declination (angular distance north or south of the celestial equator). String functions are used to prepare the data for display to an appropriate number of decimal places.

As with earlier programs, you may need to modify HOME statements and use the ENTER key instead of the RETURN key. The statements that may require modification are: 60, 180, 270, 280, 810, and 1100.

The listing of the RADEC program follows.

RADEC

```
10   REM   RA AND DEC OF ALL PLANETS
20   DEF   FN ASN(X) =  ATN (X /  SQR ( - X * X + 1)
30   DEF   FN ACO(X) =  -  ATN (X /  SQR ( - X * X + 1)) + 1.5707963
40   DEF   FN RAD(X) = .01745328 * (X)
50   DEF   FN DEG(X) = 57.29578 * (X)
60   HOME : PRINT : PRINT : PRINT : PRINT
70   PRINT  TAB( 11)"ASTRONOMY PROGRAM"
80   PRINT : PRINT
90   PRINT  TAB( 11)"-----------------"
100  PRINT  TAB( 11)"I    RADEC     I"
110  PRINT  TAB( 11)"-----------------"
120  PRINT : PRINT
130  PRINT  TAB( 8)"BY ERIC BURGESS F.R.A.S."
140  PRINT : PRINT
150  PRINT  TAB( 9)"ALL RIGHTS RESERVED BY"
160  PRINT  TAB( 9)"S & T SOFTWARE SERVICE"
170  FOR J = 3000 TO 1 STEP  - 1: NEXT
180  HOME : PRINT : PRINT : PRINT
190  PRINT : PRINT : PRINT : PRINT
200  PRINT  TAB( 2)"THIS PROGRAM GIVES RIGHT ASCENSION"
210  PRINT  TAB( 11)"AND DECLINATIONS"
220  PRINT  TAB( 3)"HELIOCENTRIC LONGITUDES AND THE"
230  PRINT  TAB( 3)"DISTANCES FROM EARTH OF ALL THE"
240  PRINT  TAB( 12)"PLANETS": PRINT : PRINT
250  PRINT  TAB( 4)"FOR THE DATE WHICH YOU INPUT"
260  PRINT : PRINT : PRINT : PRINT
270  INPUT "TO CONTINUE PRESS RETURN";A$
280  HOME : PRINT : PRINT : PRINT : PRINT
290  PRINT "ENTER THE DATE": PRINT
300  FL = 2
310  INPUT "THE YEAR ";YD$:Y =  VAL (YD$)
320  IF Y = 0 THEN  PRINT "INVALID RESPONSE": PRINT : GOTO 310
330  IF Y > 1800 GOTO 370
340  PRINT "IS ";Y;" THE CORRECT YEAR? ": INPUT Y$
350  IF Y$ = "Y" THEN 370
360  IF Y$ <  > "N" THEN  PRINT "INVALID RESPONSE": PRINT : GOTO 340
370  PRINT : INPUT "THE MONTH ";MD$:M =  VAL (MD$)
380  IF M = 0 OR M > 12 THEN  PRINT "INVALID RESPONSE": PRINT : GOTO 370
390  PRINT : INPUT "THE DAY ";DD$:D =  VAL (DD$)
400  IF D = 0 OR D > 31 THEN  PRINT "INVALID RESPONSE": PRINT : GOTO 390
410  IF M = 2 AND D > 29 THEN  PRINT "INVALID RESPONSE": PRINT : GOTO 390
420  REM   CALC GREG. DAYS TO DATE
430  REM   FROM EPOCH 1960,1,1
440  DG = 365 * Y + D + ((M - 1) * 31)
450  IF M >  = 3 GOTO 490
460  REM   CALC FOR JAN AND FEB
470  DG = DG +  INT ((Y - 1) / 4) -  INT ((.75) *  INT ((Y - 1) / 100 + 1))
480  GOTO 510
490  REM   CALC FOR MAR THRU DEC
500  DG = DG -  INT (M * .4 + 2.3) +  INT (Y / 4)
     -  INT ((.75) *  INT ((Y / 100) + 1))
510  NI = DG - 715875
520  REM   PLANETARY DATA FOR EPOCH 1960
```

```
530    IF F = 1 GOTO 790
540    RESTORE
550    DIM PD(9,10)
560    FOR YY = 0 TO 8: FOR XX = 0 TO 8
570    READ PD(YY,XX)
580    NEXT XX,YY
590    REM   MERCURY
600    DATA .071422,3.8484,.388301,1.34041,.3871,.07974,2.73514,.122173,.836013
610    REM   VENUS
620    DATA .027962,3.02812,.013195,2.28638,.7233,.00506,3.85017,.059341,
       1.33168
630    REM   EARTH
640    DATA .017202,1.74022,.032044,1.78547,1,.017,3.33929,0,0
650    REM   MARS
660    DATA .009146,4.51234,.175301,5.85209,1.5237,.141704,1.04656,.03142,
       .858702
670    REM   JUPITER
680    DATA .00145,4.53364,.090478,.23911,5.2028,.249374,1.76188,.01972,
       1.74533
690    REM   SATURN
700    DATA .000584,4.89884,.105558,1.61094,9.5385,.534156,3.1257,.043633,
       1.977458
710    REM   URANUS
720    DATA .000205,2.46615,.088593,2.96706,19.182,.901554,4.49084,.01396,
       1.28805
730    REM   NEPTUNE
740    DATA .000104,3.78556,.016965,.773181,30.06,.27054,2.33498,.031416,
       2.29162
750    REM   PLUTO
760    DATA .000069,3.16948,.471239,3.91303,39.44,9.86,5.23114,.300197,1.91812
770    FOR I9 = 1 TO 9: READ P$(I9): NEXT I9
780    DATA   MERCURY,VENUS,EARTH,MARS,JUPITER,SATURN,URANUS,NEPTUNE,PLUTO
790 F = 1
800    REM   CALC DATA FOR PLANETS
810    HOME : PRINT
820    PRINT "PLANETARY DATA FOR.. ";Y;" ";M;" ";D
830    PRINT "WHICH IS ";NI;" DAYS FROM EPOCH 1960"
840    PRINT "-:-:-:-:-:-:-:-:-:-:-:-:-:-:-:-:"
850    PRINT "         HELIO    DIST     R.A.    DEC"
860    PRINT "         LONG  TO PLANET   HRS     DEG"
870    PRINT "-:-:-:-:-:-:-:-:-:-:-:-:-:-:-:-:"
880    PRINT :I = 1
890    FOR J = 0 TO 8: GOSUB 1120
900 A(I) = A:D(I) = D:L(I) = L
910    I = I + 1: NEXT
920    FOR I = 1 TO 9
930    REM   SKIP EARTH
940    IF I = 3 THEN  NEXT
950    GOSUB 1260
960 Q(I) = Q:R(I) = R:V(I) = V
970    NEXT
980    FOR I = 1 TO 9:A(I) =  FN DEG(A(I))
990    IF I = 3 THEN  NEXT
1000   PRINT P$(I);
1010   PRINT  TAB( 10); VAL ( LEFT$ ( STR$ (A(I)),5));
```

RADEC *(continued)*

```
1020   PRINT  TAB( 18); VAL ( LEFT$ ( STR$ (Q(I)),5));
1030   PRINT  TAB( 27); VAL ( LEFT$ ( STR$ (R(I)),5));
1040   PRINT  TAB( 34); VAL ( LEFT$ ( STR$ (V(I)),5))
1050   PRINT
1060   NEXT
1070   INPUT "DO YOU WANT ANOTHER DATE? ";A$
1080   IF A$ = "Y" GOTO 280
1090   IF A$ < > "N" THEN  PRINT "INVALID RESPONSE": PRINT : GOTO 1070
1100   HOME
1110   END
1120   REM  CALC A,D,AND L
1130   REM  CALC HELIOCENTRIC LONG A
1140 A = NI * PD(J,0) + PD(J,1)
1150   IF A > 6.28318 THEN A = ((A / 6.28318) -  INT (A / 6.28318)) * 6.28318
1160   IF A < 0 THEN A = A + 6.28318: GOTO 1160
1170 C = PD(J,2) *  SIN (A - PD(J,3))
1180 A = A + C
1190   IF A > 6.28318 THEN A = A - 6.28318
1200   IF A < 0 THEN A = A + 6.28318: GOTO 1200
1210   REM  CALC DIST OF PLANET FROM SUN D
1220 D = PD(J,4) + PD(J,5) *  SIN (A - PD(J,6))
1230   REM  CALC DIST. OF PLANET FROM ECLIPTIC L
1240 L = PD(J,7) *  SIN (A - PD(J,8))
1250   RETURN
1260   REM  CALC RA AND DEC R, V
1270   REM  AND DISTANCE FROM EARTH Q
1280 Z = A(3) - A(I)
1290   IF  ABS (Z) > 3.14159 AND Z < 0 THEN Z = Z + 6.28318
1300   IF  ABS (Z) > 3.14159 AND Z > 0 THEN Z = Z - 6.28318
1310   REM  DISTANCE FROM EARTH
1320 Q =  SQR (D(I) ^ 2 + D(3) ^ 2 - 2 * D(I) * D(3) *  COS (Z))
1330   REM  CALC ANG DISTANCE FROM SUN
1340 P = (D(I) + D(3) + Q) / 2
1350 X = 2 *  FN ACO( SQR (((P * (P - D(I))) / (D(3) * Q))))
1360   REM  CALC RA
1370   IF Z < 0 THEN R =  FN DEG(A(3) + 3.14159 - X) / 15
1380   IF Z > 0 THEN R =  FN DEG(A(3) + 3.14159 + X) / 15
1390   IF R > 24 THEN R = R - 24: GOTO 1390
1400   IF R <  - 24 THEN R = R + 24: GOTO 1400
1410   IF R < 0 THEN R = R + 24: GOTO 1410
1420   REM  CALC DEC.
1430   IF Z < 0 THEN V =  SIN (A(3) + 3.14159 - X) * 23.44194 + FN DEG(L(I))
1440   IF Z > 0 THEN V =  SIN (A(3) + 3.14159 + X) * 23.44194 + FN DEG(L(I))
1450 X =  FN DEG(X)
1460   RETURN
```

Mars, known as the red planet, has intrigued mankind from time immemorial. Long associated with the god of war, the baleful planet until recently elicited images of Martians and invaders from space. Although the Viking expeditions to Mars threw considerable doubt on the possibility of there being life on Mars, it is still an intriguing planet. In this photograph Mars appears somewhat Earth-like, with clouds, volcanoes, and vast plains. Some parts of Mars are heavily cratered. The dark spots in the photograph are the huge volcanoes of the Tharsis region. In general Mars is a cold and icy planet.

Program 12: MARSP

Angular Diameter and Distance of Mars for Any Date, and Next Opposition

Because Mars orbits the Sun outside the orbit of Earth, it is referred to as a superior planet. The planet takes 687 Earth days to revolve completely around the Sun, compared with Earth's 365.25 days. Although Mars approaches relatively close to Earth compared with other outer planets, its small size makes it a difficult object to observe. The orbit of Mars is very elliptical compared with that of Earth, and consequently there are optimal times for observing the planet through any telescope. At closest approach (perihelion) Mars is 128.41 million miles from the Sun, while at its greatest distance (aphelion) the planet is 154.86 million miles from the Sun.

When Mars and Earth are aligned on the same side of the Sun, Mars is said to be in opposition (that is, opposite to the Sun in the terrestrial sky), and it is then at its closest to Earth and at its best position for telescopic observation. However, because of the elliptical Martian orbit, some oppositions are good (August and September) while others are bad (February and March). At a bad opposition Mars has an angular diameter of only about half that at a good opposition. The resolution of surface details is accordingly quite poor at a bad opposition. Consequently, favorable oppositions are welcomed by amateur astronomers.

On the average, oppositions of Mars take place once every two years and two months (780 days). When oppositions occur with Mars close to aphelion, the oppositions are only about 764 days apart, while perihelic oppositions are about 811 days apart. The best (perihelic) oppositions occur approximately every 16 years, and at those times Mars approaches within 34.9 million miles of Earth. At aphelic oppositions Mars is some 63.2 million miles from Earth. Unfortunately for observers located in the Northern Hemisphere, the August oppositions take place when Mars is on that part of the ecliptic which dips south of the celestial equator; the planet is low in the southern sky even at midnight. Good oppositions of Mars are best observed in the Southern Hemisphere. You will see why this is so if you use SYKSET/SKYPLT (Program 16) to display the midnight sky for an August and a February opposition of Mars as seen from a location in the Northern Hemisphere and then from a location in the Southern Hemisphere.

This program offers two options: you can find the angular diameter of Mars at any date, or you can find the next opposition following any date. Both result in a similar display of the angular diameter of Mars, its distance from Earth, its angular distance from the Sun, and a comparison of its angular diameter with the greatest and smallest possible angular diameters. Figure 12.1 shows the detailed display of information given by the program.

When you choose to find the next opposition following any date, the program counts down dates toward the opposition. While doing so it

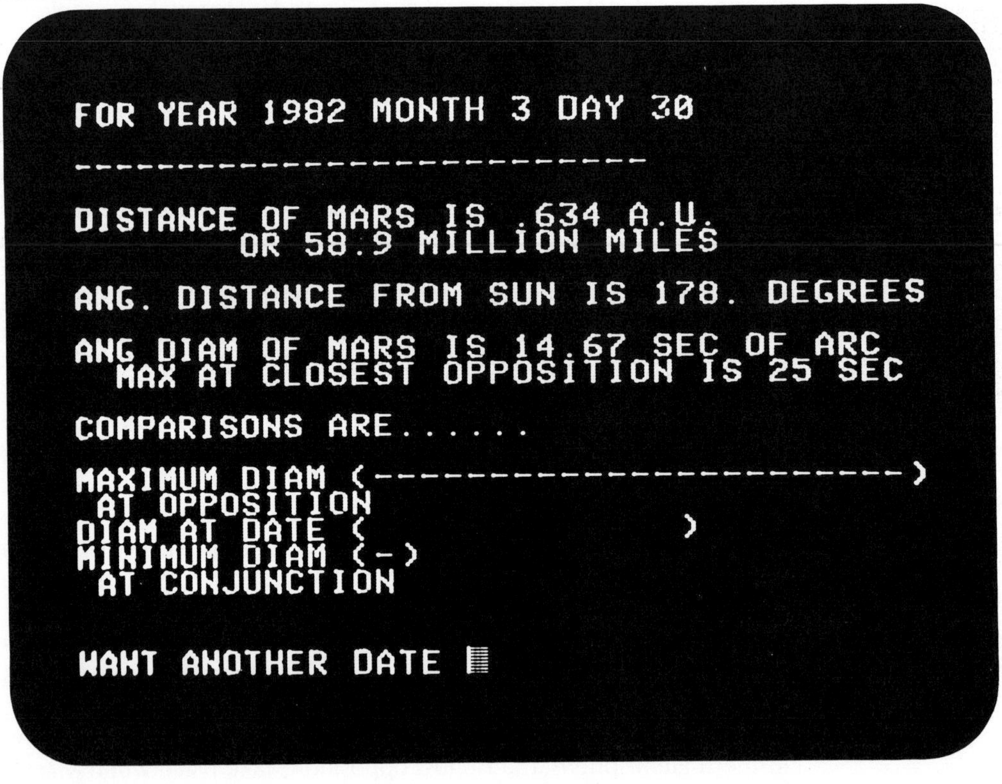

```
FOR YEAR 1982 MONTH 3 DAY 30
------------------------------

DISTANCE OF MARS IS .634 A.U.
      OR 58.9 MILLION MILES

ANG. DISTANCE FROM SUN IS 178. DEGREES

ANG DIAM OF MARS IS 14.67 SEC OF ARC
   MAX AT CLOSEST OPPOSITION IS 25 SEC

COMPARISONS ARE......

MAXIMUM DIAM <---------------------->
 AT OPPOSITION
DIAM AT DATE <               >
MINIMUM DIAM <->
 AT CONJUNCTION


WANT ANOTHER DATE ▓
```

The MARSP program will display the information shown here for any date, or it will find the next opposition of Mars and display the information for that date.

Figure 12.1

displays a graphic plot showing the Sun, Earth, and Mars relative to the direction of the vernal equinox (first point of Aries). It also shows the movement of the two planets around the Sun toward the approaching opposition, to the point where they become aligned on the same side of the Sun at the opposition.

To accomplish this the program first calculates the positions of Mars and Earth at the beginning date. The calculations are made in terms of heliocentric longitude, which is the angle between the planet and the first point of Aries, measured at the Sun. Using high resolution graphics it plots the Sun, the direction of the first point of Aries, Mars, and Earth on the screen (subroutine 1980). The plot positions are derived (instructions 2130 and 2140) from the heliocentric longitudes of Earth and Mars (AE and MA), and they are converted to a suitable scale for the monitor screen (instructions 1990 through 2030). The display appears as though seen from high above the Sun.

The program next compares the heliocentric longitudes of Mars and Earth. If they differ by more than 120 degrees it jumps a month, recalculates the positions, and replots, erasing the first plot. When the difference in longitude of the two planets is less than 120 degrees, the increments of time are reduced to five days. As the angle between the planets narrows still further, the increments are reduced to two days, and finally to one day. When the two longitudes are within two degrees of each other the program declares an opposition and calculates the right ascension of Mars. The program then ascertains in which zodiacal constellation the opposition occurs (instructions 1780 through 1890) and prints the date and the name of the constellation. The program then calculates more details about the opposition and displays them (see Figure 12.1).

The listing of the MARSP program follows.

```
┌─MARSP────────────────────────────────────────────────────┐
│                                                           │
│                                                           │
│    10  DEF  FN ACO(X) =  -  ATN (X / SQR ( - X * X + 1)) + 1.5707963 │
│    20  LY = 0                                             │
│    30  FL = 0                                             │
│    40  P1 = 3.14159:P2 = 6.28318                          │
│    50  HOME : PRINT : PRINT : PRINT                       │
│    60  PRINT  TAB( 10)"----------------"                  │
│    70  PRINT  TAB( 10)"I   M A R S   I"                   │
│    80  PRINT  TAB( 10)"----------------"                  │
│    90  PRINT : PRINT                                      │
│    100  PRINT  TAB( 8)"AN ASTRONOMY PROGRAM": PRINT : PRINT │
│    110  PRINT  TAB( 7)"BY ERIC BURGESS F.R.A.S.": PRINT : PRINT │
│    120  PRINT  TAB( 8)"ALL RIGHTS RESERVED BY"            │
│    130  PRINT  TAB( 8)"S & T SOFTWARE SERVICE": PRINT : PRINT │
│    140  PRINT                                             │
│    150  INPUT "DO YOU WANT INSTRUCTION (Y/N)";A$          │
│    160  IF A$ = "N" THEN  HOME : PRINT : PRINT : PRINT : GOTO 250 │
│    170  IF A$ < > "Y" THEN  PRINT "INVALID RESPONSE": PRINT : GOTO 150 │
│    180  HOME : PRINT : PRINT : PRINT                      │
│    190  PRINT "THIS PROGRAM OFFERS TWO ALTERNATIVES": PRINT │
│    200  PRINT "  1) CALCULATES APPROXIMATE ANGULAR"       │
│    210  PRINT "     DIAMETER OF MARS FOR ANY DATE"        │
│    220  PRINT "  2) PROVIDES APPROXIMATE DATE OF"         │
│    230  PRINT "     NEXT OPPOSITION AFTER ANY DATE": PRINT │
│    240  PRINT : PRINT : PRINT                             │
│    250  PRINT "SELECT 1) ANGULAR DIAMETER"               │
│    260  PRINT "       2) NEXT OPPOSITION"                 │
│    270  PRINT                                             │
│    280  INPUT B$                                          │
│    290  IF B$ = "1" THEN 320                              │
│    300  IF B$ < > "2" THEN  PRINT "INVALID RESPONSE": PRINT : GOTO 250 │
│    310  GOTO 1390                                         │
│    320  GOSUB 670: REM  GET DISTANCE FROM EARTH           │
│    330  HOME : PRINT : PRINT                              │
│    340 MD$ =  LEFT$ ( STR$ (9.31 / Q),5)                 │
│    350  PRINT "FOR YEAR ";Y;" MONTH ";M;" DAY ";D: PRINT  │
│    360  PRINT "-------------------------": PRINT          │
│    370  Q$ =  LEFT$ ( STR$ (Q),4)                        │
│    380  PRINT "DISTANCE OF MARS IS ";Q$;" A.U."           │
│    390  PRINT "         OR "; LEFT$ ( STR$ (Q * 92.96),4);" MILLION MILES" │
│    400  PRINT                                             │
│    410 MW$ =  LEFT$ ( STR$ (W * 57.29578),4)             │
│    420  PRINT "ANG. DISTANCE FROM SUN IS ";MW$;          │
│    430  PRINT " DEGREES": PRINT                           │
│    440  IF  VAL (MD$) > 25 THEN MD$ = "25"               │
│    450  PRINT "ANG DIAM OF MARS IS ";MD$;                │
│    460  PRINT " SEC OF ARC"                               │
│    470  PRINT "  MAX AT CLOSEST OPPOSITION IS 25 SEC"     │
│    480  PRINT                                             │
│    490  PRINT "COMPARISONS ARE......"                     │
│    500  PRINT                                             │
│    510  PRINT "MAXIMUM DIAM (----------------------)"     │
│    520  PRINT " AT OPPOSITION"                            │
│    530  IF 9.36 / Q > = 25 THEN OP = 24: GOTO 550         │
│    540 OP =  INT (9.36 / Q)                               │
│                                                           │
└───────────────────────────────────────────────────────────┘
```

```
550   PRINT "DIAM AT DATE (";
560   PRINT  TAB( 14 + OP)")"
570   PRINT "MINIMUM DIAM (-)
580   PRINT " AT CONJUNCTION"
590   PRINT : PRINT
600   INPUT "WANT ANOTHER DATE ";A$
610   IF A$ = "N" THEN  HOME : GOTO 660
620   IF A$ <  > "Y" THEN  PRINT "INVALID RESPONSE": PRINT : GOTO 600
630   HOME :FL = 0: PRINT : PRINT : PRINT
640   IF B$ = "1" THEN  GOTO 320
650   GOTO 1390
660   END
670   REM   ROUTINE FOR ANGULAR DIAMETER
680   GOSUB 930: REM   ENTER DATE
690   GOSUB 1180
700   REM   CALC DISTANCE FROM EARTH
710   REM   AND ANGULAR DISTANCE FROM SUN
720   AE = NI * .017202 + 1.74022
730   IF AE > P2 THEN AE = ((AE / P2) -  INT (AE / P2)) * P2
740   IF AE < 0 THEN AE = AE + P2: GOTO 740
750   CE = .032044 *  SIN (AE - 1.78547)
760   AE = AE + CE
770   IF AE > P2 THEN AE = AE - P2
780   IF AE < 0 THEN AE = AE + P2: GOTO 780
790   DE = 1 + .017 *  SIN (AE - 3.33926)
800   GOSUB 1290
810   Z = AE - MA
820   IF  ABS (Z) > P1 AND Z < 0 THEN Z = Z + P2
830   IF  ABS (Z) > P1 AND Z > 0 THEN Z = Z - P2
840   IF Z = 0 THEN Q = MD - DE:W = P1: GOTO 920
850   IF Z = P1 THEN Q = MD + DE:W = 0: GOTO 920
860   REM   CALC DISTANCE (Q) FROM EARTH (AU)
870   Q =  SQR (MD ^ 2 + DE ^ 2 - 2 * MD * DE *  COS (Z))
880   P = (MD + DE + Q) / 2
890   REM   CALC ANGULAR DISTANCE (W) FROM SUN
900   V =  SQR ((P * (P - MD)) / (DE * Q))
910   W = 2 *  FN ACO(V)
920   RETURN
930   REM   ENTER DATE
940   IF B$ = "2" AND FL = 1 THEN GOTO 1160
950   PRINT : PRINT
960   PRINT "ENTER THE DATE": PRINT
970   INPUT " THE YEAR ";Y$:Y =  VAL (Y$)
980   IF Y = 0 THEN  PRINT "INVALID RESPONSE": PRINT : GOTO 970
990   IF Y > 1800 GOTO 1040
1000  PRINT "IS ";Y;" THE CORRECT YEAR": INPUT A$
1010  IF A$ = "Y" THEN 1040
1020  IF A$ <  > "N" THEN  PRINT "INVALID RESPONSE": PRINT : GOTO 1000
1030  IF A$ = "N" THEN 970
1040  REM   CHECK FOR LEAP YEAR
1050  GOSUB 1070
1060  GOTO 1100
1070  Y4 = Y / 4:Y5 = Y / 100
1080  IF Y4 -  INT (Y4) = 0 AND Y5 -  INT (Y5) <  > 0 THEN LY = 1
1090  RETURN
```

MARSP (continued)

```
1100   PRINT : INPUT " THE MONTH ";M$:M =  VAL (M$)
1110   IF M = 0 OR M > 12 THEN  PRINT "INVALID RESPONSE": PRINT : GOTO 1070
1120   PRINT : INPUT " THE DAY ";D$:D =  VAL (D$)
1130   IF D = 0 OR D > 31 THEN  PRINT "INVALID RESPONSE": PRINT : GOTO 1120
1140   IF M = 2 AND D > 29 THEN  PRINT "INVALID RESPONSE": PRINT : GOTO 1120
1150   REM  STORE INITIAL DATE
1160 YI = Y:MJ = M:DI = D
1170   RETURN
1180   REM  CALC DAYS FROM EPOCH
1190 DG = 365 * Y + D + ((M - 1) * 31)
1200   IF M >  = 3 GOTO 1240
1210   REM  CALC IF JAN OR FEB
1220 DG = DG +  INT ((Y - 1) / 4) -  INT ((.75) *  INT ((Y - 1) / 100 + 1))
1230   GOTO 1260
1240   REM  CALC FOR MAR THRU DEC
1250 DG = DG -  INT (M * .4 + 2.3) +  INT (Y / 4)
       -  INT ((.75) *  INT ((Y / 100) + 1))
1260 NI = DG - 715875
1270   RETURN
1280   GOTO 320
1290   REM  CALC MARS'S POSITION
1300 MA = NI * .009146 + 4.51234
1310   IF MA > P2 THEN MA = ((MA / P2) -  INT (MA / P2)) * P2
1320   IF MA < 0 THEN MA = MA + P2: GOTO 1320
1330 MC = .175301 *  SIN (MA - 5.85209)
1340 MA = MA + MC
1350   IF MA > P2 THEN MA = MA - P2
1360   IF MA < 0 THEN MA = MA + P2: GOTO 1360
1370 MD = 1.5237 + .141704 *  SIN (MA - 1.04656)
1380   RETURN
1390   GOSUB 930: GOSUB 1180
1400   GOSUB 710
1410   PRINT : PRINT
1420   HOME : PRINT : PRINT : PRINT
1430   VTAB 23
1440   PRINT YI;" ";MI;" ";DI
1450   GOSUB 1980: REM  TO PLOT PLANETS
1460 FL = 1
1470 DS = W * 57.29578
1480   IF DS > 120 THEN  GOTO 1540
1490 FL = 1
1500 M = M + 1
1510   IF M > 12 THEN M = 1:Y = Y + 1
1520   GOSUB 1070
1530   GOTO 1390
1540   REM  COMPUTE FOR DAYS
1550   IF DS < 160 THEN D = D + 5: GOTO 1580
1560   IF DS < 170 THEN D = D + 2: GOTO 1580
1570 D = D + 1
1580   GOSUB 1630
1590 D2 = DS
1600   IF DS > 178 THEN  GOTO 1710
1610   IF D2 < DS THEN D = D + 1: GOSUB 1630
1620   GOTO 1700
1630   REM  CORRECT FOR MONTH AND YEAR ENDS
1640   IF (LY = 1 AND M = 2 AND D > 29) THEN M = 3:D = D - 29: GOTO 1640
```

```
1650   IF (LY = 0 AND M = 2 AND D > 28) THEN M = 3:D = D - 28: GOTO 1650
1660   IF D < 31 THEN   GOTO 1690
1670   IF (M = 4 OR M = 6 OR M = 9 OR M = 11) THEN M = M + 1:D = D - 30:
       GOTO 1690
1680   IF D > 31 THEN M = M + 1:D = D - 31
1690   RETURN
1700   GOTO 1390
1710   REM  COMPUTE RA FOR YI,MI,DI
1720   IF Z < 0 THEN R = AE + P1 - W
1730   IF Z > 0 THEN R = AE + P1 + W
1740 R = 57.29578 * R / 15
1750   IF R > 24 THEN R = R - 24: GOTO 1750
1760   IF R < - 24 THEN R = R + 24: GOTO 1760
1770   IF R < 0 THEN R = R + 24: GOTO 1770
1780   IF R > 0 AND R < 2 THEN C$ = "PISCES": GOTO 1900
1790   IF R > 2 AND R < 4 THEN C$ = "ARIES": GOTO 1900
1800   IF R > 4 AND R < 6 THEN C$ = "TAURUS": GOTO 1900
1810   IF R > 6 AND R < 8 THEN C$ = "GEMINI": GOTO 1900
1820   IF R > 8 AND R < 10 THEN C$ = "CANCER": GOTO 1900
1830   IF R > 10 AND R < 12 THEN C$ = "LEO": GOTO 1900
1840   IF R > 12 AND R < 14 THEN C$ = "VIRGO": GOTO 1900
1850   IF R > 14 AND R < 16 THEN C$ = "LIBRA": GOTO 1900
1860   IF R > 16 AND R < 18 THEN C$ = "SCORPIO": GOTO 1900
1870   IF R > 18 AND R < 20 THEN C$ = "SAGITTARIUS": GOTO 1900
1880   IF R > 20 AND R < 22 THEN C$ = "CAPRICORNUS": GOTO 1900
1890   IF R > 22 AND R < 24 THEN C$ = "AQUARIUS"
1900   PRINT
1910   TEXT
1920   PRINT "NEXT OPPOSITION OF MARS IS"
1930   PRINT "      "YI;" ";MI;" ";DI
1940   PRINT "IN THE CONSTELLATION OF ";C$
1950   PRINT
1960   INPUT "FOR FURTHER DETAILS PRESS RETURN KEY";Q$
1970   Y = YI:M = MI:D = DI: GOTO 320
1980   REM  SUB TO PLOT PLANETS
1990   AD = 50:DM = 75
2000   EY = AD * SIN (AE)
2010   EX = 1.5 * AD * COS (AE)
2020   MY = DM * SIN (MA)
2030   MX = 1.5 * DM * COS (MA)
2040   HGR
2050   HPLOT 138,78: HPLOT 139,79: HPLOT 137,77
2060   HPLOT 136,76: HPLOT 140,80: HPLOT 139,81
2070   HPLOT 138,81: HPLOT 137,81: HPLOT 136,80
2080   HPLOT 138,75: HPLOT 137,75: HPLOT 139,75
2090   HPLOT 140,76: HPLOT 139,75: HPLOT 140,76
2100   FOR J = 150 TO 250 STEP 10
2110   HPLOT J,78 TO J + 5,78
2120   NEXT J
2130   HPLOT (138 + EX),(78 - EY)
2140   HPLOT (138 + MX),(78 - MY)
2150   HPLOT 269,76: HPLOT 270,79: HPLOT 270,78
2160   HPLOT 270,77: HPLOT 269,76: HPLOT 268,76
2170   HPLOT 267,77: HPLOT 271,76: HPLOT 272,76
2180   HPLOT 273,77
2190   RETURN
```

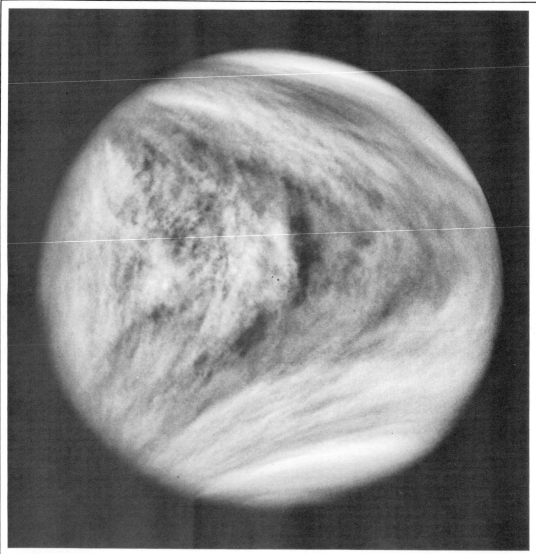

By contrast with Mars, Venus is a hot planet whose surface is obscured completely by its cloud system. Venus is larger than Mars and almost the same size as Earth. A question we still have not been able to answer is why Venus should have evolved so differently from Earth. Radar carried by spacecraft has penetrated the clouds to reveal highlands, plains, and what may be two large volcanic regions through which heat escapes from the interior. This picture is a view of the cloud-shrouded planet obtained by the Pioneer Venus *orbiting spacecraft.*

Program 13: MVENC and MERVE

Elongations, Phases, Angular Diameters, and Distances of Mercury and Venus, and Next Elongations_____

Across the Aegean Sea in the dawn glow, the ancient Greeks periodically observed a dull white star rising just before sunrise. They called it Apollo. They often saw another dull white star following the Sun to set in the Ionian Sea, and this they named Mercury, the messenger of the gods. It was not until the time of Plato that these two celestial objects were recognized as being one: the planet we now know as Mercury, innermost planet of our solar system.

Except in very clear skies, Mercury is an elusive object to find because of its proximity to the solar glare. Venus, however, the next planet in terms of distance from the Sun, is the most brilliant object in the sky after the Sun and the Moon. Venus moves much farther from the Sun than Mercury in the morning and evening skies, and it is so brilliant that the Babylonians referred to Venus as ''mistress of the heavens.''

Because Mercury and Venus follow orbits inside Earth's orbit, they are known as inferior planets. They always appear relatively close to the Sun in the sky. They are seen as morning or evening stars depending on whether they appear to the right or the left of the Sun, respectively.

The Mayas paid great attention to observing Venus and establishing calendars based on the apparitions of the planet. Venus was called *noh ek*, Great Star, and at the time of the Spanish conquest the accuracy of observations was such that the Mayas never made mistakes in predicting the apparitions of the planet. Pages 46–50 of the Dresden Codex contain information about the cycles of Venus in a dot-bar notation to base 20, a vigesimal numbering system. Much Mayan architecture is also oriented toward Venus's rising and setting points on the horizon. Sacrifices to Venus formed a prominent ritual of Mayan religious life.

As seen from Earth, both Mercury and Venus oscillate to either side of the Sun. The maximum distance east or west of the Sun is termed an elongation. At eastern elongation the planets are visible in the evening sky—they appear to follow the Sun in its apparent daily motion across Earth's sky. At western elongation the planets are ahead of the Sun and are seen as morning stars before sunrise (see Figure 13.1).

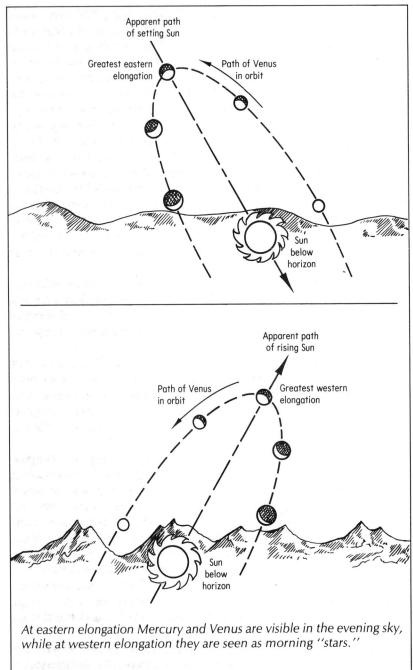

Apparent path
of setting Sun

Greatest eastern
elongation

Path of Venus
in orbit

Sun
below
horizon

Apparent path
of rising Sun

Path of Venus
in orbit

Greatest western
elongation

Sun
below
horizon

At eastern elongation Mercury and Venus are visible in the evening sky, while at western elongation they are seen as morning "stars."

Figure 13.1

Venus repeats its apparitions approximately every 584 days, while Mercury repeats every 116 days. Mercury's orbit is quite elliptical, so its apparitions vary greatly; its elongation from the Sun can be as little as 18 degrees and as much as 27 degrees. Mercury can be seen for a maximum of only 2.5 hours before sunrise or after sunset at a favorable elongation. By contrast, Venus moves about 47 degrees from the Sun at its maximum elongation and can be observed for a much greater period. On the average, Venus passes from greatest eastern elongation as an evening star to greatest western elongation as a morning star in about 140 days, and from a morning star back to an evening star in about 430 days. Venus is unobservable with the unaided eye for about eight days as it passes between Earth and the Sun, except on those very rare occasions when the planet passes in transit across the face of the Sun. When Venus swings behind the Sun it is hidden in the solar glare for about seven weeks. Thus Venus has four distinct periods to its cycle: two unequal periods of invisibility, and two periods, each of about nine months' visibility, as an evening or morning "star."

This program offers two alternatives. It will provide information on the two planets' distances from Earth, angular sizes, and phases for any date and graphically display the positions of both planets relative to the Sun (see Figure 13.2). It will also calculate the date of the next elongation of Mercury or Venus following any date.

When you select the alternative to seek the next elongation, the display counts down in dates toward the elongation and shows the positions of Mercury and Venus relative to the Sun. On arrival at the elongation, you can then ask for more details, such as distance from Earth, phase, angular distance from the Sun, and angular diameter of the planet at the date of elongation (see Figure 13.3).

Using similar routines to those of earlier planetary programs, the program determines the angular distance of the planets from the Sun (starting with instruction 930). It finds the angular distance of the planet for which you seek the next elongation, repeats the calculation for a later date, and searches for a maximum angular distance from the Sun by repeated comparisons following increments of the date. The next elongation is identified as the first maximum of angular distance to be encountered after the start date. Next the program determines if this is east or west of the Sun. For the date of the elongation it then calculates the other parameters for the display. Through a simple sort routine (instructions 1790 through 1960) the program develops a graphic display of the Sun with the planets configured alongside as they appear from Earth.

Another version of this program (MERVE) plots the positions of Mercury and Venus relative to the Sun for a number of intervals. This version gives a graphic display of the planets moving from side to side of the Sun. It allows the display to overlay earlier displays or to scroll, to show several configurations on the screen at once. When all the intervals have been plotted, the program allows you to select details about either or both planets—such as distance, phase, and angular diameter. It uses the same basic algorithms and sort routine for the graphic display as the MVENC program uses, but it does not seek an elongation. Instead it counts displays to the number of intervals requested. An example of the display provided by the MERVE program is shown in Figure 13.4.

The listings for the MVENC and MERVE programs follow.

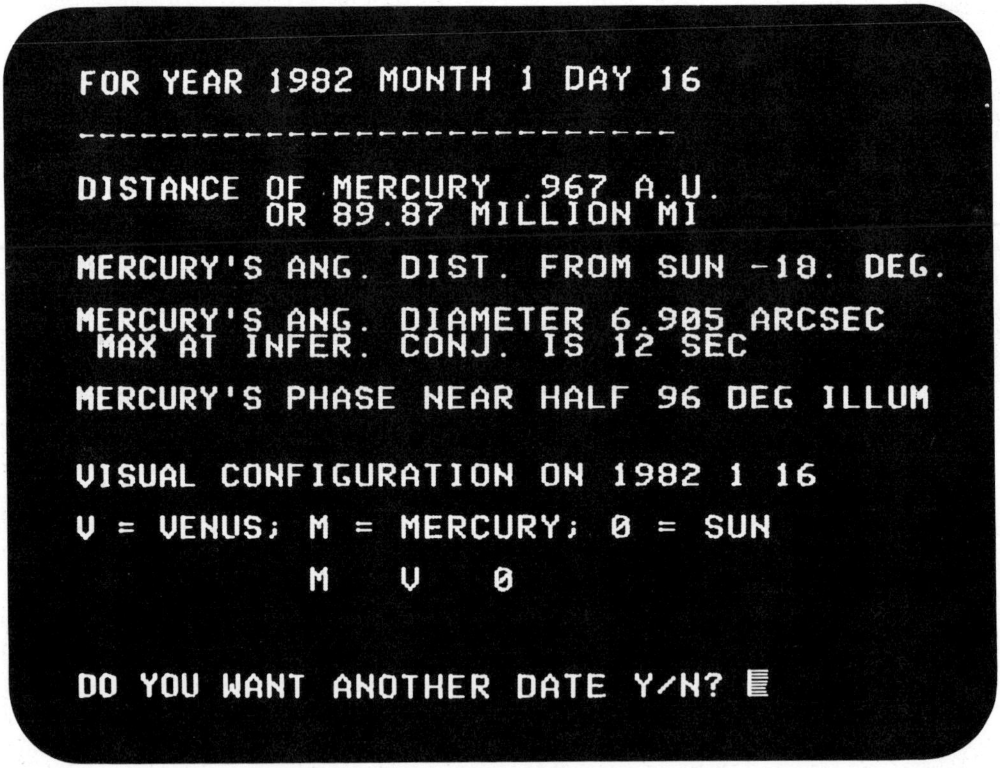

```
FOR YEAR 1982 MONTH 1 DAY 16
------------------------------

DISTANCE OF MERCURY .967 A.U.
        OR 89.87 MILLION MI

MERCURY'S ANG. DIST. FROM SUN -18. DEG.

MERCURY'S ANG. DIAMETER 6.905 ARCSEC
  MAX AT INFER. CONJ. IS 12 SEC

MERCURY'S PHASE NEAR HALF 96 DEG ILLUM


VISUAL CONFIGURATION ON 1982 1 16

V = VENUS; M = MERCURY; 0 = SUN

        M    V    0




DO YOU WANT ANOTHER DATE Y/N?
```

Details of either or both planets can be displayed by the MVENC and MERVE programs.

Figure 13.2

```
NEXT ELONGATION OF MERCURY 1982 1 16

AND THE PLANET IS....
     EAST OF SUN (AN EVENING STAR)
VISUAL CONFIGURATION ON 1982 1 16
V = VENUS; M = MERCURY; 0 = SUN
          M    V    0


DO YOU WANT MORE DETAILS Y/N? ▊
```

Information on the next elongation of Mercury or Venus is displayed by the MVENC program.

— Figure 13.3 —

```
DISPLAY 1 IS DATE 1982 1 1

        V        M        0

DISPLAY 2 IS DATE 1982 1 31

                      M 0      V

DISPLAY 3 IS DATE 1982 3 2

                 0              M      V



SELECT DETAILS FOR DISPLAY
MERCURY (1); VENUS (2); OR BOTH (3) ▤
```

Positions of Mercury (M) and Venus (V) are shown relative to the Sun (0) for a series of intervals in this typical display generated by the MERVE program.

— Figure 13.4 —

```
┌─ MVENC ─────────────────────────────────────────────────────────────┐
│                                                                       │
│   10   REM   INNER PLANETS' ELONGATIONS                               │
│   20   DEF   FN ACO(X) =  -  ATN (X / SQR ( - X * X + 1)) + 1.5707963 │
│   30   P1 = 3.14159:P2 = 6.28318                                      │
│   40   HOME : PRINT : PRINT : PRINT                                   │
│   50   PRINT  TAB( 10)"AN ASTRONOMY PROGRAM"                          │
│   60   PRINT : PRINT                                                  │
│   70   PRINT  TAB( 10)"-------------------"                           │
│   80   PRINT  TAB( 10)"I  INNER   PLANETS  I"                         │
│   90   PRINT  TAB( 10)"-------------------"                           │
│   100  PRINT : PRINT                                                  │
│   110  PRINT  TAB( 9)"BY ERIC BURGESS F.R.A.S."                       │
│   120  PRINT : PRINT                                                  │
│   130  PRINT  TAB( 9)"ALL RIGHTS RESERVED BY"                         │
│   140  PRINT  TAB( 9)"S & T SOFTWARE SERVICE"                         │
│   150  PRINT : PRINT : PRINT                                          │
│   160  INPUT "DO YOU WANT INSTRUCTIONS Y/N? ";A$                      │
│   170  IF A$ = "N" THEN  HOME : PRINT : PRINT : PRINT : GOTO 350      │
│   180  IF A$ < > "Y" THEN  PRINT "INVALID RESPONSE:PRINT:GOTO140      │
│   190  HOME : PRINT : PRINT : PRINT                                   │
│   200  PRINT "THIS PROGRAM OFFERS TWO ALTERNATIVES"                   │
│   210  PRINT : PRINT                                                  │
│   220  PRINT "   1) CALCULATES APPROXIMATE ANGULAR"                   │
│   230  PRINT "      DISTANCES OF MERCURY AND VENUS"                   │
│   240  PRINT "      FROM THE SUN, AND APPROXIMATE"                    │
│   250  PRINT "      DISTANCES FROM EARTH, AND"                        │
│   260  PRINT "      PROVIDES APPROXIMATE ANGULAR"                     │
│   270  PRINT "      DIAMETER AND PHASE FOR ANY DATE"                  │
│   280  PRINT                                                          │
│   290  PRINT "   2) PROVIDES DATE OF THE NEXT"                        │
│   300  PRINT "      ELONGATION OF MERCURY OR VENUS"                   │
│   310  PRINT "           AFTER ANY DATE"                              │
│   320  PRINT : PRINT                                                  │
│   330  INPUT "TO CONTINUE PRESS RETURN ";A$                          │
│   340  HOME : PRINT : PRINT : PRINT                                   │
│   350  PRINT "SELECT 1) ANGULAR DISTANCES"                           │
│   360  PRINT "       2) NEXT ELONGATION"                             │
│   370  PRINT : INPUT B$                                              │
│   380  IF B$ = "1" THEN 410                                          │
│   390  IF B$ < > "2" THEN  PRINT "INVALID RESPONSE": PRINT : GOTO 340 │
│   400  GOTO 2110                                                      │
│   410  GOSUB 890: REM  GET DISTANCE FROM EARTH                        │
│   420  HOME : PRINT : PRINT                                           │
│   430  GOSUB 440: GOTO 820                                            │
│   440 MD$ =  LEFT$ ( STR$ (6.68 / MQ),5)                              │
│   450 VD$ =  LEFT$ ( STR$ (16.82 / VQ),5)                             │
│   460  PRINT "FOR YEAR ";Y;" MONTH ";M;" DAY ";D: PRINT               │
│   470  PRINT "-------------------------": PRINT                       │
│   480 MQ$ =  LEFT$ ( STR$ (MQ),4)                                     │
│   490 VQ$ =  LEFT$ ( STR$ (VQ),4)                                     │
│   500  IF E$ = "2" THEN  GOTO 540                                     │
│   510  PRINT "DISTANCE OF MERCURY ";MQ$;" A.U."                       │
│   520  PRINT  TAB( 10)"OR "; VAL ( LEFT$ ( STR$ (MQ * 92.9),5));" MILLION MI" │
│   530  IF E$ = "1" THEN  GOTO 560                                     │
│   540  PRINT "DISTANCE OF VENUS ";VQ$;" A.U."                         │
│                                                                       │
└───────────────────────────────────────────────────────────────────┘
```

─────── **MVENC** *(continued)* ───────

```
550   PRINT   TAB( 10)"OR "; VAL ( LEFT$ ( STR$ (VQ * 92.9),5));" MILLION MI"
560   PRINT
570   MW$ =  LEFT$ ( STR$ (MW * 57.29578),4)
580   VW$ =  LEFT$ ( STR$ (VW * 57.29578),4)
590   IF E$ = "2" THEN  GOTO 620
600   PRINT "MERCURY'S ANG. DIST. FROM SUN ";MW$;" DEG."
610   IF E$ = "1" THEN  PRINT : GOTO 640
620   PRINT "VENUS'S ANGULAR DIST. FROM SUN ";VW$;" DEG."
630   PRINT
640   IF E$ = "2" THEN  GOTO 690
650   PRINT "MERCURY'S ANG. DIAMETER ";MD$;" ARCSEC"
660   PRINT " MAX AT INFER. CONJ. IS 12 SEC"
670   IF E$ = "1" THEN  GOTO 710
680   PRINT
690   PRINT "VENUS'S ANG. DIAMETER ";VD$;" ARCSEC"
700   PRINT " MAX AT INFER. CONJ. IS 61 SEC"
710   PRINT
720   PH =  ABS (MW) +  ABS (MZ): GOSUB 2880
730   IF E$ = "2" THEN  GOTO 760
740   PRINT "MERCURY'S PHASE ";PH$;" ";PH;" DEG ILLUM"
750   IF E$ = "1" THEN  GOTO 820
760   PH =  ABS (VW) +  ABS (ZV): GOSUB 2880
770   PRINT "PHASE OF VENUS   ";PH$;" ";PH;" DEG ILLUM"
780   :
790   IF E$ = "2" THEN 820
800   GOSUB 1790
810   GOSUB 1970
820   :
830   IF B$ = "2" THEN  RETURN
840   INPUT "WANT ANOTHER DATE Y/N ? ";A$
850   IF A$ = "N" THEN  HOME : GOTO 880
860   IF A$ < > "Y" THEN  PRINT "INVALID RESPONSE": PRINT : GOTO 840
870   HOME : PRINT : PRINT : PRINT : GOTO 410
880   END
890   REM  ROUTINE FOR ANG. DIAMETER
900   HOME : PRINT : PRINT : PRINT
910   GOSUB 1180: REM  ENTER DATE
920   GOSUB 1360
930   REM  CALC DIST. FROM EARTH
940   REM  AND ANG. DISTANCE FROM SUN
950   AE = NI * .017202 + 1.74022
960   IF AE > P2 THEN AE = ((AE / P2) -  INT (AE / P2)) * P2
970   IF AE < 0 THEN AE = AE + P2: GOTO 970
980   CE = .032044 *  SIN (AE - 1.78547)
990   AE = AE + CE
1000  IF AE > P2 THEN AE = AE - P2
1010  IF AE < 0 THEN AE = AE + P2: GOTO 1010
1020  DE = 1 + .017 *  SIN (AE - 3.33926)
1030  GOSUB 1470
1040  MZ = AE - MA
1050  ZM = MZ
1060  IF  ABS (MZ) > P1 AND MZ < 0 THEN MZ = MZ + P2
1070  IF  ABS (MZ) > P1 AND MZ > 0 THEN MZ = MZ - P2
1080  REM  CALC DISTANCE FROM EARTH IN AU
1090  MQ =  SQR (MD ^ 2 + DE ^ 2 - 2 * MD * DE *  COS (MZ))
```

MVENC *(continued)*

```
1100 MP = (MD + DE + MQ) / 2
1110  REM  CALC ANG. DISTANCE FROM SUN
1120 MV =  SQR ((MP * (MP - MD)) / (DE * MQ))
1130 MW = 2 *  FN ACO(MV)
1140  IF  ABS (ZM) > P1 THEN ZM = ZM + P2
1150  IF ZM > 0 THEN MW =  - MW
1160  GOSUB 1570
1170  RETURN
1180  REM  ENTER DATE
1190  PRINT : PRINT
1200  PRINT : PRINT "ENTER THE DATE ": PRINT
1210  INPUT "THE YEAR ";Y$:Y =  VAL (Y$)
1220  IF Y = 0 THEN  PRINT "INVALID RESPONSE"; PRINT : GOTO 1210
1230  IF Y > 1800 GOTO 1280
1240  PRINT "IS ";Y;" THE CORRECT YEAR": INPUT A$
1250  IF A$ = "Y" THEN 1280
1260  IF A$ < > "N" THEN  PRINT "INVALID RESPONSE": PRINT : GOTO 1240
1270  IF A$ = "N" THEN 1210
1280  PRINT : INPUT "THE MONTH ";M$:M =  VAL (M$)
1290  IF M = 0 OR M > 12 THEN  PRINT "INVALID RESPONSE": PRINT : GOTO 1280
1300  PRINT : INPUT "THE DAY ";D$:D =  VAL (D$)
1310  IF D = 0 OR D > 31 THEN  PRINT "INVALID RESPONSE": PRINT : GOTO 1300
1320  IF M = 2 AND M > 29 THEN  PRINT "INVALID RESPONSE": PRINT : GOTO 1300
1330  REM  STORE INITIAL DATE AND TIME
1340 YI = Y:MI = M:DI = D
1350  RETURN
1360  REM  CALC DAYS FROM EPOCH 1960
1370 DG = 365 * Y + D + ((M - 1) * 31)
1380  IF M > = 3 GOTO 1420
1390  REM  CALC IF JAN OR FEB
1400 DG = DG +  INT ((Y - 1) / 4) -  INT ((.75) *  INT ((Y - 1) / 100 + 1))
1410  GOTO 1440
1420  REM  CALC FOR MAR THRU DEC
1430 DG = DG -  INT (M * .4 + 2.3) +  INT (Y / 4)
      -  INT ((.75) *  INT ((Y / 100) + 1))
1440 NI = DG - 715875
1450  RETURN
1460  GOTO 410
1470  REM  CALC MERCURY'S POSITION
1480 MA = NI * .071425 + 3.8494
1490  IF MA > P2 THEN MA = ((MA / P2) -  INT (MA / P2)) * P2
1500  IF MA < 0 THEN MA = MA + P2: GOTO 1500
1510 MC = .388301 *  SIN (MA - 1.34041)
1520 MA = MA + MC
1530  IF MA > P2 THEN MA = MA - P2
1540  IF MA < 0 THEN MA = MA + P2: GOTO 1540
1550 MD = .3871 + .079744 *  SIN (MA - 2.73514)
1560  RETURN
1570  REM  CALC VENUS'S POSITION
1580 VA = NI * .027962 + 3.02812
1590  IF VA > P2 THEN VA = ((VA / P2) -  INT (VA / P2)) * P2
1600  IF VA < 0 THEN VA = VA + P2: GOTO 1600
1610 VC = .013195 *  SIN (VA - 2.28638)
1620 VA = VA + VC
1630  IF VA > P2 THEN VA = VA - P2
```

─── MVENC *(continued)* ───

```
1640  IF VA < 0 THEN VA = VA + P2: GOTO 1640
1650  VD = .7233 + .00506 *  SIN (VA - 3.85017)
1660  ZV = AE - VA
1670  VZ = ZV
1680   IF  ABS (ZV) > P1 AND ZV < 0 THEN ZV = ZV + P2
1690   IF  ABS (ZV) > P1 AND ZV > 0 THEN ZV = ZV - P2
1700  REM   CALC DISTANCE FROM EARTH IN AU
1710  VQ =  SQR (VD ^ 2 + DE ^ 2 - 2 * VD * DE *  COS (ZV))
1720  VP = (VD + DE + VQ) / 2
1730  REM   CALC ANG. DISTANCE FROM SUN
1740  VV =  SQR ((VP * (VP - VD)) / (DE * VQ))
1750  VW = 2 *  FN ACO(VV)
1760   IF  ABS (VZ) > P1 THEN VZ = VZ + P2
1770   IF VZ > 0 THEN VW =  - VW
1780  RETURN
1790  REM   SHOW SUN AND PLANETS
1800  Z$(0) = MW$ + "M":Z$(1) = VW$ + "V"
1810  Z$(2) =  STR$ (0) + ""
1820  FOR I = 0 TO 2
1830  E$(I) = "XXX"
1840  NEXT I
1850  K = 0
1860  FOR I = 0 TO 2
1870  IF Z$(I) = "XXX" GOTO 1950
1880  FOR J = I TO 2
1890  IF Z$(J) = "XXX" GOTO 1910
1900  IF  VAL (Z$(J)) <  VAL (Z$(I)) THEN I = J: GOTO 1910
1910  NEXT J
1920  E$(K) = Z$(I):Z$(I) = "XXX"
1930  K = K + 1
1940  I =  - 1
1950  NEXT I
1960  RETURN
1970  INPUT "FOR VISUAL CONFIGURATION PRESS RETURN";A$
1980  PRINT
1990  PRINT "VISUAL CONFIGURATION ON ";YI;" ";MI;" ";DI
2000  PRINT
2010  PRINT "V = VENUS; M = MERCURY; 0 = SUN"
2020  PRINT
2030  M1 = 20 + .8 *  VAL ( LEFT$ (E$(0),5)) / 2
2040  PRINT  TAB( M1); RIGHT$ (E$(0),1);
2050  M2 = 20 + .8 *  VAL ( LEFT$ (E$(1),5)) / 2
2060  PRINT  TAB( M2); RIGHT$ (E$(1),1);
2070  M3 = 20 + .8 *  VAL ( LEFT$ (E$(2),5)) / 2
2080  PRINT  TAB( M3); RIGHT$ (E$(2),1)
2090  PRINT : PRINT
2100  RETURN
2110  HOME : REM  ROUTINE FOR ELONGATIONS
2120  PRINT : PRINT : PRINT
2130  PRINT "SELECT MERCURY  (1)"
2140  INPUT "      OR VENUS  (2) ";E$
2150  IF E$ = "1" THEN  GOSUB 2170: GOTO 2430
2160  GOSUB 2620
2170  HOME : GOSUB 1200: REM  GET DATE
2180  FL = 0
```

MVENC *(continued)*

```
2190   GOSUB 1360: REM   CALC DAYS FROM EPOCH
2200   GOSUB 930: REM   CALC FOR EARTH AND PLANETS
2210 VW$ =  STR$ (VW * 57.29578)
2220   HOME
2230   PRINT
2240   PRINT YI;" ";MI;" ";DI
2250   PRINT : PRINT : PRINT : PRINT : PRINT
2260 MW$ =  STR$ (MW * 57.29578): GOSUB 1790: GOSUB 2030
2270   PRINT : PRINT : PRINT
2280   PRINT "M = MERCURY, V = VENUS, O = SUN"
2290 DS = MW * 57.29578
2300   IF FL = 0 AND  ABS (DS) < 20 THEN D = D + 5: GOTO 2350
2310 CK =  ABS (DS) -  ABS (D2)
2320   IF CK < .05 AND CK >  - .05 THEN 2400
2330   IF  ABS (DS) < 15 THEN D = D + 5: GOTO 2350
2340   IF CK < .1 OR CK >  - .1 THEN D = D + 1: GOTO 2350
2350   IF D > 30 THEN M = M + 1:D = D - 30
2360   IF M > 12 THEN M = 1:Y = Y + 1
2370 D2 = DS:FL = 1
2380   GOSUB 1340
2390   GOTO 2190
2400   HOME : PRINT : PRINT : PRINT
2410   PRINT "NEXT ELONGATION OF MERCURY ";Y;" ";M;" ";D
2420   PRINT : PRINT
2430   IF MW < 0 THEN P$ = "EAST OF SUN (AN EVENING STAR)": GOTO 2450
2440 P$ = "WEST OF SUN (A MORNING STAR)"
2450   PRINT "AND THE PLANET IS...."
2460   PRINT
2470   PRINT  TAB( 5);P$
2480   GOSUB 1980
2490   INPUT "DO YOU WANT MORE DETAILS Y/N? ";A$
2500   IF A$ = "N" THEN  HOME : PRINT : GOTO 2530
2510   IF A$ = "Y" THEN  HOME : PRINT : GOSUB 440
2520   PRINT : GOSUB 1980: PRINT
2530   INPUT "DO YOU WANT ANOTHER DATE Y/N? ";A$
2540   IF E$ = "2" AND A$ = "Y" THEN GOTO 2620
2550   IF E$ = "1" AND A$ = "Y" THEN GOTO 2170
2560   IF A$ <  > "N" THEN  PRINT "INVALID RESPONSE": PRINT : GOTO 2530
2570   PRINT : PRINT
2580   INPUT "DO YOU WANT ANOTHER PLANET Y/N? ";A$
2590   IF A$ = "Y" THEN  GOTO 2110
2600   IF A$ <  > "N" THEN  PRINT "INVALID RESPONSE": PRINT : GOTO 2580
2610   GOTO 880
2620   HOME : GOSUB 1200: REM   GET DATE
2630 FL = 0
2640   GOSUB 1360: REM   CALC DAYS FROM EPOCH
2650   GOSUB 930: REM   CALC FOR EARTH AND PLANETS
2660   HOME
2670 MW$ =  STR$ (MW * 57.29578)
2680   PRINT
2690   PRINT YI;" ";MI;" ";DI
2700   PRINT : PRINT : PRINT : PRINT : PRINT
2710 VW$ =  STR$ (VW * 57.29578): GOSUB 1790: GOSUB 2030
2720   PRINT : PRINT : PRINT : PRINT
2730   PRINT "V= VENUS, M = MERCURY, S = SUN"
```

MVENC *(continued)*

```
2740 VS = VW * 57.29578
2750  IF FL = 0 AND  ABS (VS) < 45 THEN D = D + 20: GOTO 2800
2760 CK =  ABS (VS) -  ABS (V2)
2770  IF CK < .05 AND CK >  - .05 THEN 2850
2780  IF  ABS (VS) < 40 THEN D = D + 10: GOTO 2800
2790  IF CK < .1 OR CK >  - .1 THEN D = D + 1: GOTO 2800
2800  IF D > 30 THEN M = M + 1:D = D - 30
2810  IF M > 12 THEN M = 1:Y = Y + 1
2820 V2 = VS:FL = 1
2830  GOSUB 1340
2840  GOTO 2640
2850  HOME : PRINT : PRINT : PRINT
2860  PRINT "NEXT ELONGATION OF VENUS IS ";Y;" ";M;" ";D
2870 MW = VW: PRINT : GOTO 2430
2880  REM  SUB FOR PHASES
2890 PH = 180 - PH * 57.2958
2900  IF PH > 150 THEN PH$ = "THIN CRESCENT"
2910  IF PH > 120 AND PH < 151 THEN PH$ = "FAT CRESCENT"
2920  IF PH > 70 AND PH < 121 THEN PH$ = "NEAR HALF"
2930  IF PH > 29 AND PH < 71 THEN PH$ = "GIBBOUS"
2940  IF PH < 21 THEN PH$ = "NEAR FULL"
2950 PH =  INT (180 - PH)
2960  RETURN
```

```
┌─ MERVE ─────────────────────────────────────────────────────────┐
│                                                                   │
│  10  HOME                                                         │
│  20  DEF  FN ACO(X) =  -  ATN (X / SQR ( - X * X + 1)) + 1.570793 │
│  30  P1 = 3.14159:P2 = 6.28318                                    │
│  40  PRINT : PRINT                                                │
│  50  PRINT  TAB( 5)"----------------------------"                 │
│  60  PRINT  TAB( 5)"I     MERCURY AND VENUS      I"               │
│  70  PRINT  TAB( 5)"----------------------------"                 │
│  80  PRINT : PRINT                                                │
│  90  PRINT  TAB( 11)"AN ASTRONOMY PROGRAM": PRINT : PRINT         │
│  100  PRINT  TAB( 10)"BY ERIC BURGESS F.R.A.S.": PRINT : PRINT    │
│  110  PRINT  TAB( 11)"ALL RIGHTS RESERVED BY"                     │
│  120  PRINT  TAB( 11)"S $ T SOFTWARE SERVICE"                     │
│  130  PRINT : PRINT                                               │
│  140  PRINT  TAB( 13)"JAN. 82 VERSION": PRINT : PRINT             │
│  150  PRINT  TAB( 5)"";                                           │
│  160  INPUT "DO YOU WANT INSTRUCTIONS? Y/N ";A$                   │
│  170  IF A$ = "N" THEN  HOME : PRINT : PRINT : PRINT : GOTO 320   │
│  180  IF A$ < > "Y" THEN  PRINT "INVALID RESPONSE": PRINT : GOTO 160 │
│  190  HOME : PRINT : PRINT : PRINT                                │
│  200  PRINT : PRINT : PRINT                                       │
│  210  PRINT  TAB( 5)"THIS PROGRAM PLOTS THE POSITIONS"            │
│  220  PRINT  TAB( 5)"OF MERCURY AND VENUS RELATIVE"               │
│  230  PRINT  TAB( 5)"TO THE SUN AT INTERVALS CHOSEN BY"           │
│  240  PRINT  TAB( 5)"YOU IN UNITS OF DAYS FOR AS MANY"            │
│  250  PRINT  TAB( 5)"INTERVALS AS YOU REQUEST."                   │
│  260  PRINT : PRINT                                               │
│  270  PRINT  TAB( 5)"YOU CAN THEN ASK TO SEE DATA ABOUT"          │
│  280  PRINT  TAB( 5)"EITHER OR BOTH PLANETS, SUCH AS"             │
│  290  PRINT  TAB( 5)"DISTANCE, APPARENT SIZE,& PHASE."            │
│  300  PRINT : PRINT : PRINT                                       │
│  310  INPUT "TO CONTINUE PRESS RETURN ";A$                        │
│  320  FL = 0                                                      │
│  330  HOME : PRINT : PRINT : PRINT                                │
│  340  F9 = 0:NT = 1                                               │
│  350  INPUT "SELECT INTERVAL (DAYS) ";ND                          │
│  360  PRINT : PRINT : PRINT                                       │
│  370  INPUT "SCROLLED DISPLAY? Y/N ";FL$                          │
│  380  IF FL$ = "N" THEN F9 = 1                                    │
│  390  PRINT : PRINT                                               │
│  400  INPUT "HOW MANY INTERVALS? ";NS                             │
│  410  GOTO 2200                                                   │
│  420  GOSUB 990: REM  GET DISTANCE FROM EARTH                     │
│  430  HOME : PRINT : PRINT                                        │
│  440  GOSUB 450: GOTO 930                                         │
│  450  HOME : PRINT : PRINT                                        │
│  460  MD$ =  LEFT$ ( STR$ (6.68 / MQ),5)                          │
│  470  VD$ =  LEFT$ ( STR$ (16.82 / VQ),5)                         │
│  480  PRINT "FOR YEAR ";Y;" MONTH ";M;" DAY ";D: PRINT            │
│  490  PRINT "------------------------": PRINT                     │
│  500  MQ$ =  LEFT$ ( STR$ (MQ),4)                                 │
│  510  VQ$ =  LEFT$ ( STR$ (VQ),4)                                 │
│  520  IF E$ = "2" THEN  GOTO 560                                  │
│  530  PRINT "DISTANCE OF MERCURY IS ";MQ$;" A.U."                 │
│  540  PRINT  TAB( 15)"OR "; VAL ( LEFT$ ( STR$ (MQ * 92.9),5));" MILLION │
│       MILES"                                                      │
│                                                                   │
└───────────────────────────────────────────────────────────────┘
```

─────── **MERVE** *(continued)* ───────

```
550   IF E$ = "1" THEN  GOTO 580
560   PRINT "DISTANCE OF VENUS IS ";VQ$;" A.U."
570   PRINT  TAB( 15)"OR "; VAL ( LEFT$ ( STR$ (VQ * 92.9),5));" MILLION
      MILES"
580   PRINT
590   MW$ =  LEFT$ ( STR$ (MW * 57.29578),4)
600   VW$ =  LEFT$ ( STR$ (VW * 57.29578),4)
610   IF E$ = "2" THEN  GOTO 670
620   PRINT "MERCURY ANG.DIST. FROM SUN IS ";MW$;
630   PRINT " DEG"
640   GOSUB 2710
650   PRINT  TAB( 2)"(LOCATED ";AP$;" FOR VIEWING)"
660   IF E$ = "1" THEN  PRINT : GOTO 700
670   PRINT "ANG. DIST. OF VENUS FROM SUN IS ";VW$;
680   PRINT " DEG"
690   IF E$ = "3" OR  VAL (SD$) = 3 THEN  INPUT "PRESS RETURN TO
      CONTINUE ";A$: HOME : PRINT
700   IF E$ = "2" THEN  GOTO 760
710   PRINT "ANG. DIAM. OF MERCURY IS ";MD$;
720   PRINT " ARCSEC"
730   PRINT " (AT INFER.CONJ. DIAM. IS 12 ARCSEC)"
740   IF E$ = "1" THEN  GOTO 790
750   PRINT
760   PRINT "ANG. DIAM. OF VENUS IS ";VD$;
770   PRINT " ARCSEC"
780   PRINT "  (AT INFER. CONJ. DIAM. IS 61 ARCSEC)"
790   PRINT
800   PH =  ABS (MW) +  ABS (MZ): GOSUB 2520
810   IF E$ = "2" THEN  GOTO 860
820   PRINT "MERCURY'S PHASE IS ";PH$
830   PRINT  TAB( 5)PH;" DEG. ILLUMINATED"
840   PRINT
850   IF E$ = "1" THEN  GOTO 890
860   PH =  ABS (VW) +  ABS (ZV): GOSUB 2520
870   PRINT "PHASE OF VENUS IS ";PH$
880   PRINT  TAB( 5);PH;" DEG. ILLUMINATED"
890   PRINT
900   GOSUB 1920
910   GOSUB 2100
920   RETURN
930   PRINT : INPUT "WANT ANOTHER DATE? ";A$
940   IF A$ = "N" THEN  HOME : GOTO 980
950   IF A$ <  > "Y" THEN  PRINT "INVALID RESPONSE": PRINT : GOTO 930
960   HOME : PRINT : PRINT : PRINT
970   GOTO 320
980   END
990   REM   FIND ANGULAR DIAM
1000   HOME : PRINT : PRINT : PRINT
1010   GOSUB 1290: REM   ENTER DATE
1020   GOSUB 1480: REM   CALC DAYS FROM EPOCH
1030   REM   CALC DIST. FROM EARTH
1040   REM   AND ANG. DIST. FROM SUN
1050   AE = NI * .017202 + 1.74022
1060   IF AE > P2 THEN AE = ((AE / P2) -  INT (AE / P2)) * P2
1070   IF AE < 0 THEN AE = AE + P2: GOTO 1070
```

```
-- MERVE (continued) --

1080 CE = .032044 *  SIN (AE - 1.78547)
1090 AE = AE + CE
1100  IF AE > P2 THEN AE = AE - P2
1110  IF AE < 0 THEN AE = AE + P2: GOTO 1110
1120 DE = 1 + .017 *  SIN (AE - 3.33926)
1130  GOSUB 1590: REM  CALC MERCURY'S POSITION
1140 MZ = AE - MA
1150 ZM = MZ
1160  IF  ABS (MZ) > P1 AND MZ < 0 THEN MZ = MZ + P2
1170  IF  ABS (MZ) > P1 AND MZ > 0 THEN MZ = MZ - P2
1180  REM  CALC DIST FROM EARTH
1190 MQ =  SQR (MD ^ 2 + DE ^ 2 - 2 * MD * DE *  COS (MZ))
1200 MP = (MD + DE + MQ) / 2
1210  REM  CALC ANG DIST. FROM SUN
1220 MV =  SQR ((MP * (MP - MD)) / (DE * MQ))
1230 MW = 2 *  FN ACO(MV)
1240  IF  ABS (ZM) > P1 AND ZM < 0 THEN ZM = ZM + P2
1250  IF  ABS (ZM) > P1 AND ZM > 0 THEN ZM = ZM - P2
1260  IF ZM > 0 THEN MW =  - MW
1270  GOSUB 1690: REM  CALC VENUS'S POSITION
1280  RETURN
1290  REM  ENTER DATE
1300  PRINT : PRINT
1310  PRINT "ENTER THE DATE": PRINT
1320  INPUT "   THE YEAR ";Y$:Y = VAL (Y$)
1330  IF Y = 0 THEN  PRINT "INVALID RESPONSE": PRINT : GOTO 1320
1340  IF Y > 1800 THEN  GOTO 1390
1350  PRINT " ";Y;" THE CORRECT YEAR? ": INPUT A$
1360  IF A$ = "Y" THEN  GOTO 1390
1370  IF A$ < > "N" THEN  PRINT "INVALID RESPONSE": PRINT : GOTO 1350
1380  IF A$ = "N" THEN  GOTO 1320
1390  PRINT : INPUT "   THE MONTH ";M$:M =  VAL (M$)
1400  IF M = 0 OR M > 12 THEN  PRINT "INVALID RESPONSE": PRINT : GOTO 1390
1410  PRINT : INPUT "   THE DAY ";D$:D =  VAL (D$)
1420  IF D = 0 OR D > 31 THEN  PRINT "INVALID RESPONSE": PRINT : GOTO 1410
1430  IF M = 2 AND D > 29 THEN  PRINT "INVALID RESPONSE": PRINT : GOTO 1410
1440  HOME
1450  REM  STORE INITIAL DATE
1460 YI = Y:MI = M:DI = D
1470  RETURN
1480  REM  CALC DAYS FROM EPOCH
1490 DG = 365 * Y + D + ((M - 1) * 31)
1500  IF M > = 3 THEN  GOTO 1540
1510  REM  CALC FOR JAN OR FEB
1520 DG = DG +  INT ((Y - 1) / 4) -  INT ((.75) *  INT ((Y - 1) / 100 + 1))
1530  GOTO 1560
1540  REM  CALC FOR MAR THRU DEC
1550 DG = DG -  INT (M * .4 + 2.3) +  INT (Y / 4)
      -  INT ((.75) *  INT ((Y / 100) + 1))
1560 NI = DG - 715875
1570  RETURN
1580  GOTO 420
1590  REM  CALC MERCURY'S POSITION
1600 MA = NI * .071425 + 3.8494
1610  IF MA > P2 THEN MA = ((MA / P2) -  INT (MA / P2)) * P2
```

────────────────────── **MERVE** (continued) ──────────────────────

```
1620  IF MA < 0 THEN MA = MA + P2: GOTO 1620
1630 MC = .388301 *  SIN (MA - 1.34041)
1640 MA = MA + MC
1650  IF MA > P2 THEN MA = MA - P2
1660  IF MA < 0 THEN MA = MA + P2: GOTO 1660
1670 MD = .3871 + .079744 *  SIN (MA - 2.73514)
1680  RETURN
1690  REM   CALC VENUS'S POSITION
1700 VA = NI * .027962 + 3.02812
1710  IF VA > P2 THEN VA = ((VA / P2) -  INT (VA / P2)) * P2
1720  IF VA < 0 THEN VA = VA + P2: GOTO 1720
1730 VC = .013195 *  SIN (VA - 2.28638)
1740 VA = VA + VC
1750  IF VA > P2 THEN VA = VA - P2
1760  IF VA < 0 THEN VA = VA + P2: GOTO 1760
1770 VD = .7233 + .00506 *  SIN (VA - 3.85017)
1780 ZV = AE - VA
1790 VZ = ZV
1800  IF  ABS (ZV) > P1 AND ZV < 0 THEN ZV = ZV + P2
1810  IF  ABS (ZV) > P1 AND ZV > 0 THEN ZV = ZV - P2
1820  REM   CALC DIST FROM EARTH
1830 VQ =  SQR (VD ^ 2 + DE ^ 2 - 2 * VD * DE *  COS (ZV))
1840 VP = (VD + DE + VQ) / 2
1850  REM   CALC ANG DIST FROM SUN
1860 VV =  SQR ((VP * (VP - VD)) / (DE * VQ))
1870 VW = 2 *  FN ACO(VV)
1880  IF  ABS (VZ) > P1 AND VZ < 0 THEN VZ = VZ + P2
1890  IF  ABS (VZ) > P1 AND VZ > 0 THEN VZ = VZ - P2
1900  IF VZ > 0 THEN VW = - VW
1910  RETURN
1920  REM   SHOW SUN AND PLANETS
1930 Z$(0) = MW$ + "M":Z$(1) = VW$ + "V"
1940 Z$(2) =  STR$ (0) + "O"
1950  FOR I = 0 TO 2
1960 U$(I) = "999"
1970  NEXT I
1980 K = 0
1990  FOR I = 0 TO 2
2000  IF Z$(I) = "999" THEN  GOTO 2080
2010  FOR J = I TO 2
2020  IF Z$(J) = "999" THEN  GOTO 2040
2030  IF  VAL (Z$(J)) <  VAL (Z$(I)) THEN I = J: GOTO 2040
2040  NEXT J
2050 U$(K) = Z$(I):Z$(I) = "999"
2060 K = K + 1
2070 I = - 1
2080  NEXT I
2090  RETURN
2100  PRINT "VISUAL CONFIG. RELATIVE TO SUN"
2110  PRINT
2120 M1 = 20 + .8 *  VAL ( LEFT$ (U$(0),5)) / 2
2130  PRINT  TAB( M1); RIGHT$ (U$(0),1);
2140 M2 = 20 + .8 *  VAL ( LEFT$ (U$(1),5)) / 2
2150  PRINT  TAB( M2); RIGHT$ (U$(1),1);
2160 M3 = 20 + .8 *  VAL ( LEFT$ (U$(2),5)) / 2
```

MERVE *(continued)*

```
2170   PRINT  TAB( M3); RIGHT$ (U$(2),1)
2180   PRINT
2190   RETURN
2200   HOME : REM   ROUTINE FOR ELONGATIONS
2210   GOSUB 2220
2220   HOME : GOSUB 1290: REM   GET DATE
2230   FL = 0
2240   GOSUB 1480: REM   CALC DAYS FROM EPOCH
2250   :
2260   GOSUB 1030:VW$ =  STR$ (VW * 57.29578)
2270   MW$ =  STR$ (MW * 57.29578): GOSUB 1920
2280   IF F9 = 1 THEN  HOME : PRINT : PRINT : PRINT
2290   PRINT "DISPLAY ";NT;" IS DATE ";YI;" ";MI;" ";DI: PRINT : GOSUB 2110
2300   IF NT = NS THEN  PRINT : PRINT
2310   IF NT = NS THEN  GOSUB 2800
2320   IF NT = NS THEN  HOME : GOTO 2390
2330   PRINT
2340   D = D + ND
2350   GOSUB 2610
2360   YI = Y:DI = D:MI = M
2370   FL = 1:NT = NT + 1
2380   GOTO 2240
2390   GOSUB 450
2400   IF SD$ = "3" THEN E$ = "":SD$ = "": PRINT : GOTO 2480
2410   IF OP = 1 THEN OP = 0: GOTO 2480
2420   INPUT "WANT THE OTHER PLANET? Y/N ";A$
2430   IF A$ = "Y" AND E$ = "1" THEN E$ = "2":A$ = "": GOSUB 450: GOTO 2460
2440   IF A$ = "Y" AND E$ = "2" THEN E$ = "1":A$ = "": GOSUB 450: GOTO 2460
2450   IF A$ < > "N" THEN  PRINT "INVALID RESPONSE": PRINT : GOTO 2420
2460   PRINT : INPUT "WANT BOTH PLANETS? Y/N ";A$
2470   IF A$ = "Y" THEN E$ = "3":OP = 1: GOSUB 450:E$ = "":OP = 0
2480   INPUT "DO YOU WANT ANOTHER DATE? Y/N ";A$
2490   IF A$ = "Y" THEN  GOTO 320
2500   IF A$ < > "N" THEN  PRINT "INVALID RESPONSE": PRINT : GOTO 2480
2510   HOME : GOTO 980
2520   REM   SUB FOR PHASES
2530   PH = 180 - PH * 57.2958
2540   IF PH > 150 THEN PH$ = "THIN CRESCENT"
2550   IF PH > 120 AND PH < 151 THEN PH$ = "FAT CRESCENT"
2560   IF PH > 70 AND PH < 121 THEN PH$ = "NEAR HALF"
2570   IF PH > 29 AND PH < 71 THEN PH$ = "GIBBOUS"
2580   IF PH < 30 THEN PH$ = "NEAR FULL"
2590   PH =  INT (180 - PH)
2600   RETURN
2610   REM   SUB FOR MONTH END ADJUSTS
2620   IF Y / 4 -  INT (Y / 4) = 0 AND Y / 100 -  INT (Y / 100) < > 0 THEN
       LY = 1
2630   IF M > 12 THEN M = M - 12:Y = Y + 1
2640   IF LY = 1 AND M = 2 AND D > 29 THEN M = 3:D = D - 29: GOTO 2660
2650   IF LY = 0 AND M = 2 AND D > 28 THEN M = 3:D = D - 28: GOTO 2660
2660   IF D < 31 GOTO 2700
2670   IF (M = 4 OR M = 6 OR M = 9 OR M = 11) THEN M = M + 1:D = D - 30:
       GOTO 2630
2680   IF D > 31 THEN M = M + 1:D = D - 31: GOTO 2630
```

─────── **MERVE** *(continued)* ───────

```
2690   IF M > 12 THEN M = M - 12:Y = Y + 1
2700   RETURN
2710   REM   SUB FOR VIEWING MERCURY
2720 MX =   VAL (MW$)
2730 AP$ = ""
2740   IF MX > 13 AND M > 7 AND M < 12 THEN AP$ = "FAVORABLY": GOTO 2790
2750   IF MX > 18 AND M > 7 AND M < 12 THEN AP$ = "VERY FAVORABLY": GOTO 2790
2760   IF MX <  - 13 AND M > 1 AND M < 7 THEN AP$ = "FAVORABLY": GOTO 2790
2770   IF MX <  - 18 AND M > 1 AND M < 7 THEN AP$ = "VERY FAVORABLY":
       GOTO 2790
2780 AP$ = "UNFAVORABLY"
2790   RETURN
2800   PRINT "SELECT DETAILS FOR DISPLAY"
2810   INPUT "MERCURY (1); VENUS (2); OR BOTH (3) ";SD$
2820   IF   VAL (SD$) = 3 THEN   RETURN
2830 E$ = SD$: RETURN
2840   PRINT "INVALID RESPONSE": PRINT : GOTO 2800
```

Photo Credit: NASA/Jet Propulsion Laboratory

The closest planet to the Sun, Mercury has a Moon-like, heavily-cratered surface, first revealed by the spacecraft Mariner 10. Mercury is so difficult to observe from Earth that prior to the space mission no definite details had been identified on its surface. This photograph shows one hemisphere of the planet with the large impact basin Caloris half illuminated by the Sun. The feature is just below the middle of the curved terminator on the right of the image. The terminator is the boundary between day and night on the planet.

Program 14: PRISE

Times of Rising and Setting of Mercury and Venus before or after the Sun for Any Date_____

Because of the inclination of the ecliptic to the celestial equator, there are times of the terrestrial year when Venus and Mercury are configured more easily for observation; that is, when they rise before the Sun or set after the Sun with sufficient time for observation in a dark sky. In the Northern Hemisphere, spring is the best time to observe either of these planets as an evening star. Fall is the best time to observe them as morning stars. During spring the ecliptic makes a large angle with the western horizon after sunset. In fall it is inclined at a large angle to the eastern horizon before sunrise. Opposite conditions apply for the Southern Hemisphere.

This program calculates for Venus and Mercury the approximate number of minutes between sunset and planet set and between sunrise and planet rise for any given date, longitude, and latitude. It also indicates whether the planet is a morning or an evening star. Used with RISES and MVENC, this program helps you select the best times to observe Venus and the elusive planet Mercury. A typical display generated by this program appears in Figure 14.1.

The listing is set for a given longitude and latitude. You can change this when you key in the program by altering line 80 to your local coordinates. You can also change these coordinates during the running of the program if you wish—to a vacation observing site, for example.

The program calculates the right ascension and declination of the planets and of the Sun. From the right ascension difference (instruction 1820), the difference in declination (instruction 1810), and the latitude of the observing site (LA), it calculates the time differences at the horizon (instructions 1700 through 1830).

The listing for the PRISE program follows.

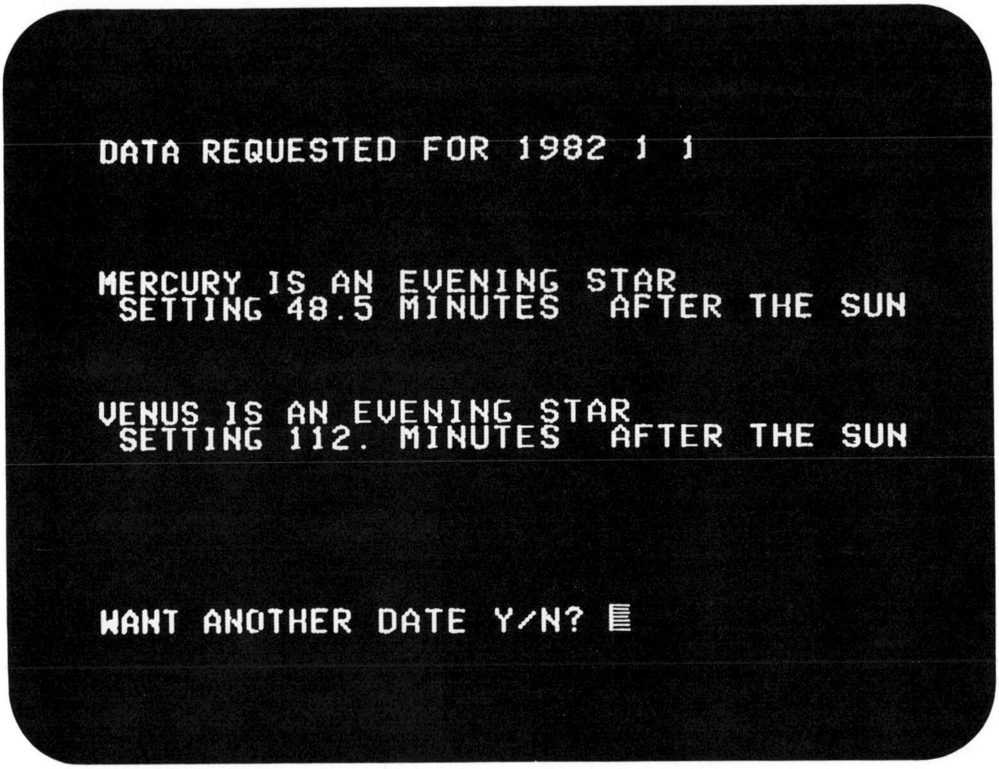

```
DATA REQUESTED FOR 1982 1 1


MERCURY IS AN EVENING STAR
   SETTING 48.5 MINUTES  AFTER THE SUN


VENUS IS AN EVENING STAR
   SETTING 112. MINUTES  AFTER THE SUN



WANT ANOTHER DATE Y/N? ▊
```

For good observation it is important to know how long after the Sun Venus or Mercury sets, or how soon they rise before the Sun. The program PRISE gives this information in the display format shown.

Figure 14.1

```
20   HOME : PRINT : PRINT
30   DEF  FN ASN(X) =  ATN (X /  SQR ( - X * X + 1)
40   DEF  FN ACO(X) =  -  ATN (X /  SQR ( - X * X + 1)) + 1.5707963
50   DEF  FN RAD(X) = .01745328 * (X)
60   DEF  FN DEG(X) = 57.29578 * (X)
70   REM  INSERT LOCAL LONGITUDE AND LATITUDE IN FOLLOWING STATEMENT
80   LO = 122.49:LA = 38.24
90   PRINT : PRINT : PRINT : PRINT
100  PRINT  TAB( 10)"ASTRONOMY PROGRAM": PRINT
110  PRINT  TAB( 9)"-------------------"
120  PRINT  TAB( 9)"I  INNER PLANETS  I"
130  PRINT  TAB( 9)"I TIMES FROM SUN  I"
140  PRINT  TAB( 9)"-------------------"
150  PRINT
160  PRINT  TAB( 7)"BY ERIC BURGESS F.R.A.S."
170  PRINT
180  PRINT  TAB( 7)"ALL RIGHTS RESERVED BY"
190  PRINT  TAB( 7)"S & T SOFTWARE SERVICE"
200  FOR J = 2000 TO 1 STEP  - 1: NEXT J
210  HOME : PRINT : PRINT : PRINT
220  PRINT : PRINT : PRINT : PRINT
230  PRINT "THIS PROGRAM GIVES APPROXIMATE"
240  PRINT "TIME IN MINUTES THAT MERCURY AND"
250  PRINT "VENUS RISE BEFORE THE SUN AS"
260  PRINT "MORNING STARS, OR SET AFTER THE"
270  PRINT "SUN AS EVENING STARS, FOR ANY DATE"
280  PRINT "       WHICH YOU INPUT"
290  PRINT : PRINT
300  PRINT "INITIAL CONDITIONS ARE SET FOR......"
310  PRINT  TAB( 5)"SEBASTOPOL, CA LONG. 122.49"
320  PRINT  TAB( 21)"LAT. 38.24": PRINT
330  PRINT : INPUT "WANT TO CHANGE THEM Y/N? ";A$
340  IF A$ <  > "Y" GOTO 380
350  HOME : PRINT : PRINT : PRINT : PRINT
360  INPUT "    GIVE LATITUDE ";LA
370  INPUT "    AND LONGITUDE ";LO
380  HOME : PRINT : PRINT : PRINT
390  PRINT "ENTER THE DATE": PRINT
400  FL = 2
410  INPUT "  THE YEAR ? ";YD$:Y = VAL (YD$)
420  IF Y = 0 THEN  PRINT "INVALID RESPONSE": PRINT : GOTO 410
430  IF Y > 1800 THEN  GOTO 490
440  PRINT "IS ";Y;" THE CORRECT YEAR? "
450  INPUT ;Y$
460  IF Y$ = "Y" THEN  GOTO 490
470  IF Y$ <  > "N" THEN  PRINT "INVALID RESPONSE": PRINT : GOTO 440
480  IF Y$ = "N" THEN  PRINT : GOTO 410
490  LY = 0
500  IF Y / 4 -  INT (Y / 4) = 0 AND Y / 100 -  INT (Y / 100) <  > 0 THEN
     LY = 1
510  PRINT : INPUT "  THE MONTH? ";MD$:M =  VAL (MD$)
520  IF M = 0 OR M > 12 THEN  PRINT "INVALID RESPONSE": PRINT : GOTO 490
```

PRISE *(continued)*

```
530   PRINT : INPUT "  THE DAY? ";DD$:D =  VAL (DD$)
540   IF D = 0 OR D > 31 THEN  PRINT "INVALID RESPONSE": PRINT : GOTO 530
550   IF M < > 2 THEN  GOTO 580
560   IF LY = 0 AND D > 28 THEN  PRINT "INVALID RESPONSE": PRINT : GOTO 530
570   IF LY = 1 AND D > 29 THEN  PRINT "INVALID RESPONSE": PRINT : GOTO 530
580   DX = D:MX = M:YX = Y
590   D = D + (LO / 15) / 24
600   IF (LY = 1 AND M = 2 AND  INT (D) > 29) THEN M = 3:D = D - 29: GOTO 660
610   IF (LY = 0 AND M = 2 AND  INT (D) > 28) THEN M = 3:D = D - 28: GOTO 660
620   IF D < 31 GOTO 660
630   IF (M = 4 OR M = 6 OR M = 9 OR M = 11) THEN M = M + 1:D = D - 30:
      GOTO 660
640   IF  INT (D) > 31 THEN M = M + 1:D = D - 31
650   IF M = 13 THEN M = 1:Y = Y + 1
660   REM  CALC DAYS TO DATE REQUESTED
670   REM  FROM 1960,1,1 EPOCH
680   DG = 365 * Y + D + (M - 1) * 31
690   IF M > = 3 THEN  GOTO 730
700   REM  CALC FOR JAN OR FEB
710   DG = DG +  INT ((Y - 1) / 4) -  INT ((.75) *  INT ((Y - 1) / 100 + 1))
720   GOTO 750
730   REM  CALC FOR MAR THRU DEC
740   DG = DG -  INT (M * .4 + 2.3) +  INT (Y / 4)
      -  INT ((.75) *  INT ((Y / 100) + 1))
750   NI = DG - 715875
760   GOSUB 1620: REM  GET SIDEREAL TIME
770   REM  INPUT OF ORBIT DATA FOR PLANETS
780   IF F = 1 THEN  GOTO 950
790   RESTORE
800   DIM PD(3,9)
810   FOR YY = 0 TO 2: FOR XX = 0 TO 8
820   READ PD(YY,XX)
830   NEXT XX,YY
840   REM  MERCURY
850   DATA .071422,3.8484,.388301,1.34041,.3871,.07974,2.73514
860   DATA .12223,.836013
870   REM  VENUS
880   DATA .027962,3.02812,.013195,2.28638,.7233,.00506,3.85017
890   DATA .059341,1.33168
900   REM  EARTH
910   DATA .017202,1.74022,.032044,1.78547,1,.017,3.33926
920   DATA 0,0
930   FOR I9 = 1 TO 3: READ P$(I9): NEXT I9
940   DATA  MERCURY,VENUS,SUN
950   F = 1
960   REM  CALC DATA FOR PLANETS
970   HOME
980   PRINT : PRINT : PRINT
990   PRINT "DATA REQUESTED FOR ";YX;" ";MX;" ";DX
1000  PRINT : PRINT : PRINT
1010  PRINT :I = 1
1020  FOR J = 0 TO 2: GOSUB 1230
1030  A(I) = A:D(I) = DS:L(I) = L
1040  I = I + 1: NEXT
1050  FOR I = 1 TO 2
```

PRISE (continued)

```
1060   REM   SKIP EARTH
1070   IF I = 3 THEN   NEXT
1080   GOSUB 1370
1090   Q(I) = Q:X(I) = X:R(I) = R:V(I) = V
1100   NEXT
1110   FOR K = 1 TO 2
1120   I = K:A(I) =  FN DEG(A(I))
1130   GOSUB 1710
1140   NEXT
1150   PRINT : PRINT : PRINT
1160   INPUT "WANT ANOTHER DATE Y/N? ";A$
1170   IF A$ = "Y" THEN   GOTO 380
1180   IF A$ <  > "N" THEN   PRINT "INVALID RESPONSE": PRINT : GOTO 1160
1190   HOME
1200   END
1210   REM   CALC POSITION DATA
1220   REM   CALC HELIOCENTRIC LONGITUDE
1230   A = NI * PD(J,0) + PD(J,1)
1240   IF A > 6.28318 THEN A = ((A / 6.28318) -  INT (A / 6.28318)) * 6.28318
1250   IF A < 0 THEN A = A + 6.28318: GOTO 1250
1260 C = PD(J,2) *  SIN (A - PD(J,3))
1270 A = A + C
1280   IF A > 6.28318 THEN A = A - 6.28318
1290   IF A < 0 THEN A = A + 6.28318: GOTO 1290
1300   REM   CALC PLANET DIST FROM SUN
1310   DS = PD(J,4) + PD(J,5) *  SIN (A - PD(J,6))
1320   REM   CALC DIST FROM ECLIPTIC
1330   L = PD(J,7) *  SIN (A - PD(J,8))
1340   RETURN
1350   REM   CALC PLANETARY DATA
1360   REM   CALC ANG DISTANCE FROM EARTH
1370 Z = A(3) - A(I)
1380   IF   ABS (Z) > 3.14159 AND Z < 0 THEN Z = Z + 6.28318
1390   IF   ABS (Z) > 3.14159 AND Z > 0 THEN Z = Z - 6.28318
1400   REM    CALC DISTANCE FROM EARTH
1410   Q =  SQR (D(I) ^ 2 + D(3) ^ 2 - 2 * D(I) * D(3) *  COS (Z))
1420   REM   CALC ANG DISTANCE FROM SUN
1430   P = (D(I) + D(3) + Q) / 2
1440   X = 2 *  FN ACO( SQR (((P * (P - D(I))) / (D(3) * Q))))
1450   REM   CALC RA OF PLANET
1460   IF Z < 0 THEN R =  FN DEG(A(3) + 3.14159 - X) / 15
1470   IF Z > 0 THEN R =  FN DEG(A(3) + 3.14159 + X) / 15
1480   IF R > 24 THEN R = R - 24: GOTO 1480
1490   IF R <  - 24 THEN R = R + 24: GOTO 1490
1500   IF R < 0 THEN R = R + 24: GOTO 1500
1510   REM   CALC DECLINATION
1520   IF Z < 0 THEN V =  SIN (A(3) + 3.14159 - X) * 23.44194 + FN DEG(L(I))
1530   IF Z > 0 THEN V =  SIN (A(3) + 3.14159 + X) * 23.44194 + FN DEG(L(I))
1540 X =  FN DEG(X)
1550 R(3) =  FN DEG(A(3) + 3.14159) / 15
1560   IF R(3) > 24 THEN R(3) = R(3) - 24: GOTO 1560
1570   IF R(3) <  - 24 THEN R(3) = R(3) + 24: GOTO 1570
1580   IF R(3) < 0 THEN R(3) = R(3) + 24
1590 V(3) =  SIN (A(3) + 3.14159) * 23.44194
1600   RETURN
```

```
   ┌─PRISE (continued)──────────────────────────────────────────────────┐
   │                                                                     │
   │ 1610   RETURN                                                       │
   │ 1620   REM   CALC SIDEREAL TIME                                     │
   │ 1630 GC = 11.927485                                                 │
   │ 1640 TC = .065711                                                   │
   │ 1650 T2 = TC * (NI - 7020) + GC                                     │
   │ 1660   IF T2 > 24 THEN T2 = T2 - 24: GOTO 1660                      │
   │ 1670   IF T2 <  - 24 THEN T2 = T2 + 24: GOTO 1670                   │
   │ 1680   IF T2 < 0 THEN T2 = T2 + 24                                  │
   │ 1690   RETURN                                                       │
   │ 1700   REM   GET TIME DIFFERENCES                                   │
   │ 1710 A1 =  SIN ( FN RAD(V(3))) * SIN ( FN RAD(LA))                  │
   │ 1720   IF A1 < 0 THEN A1 =  ABS (A1): GOTO 1740                     │
   │ 1730   IF A1 > 0 THEN A1 =  - A1                                    │
   │ 1740 A1 = (A1 - .01454) /  COS ( FN RAD(LA))                        │
   │ 1750 A1 = A1 /  COS ( FN RAD(V(3)))                                 │
   │ 1760 B1 =  SIN ( FN RAD(D(I))) * SIN ( FN RAD(LA))                  │
   │ 1770   IF B1 < 0 THEN B1 =  ABS (B1): GOTO 1790                     │
   │ 1780   IF B1 > 0 THEN B1 =  - B1                                    │
   │ 1790 B1 = (B1 - .00989) /  COS ( FN RAD(LA))                        │
   │ 1800 B1 = B1 /  COS ( FN RAD(V(I)))                                 │
   │ 1810 C1 = 4 * (A1 - B1)                                             │
   │ 1820 D1 = 60 * (R(3) - R(I))                                        │
   │ 1830 M1 = C1 + D1                                                   │
   │ 1840   REM   ADJUST FOR RA PASSING ZERO                            │
   │ 1850   IF M1 > 720 THEN M1 = M1 - 1440: GOTO 1870                   │
   │ 1860   IF M1 <  - 720 THEN M1 = M1 + 1440                           │
   │ 1870   IF M1 < 0 THEN AS$ = "AN EVENING STAR ":MS$ = "SETTING":BA$ = " AFTER" │
   │ 1880   IF M1 > 0 THEN AS$ = " A MORNING STAR ":MS$ = "RISING ":BA$ = " BEFORE" │
   │ 1890 M1 =  ABS (M1)                                                 │
   │ 1900 M1$ =  LEFT$ ( STR$ (M1),4)                                    │
   │ 1910   PRINT P$(I);" IS ";AS$                                       │
   │ 1920   PRINT " ";MS$;" ";M1$;" MINUTES ";BA$;" THE SUN"             │
   │ 1930   PRINT : PRINT : PRINT                                        │
   │ 1940   RETURN                                                       │
   │                                                                     │
   └─────────────────────────────────────────────────────────────────────┘
```

When inspected at close hand from spacecraft, all the planets of the solar system have proved to be quite different from what was anticipated. Mars has intriguing features, many of which cannot be easily explained. This picture from the Viking Orbiter shows an interesting area of hills and valleys with streamlike channels at the bottom of many of the valleys. The region has very few craters and is believed to be geologically young. It is located in the Nilosyrtis region of the planet.

Program 15: RISES

Times of Rising, Transit, and Setting of Planets, Sun, and Moon for Any Date_____

Observing the risings, transits, and settings of the Sun, Moon, and planets occupied generations of ancient astronomers. They painstakingly sighted on these objects through window slits, by stone monuments and obelisks, and in more primitive societies, by arrangements of sticks. Because of the inclination of Earth's orbit and the complex motions of the planets, such observations were not easy. It took many years to accumulate sufficient data for the priest-astronomers to make realistic predictions. Examples of the ancient horizon-scanning observatories are found worldwide, indicating the importance many civilizations attached to knowing the motions of the heavenly bodies. Their observations had religious, astrological, and agricultural significance.

Today a personal computer and this program can give you such data with reasonable accuracy almost instantaneously. The only things you have to do are provide it with a date and choose the Sun, the Moon, or a planet. This program calculates the approximate time of rising, transit, and setting of any of these bodies for any date, at any latitude and longitude. Times of actual visibility after rising and before setting depend on local horizon and atmospheric conditions as well as on the rate of rising and setting—that is, the body's position on the ecliptic relative to the latitude of the observer. The Sun and Moon, because of their brightness, are often visible sooner after rising and closer to setting than are the planets. A typical display generated by this program is shown in Figure 15.1.

The program loads data on the planetary orbits, as did Program 11, and calculates for each planet its right ascension and declination at the requested date. It displays these data for the selected planet together with its distance from Earth and its angular distance from the Sun. The program uses the calculated sidereal time for the chosen date together with the latitude to calculate the rise, transit, and set times of the first point of Aries (subroutine 2480). From this the program calculates the local times when that point in the celestial sphere with the right ascension and declination of the planet rises above the horizon, culminates, and sets below the horizon at the observer's location.

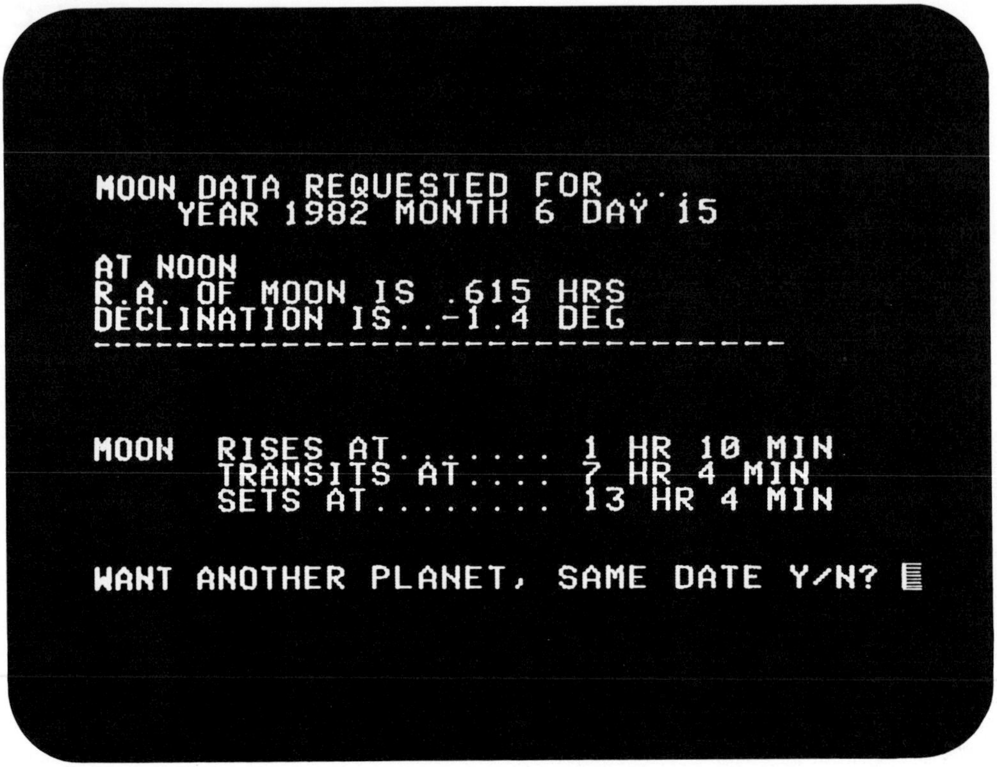

```
MOON DATA REQUESTED FOR
        YEAR 1982 MONTH 6 DAY 15

AT NOON
R.A. OF MOON IS .615 HRS
DECLINATION IS..-1.4 DEG
- - - - - - - - - - - - - - - - - - - - - - - - - - -


MOON    RISES AT.......  1 HR 10 MIN
        TRANSITS AT....  7 HR 4 MIN
        SETS AT........  13 HR 4 MIN

WANT ANOTHER PLANET, SAME DATE Y/N? ▤
```

The program RISES displays the times of rising and setting for a planet, the Sun, or the Moon on a selected date.

Figure 15.1

The basic equations are:

Sidereal Rise Time = RA − (1/15) × arccos[tan(lat) × tan(dec)]

Sidereal Transit Time = RA

Sidereal Set Time = RA + (1/15) × arccos[tan(lat) × tan(dec)]

Starting at instruction 2690, the program calculates the right ascension and declination of the Moon. It calculates the Moon's rise, culmination, and set times by using the 2480 subroutine.

The listing of the RISES program follows.

RISES

```
10    HOME : PRINT : PRINT
20    PRINT : PRINT : PRINT
30    DIM PD(9,9): DIM IP(10): DIM A(9): DIM DS(9): DIM L(9)
40    DIM Q(9): DIM X(9): DIM R(9): DIM V(9)
50    DEF  FN ASN(X) =  ATN (X / SQR ( - X * X + 1))
60    DEF  FN ACO(X) =  -  ATN (X / SQR ( - X * X + 1)) + 1.5707963
70    DEF  FN RAD(X) = .01745328 * (X)
80    DEF  FN DEG(X) = 57.299578 * (X)
90    REM   INSERT YOUR LOCAL LONGITUDE AND LATITUDE IN NEXT LINE
100   LO = 122.49:LA = 38.24
110   PRINT  TAB( 8)"ASTRONOMY PROGRAM"
120   PRINT  TAB( 5)"----------------------"
130   PRINT  TAB( 5)"I  RISE AND SET TIMES  I"
140   PRINT  TAB( 5)"----------------------"
150   PRINT
160   PRINT  TAB( 5)"BY ERIC BURGESS F.R.A.S."
170   PRINT : PRINT
180   PRINT  TAB( 5)"ALL RIGHTS RESERVED BY"
190   PRINT  TAB( 5)"S & T SOFTWARE SERVICE"
200   PRINT : PRINT
210   PRINT  TAB( 7)"VERSION APR.1982"
220   FOR J = 3000 TO 1 STEP  - 1: NEXT J
230   HOME : PRINT : PRINT : PRINT
240   PRINT : PRINT : PRINT
250   PRINT "THIS PROGRAM GIVES APPROXIMATE TIMES"
260   PRINT "OF THE RISING, TRANSIT, AND SETTING"
270   PRINT "OF A SELECTED PLANET, OR THE SUN"
280   PRINT  TAB( 13)"OR THE MOON"
290   PRINT : PRINT : PRINT
300   PRINT "INITIAL CONDITIONS ARE SET FOR"
310   PRINT
320   REM   CHANGE NEXT TWO LINES TO YOUR LOCAL LAT AND LONG.
330   PRINT  TAB( 5)"SEBASTOPOL, CALIFORNIA"
340   PRINT  TAB( 5)"LONGITUDE 122.49; LATITUDE 38.24"
350   PRINT
360   INPUT "DO YOU WANT TO CHANGE THEM Y/N? ";A$
370   IF A$ < > "Y" GOTO 430
380   HOME : PRINT : PRINT : PRINT
390   PRINT : PRINT
400   INPUT "GIVE LATITUDE ";LA
410   PRINT
420   INPUT "AND LONGITUDE ";LO
430   HOME : PRINT : PRINT : PRINT : PRINT
440   PRINT "ENTER THE DATE": PRINT
450   FL = 2
460   INPUT "THE YEAR ";YD$:Y =  VAL (YD$)
470   IF Y = 0 THEN  PRINT "INVALID RESPONSE:PRINT:GOTO410
480   IF Y > 1800 THEN  GOTO 540
490   PRINT "IS ";Y;" THE CORRECT YEAR? "
500   INPUT "Y/N ";Y$
510   IF Y$ = "Y" THEN  GOTO 540
520   IF Y$ = "N" THEN  PRINT : GOTO 460
530   PRINT "INVALID RESPONSE": PRINT : GOTO 490
```

RISES *(continued)*

```
540  LY = 0
550  IF Y / 4 -  INT (Y / 4) = 0 AND Y / 100 -  INT (Y / 100) <  > 0 THEN
     LY = 1
560  PRINT
570  INPUT "THE MONTH ";MD$:M =  VAL (MD$)
580  IF M = 0 OR M > 12 THEN  PRINT "INVALID RESPONSE": PRINT : GOTO 570
590  PRINT
600  INPUT "THE DAY ";DD$:D =  VAL (DD$)
610  IF D = 0 OR D > 31 THEN  PRINT "INVALID RESPONSE": PRINT : GOTO 600
620  IF M <  > 2 THEN  GOTO 650
630  IF LY = 0 AND D > 28 THEN  PRINT "INVALID RESPONSE": PRINT : GOTO 600
640  IF LY = 1 AND D > 29 THEN  PRINT "INVALID RESPONSE": PRINT : GOTO 600
650  GOSUB 2190: REM   TO SELECT PLANET
660  DX = D:MX = M:YX = Y
670  D = D + (LO / 15) / 24
680  IF (LY = 1 AND M = 2 AND  INT (D) > 29) THEN M = 3:D = D - 29: GOTO 740
690  IF (LY = 0 AND M = 2 AND D > 28) THEN M = 3:D = D - 28: GOTO 740
700  IF D < 31 THEN  GOTO 740
710  IF (M = 4 OR M = 6 OR M = 9 OR M = 11) THEN M = M + 1:D = D - 30:
     GOTO 740
720  IF  INT (D) > 31 THEN M = M + 1:D = D - 31
730  IF M = 13 THEN M = 1:Y = Y + 1
740  REM  CALC DAYS TO DATE REQUESTED FROM EPOCH 1960,1,1
750  DG = 365 * Y + D + ((M - 1) * 31)
760  IF M > 2 THEN  GOTO 800
770  REM  FOR JAN AND FEB
780  DG = DG +  INT ((Y - 1) / 4) - INT ((.75) *  INT ((Y - 1) / 100 + 1))
790  GOTO 820
800  REM  FOR MAR THRU DEC
810  DG = DG -  INT (M * .4 + 2.3) +  INT (Y / 4)
     -  INT ((.75) *  INT ((Y / 100) + 1))
820  NI = DG - 715875
830  GOSUB 2400: REM   TO GET SIDEREAL TIME (T2)
840  IF CI = 10 THEN  GOTO 2690
850  REM  INPUT OF PLANETARY ORBIT DATA
860  IF F = 1 THEN  GOTO 1210
870  RESTORE
880  FOR PY = 0 TO 8: FOR PX = 0 TO 8
890  READ PD(PY,PX)
900  NEXT PX,PY
910  REM  MERCURY
920  DATA .071422,3.8484,.388301,1.34041
930  DATA .3871,.07974,2.73514,.12223,.836013
940  REM   VENUS
950  DATA .027962,3.02812,.013195,2.28638
960  DATA .7233,.00506,3.85017,.059341,1.33168
970  REM   EARTH
980  DATA .017202,1.74022,.032044,1.78547,1
990  DATA .017,3.33926,0,0
1000  REM   MARS
1010  DATA .009146,4.51234,.175301,5.85209,1.5237
1020  DATA .141704,1.04656,.03142,.858702
1030  REM   JUPITER
1040  DATA .00145,4.53364,.090478,.23911,5.2028
1050  DATA .249374,1.76188,.01972,1.74533
```

RISES *(continued)*

```
1060   REM   SATURN
1070   DATA .000584,4.89884,.105558,1.61094,9.5385
1080   DATA .534156,3.1257,.043633,1.977458
1090   REM   URANUS
1100   DATA .000205,2.46615,.088593,2.96706,19.182
1110   DATA .901554,4.49084,.01396,1.28805
1120   REM   NEPTUNE
1130   DATA .000104,3.78556,.016965,.773181,30.06
1140   DATA .27054,2.33498,.031416,2.29162
1150   REM   PLUTO
1160   DATA .000069,3.16948,.471239,3.91303,39.44
1170   DATA 9.86,5.23114,.300197,1.91812
1180   FOR IP = 1 TO 10: READ P$(IP): NEXT IP
1190   DATA MERCURY,VENUS,SUN,MARS,JUPITER
1200   DATA  SATURN,URANUS,NEPTUNE,PLUTO,MOON
1210 F = 1
1220   REM   CALCULATE DATA FOR PLANETS
1230   HOME : PRINT : PRINT : PRINT
1240   PRINT "DATA REQUESTED FOR ";YX;" ";MX;" ";DX
1250   PRINT
1260   PRINT "-:-:-:-:-:-:-:-:-:-:-:-:-:-:-:-:-:-
1270   IF CI = 3 THEN  PRINT  TAB( 28)"R.A.  DEC"
1280   IF CI = 3 THEN  PRINT  TAB( 27)"(HRS) (DEG)": GOTO 1320
1290   PRINT "           DIST     ANG.DIST   R.A.   DEC"
1300   PRINT "        TO PLANET   FROM SUN  (HRS) (DEG)"
1310   PRINT  TAB( 11)"A.U."
1320   PRINT "-:-:-:-:-:-:-:-:-:-:-:-:-:-:-:-:-:-
1330   IF FL = AP THEN  GOTO 1440
1340   PRINT :I = 1
1350   FOR J = 0 TO 8: GOSUB 1780
1360   A(I) = A:DS(I) = DS:L(I) = L
1370   I = I + 1: NEXT J
1380   FOR I = 1 TO 9
1390   REM  SKIP EARTH
1400   IF I = 3 THEN  NEXT I
1410   GOSUB 1920
1420   Q(I) = Q:X(I) = X:R(I) = R:V(I) = V
1430   NEXT I
1440   I = CI:A(I) =  FN DEG(A(I))
1450   Q$(I) =  LEFT$ ( STR$ (Q(I)),5)
1460   X$(I) =  LEFT$ ( STR$ (X(I)),5)
1470   R$(I) =  LEFT$ ( STR$ (R(I)),5)
1480   V$(I) =  LEFT$ ( STR$ (V(I)),5)
1490   IF I = 3 THEN  PRINT P$(I) TAB( 27)R$(3);
1500   IF I = 3 THEN  PRINT  TAB( 33)V$(3)
1510   IF I = 3 THEN  PRINT : GOTO 1540
1520   PRINT P$(I) TAB( 9)Q$(I) TAB( 19)X$(I) TAB( 28)R$(I);
1530   PRINT  TAB( 36)V$(I)
1540   PRINT "-:-:-:-:-:-:-:-:-:-:-:-:-:-:-:-:-:-
1550   GOSUB 2480: REM  GET RISE,TRANSIT SET
1560   PRINT : PRINT
1570   TM = 60 * (TR -  INT (TR)):TR =  INT (TR)
1580   TN = 60 * (TT -  INT (TT)):TT =  INT (TT)
1590   TP = 60 * (TS -  INT (TS)):TS =  INT (TS)
1600   PRINT
```

```
┌── RISES (continued) ──────────────────────────────────────────────────────

  1610   PRINT P$(I)"  RISES AT....... ";TR;" HR "; INT (TM);" MIN"
  1620   PRINT  TAB( LEN (P$(I)));"   TRANSITS AT.... ";TT;" HR ";
  1630   PRINT  INT (TN);" MIN"
  1640   PRINT  TAB(  LEN (P$(I)))" SETS AT........ ";TS;" HR ";
  1650   PRINT  INT (TP);" MIN"
  1660   PRINT : PRINT
  1670   INPUT "WANT ANOTHER PLANET, SAME DATE Y/N? ";A$
  1680   IF A$ = "N" THEN  GOTO 1740
  1690   IF A$ < > "Y" THEN  PRINT "INVALID RESPONSE": PRINT : GOTO 1670
  1700   IF CI = 10 THEN D = DX:M = MX:Y = YX: GOTO 650
  1710   GOSUB 2200: IF CI = 3 THEN FL = AP
  1720   IF CI = 10 THEN  GOSUB 2690: GOTO 1740
  1730   GOTO 1230
  1740   PRINT : INPUT "WANT ANOTHER DATE Y/N? ";A$
  1750   IF A$ = "Y" THEN FL = 0:F = 0: RESTORE : GOTO 430
  1760   IF A$ < > "N" THEN  PRINT "INVALID RESPONSE": PRINT : GOTO 1740
  1770   HOME : GOTO 3200
  1780   REM  CALC A,DS, AND L
  1790   REM  HELIOCENTRIC LONGITUDE (A)
  1800   A = NI * PD(J,0) + PD(J,1)
  1810   IF A > 6.28318 THEN A = ((A / 6.28318) -  INT (A / 6.28318)) * 6.28318
  1820   IF A < 0 THEN A = A + 6.28318: GOTO 1820
  1830   C = PD(J,2) *  SIN (A - PD(J,3))
  1840   A = A + C
  1850   IF A > 6.28318 THEN A = A - 6.28318
  1860   IF A < 0 THEN A = A + 6.28318: GOTO 1860
  1870   REM DISTANCE OF PLANET FROM SUN (DS)
  1880   DS = PD(J,4) + PD(J,5) *  SIN (A - PD(J,6))
  1890   REM  DISTANCE FROM ECLIPTIC (L)
  1900   L = PD(J,7) *  SIN (A - PD(J,8))
  1910   RETURN
  1920   REM  CALC RA AND DEC
  1930   REM  ANG. DIST FROM SUN (Z)
  1940   Z = A(3) - A(I)
  1950   IF  ABS (Z) > 3.14159 AND Z < 0 THEN Z = Z + 6.28318
  1960   IF  ABS (Z) > 3.14159 AND Z > 0 THEN Z = Z - 6.28318
  1970   IF CI = 3 THEN X = 0: GOTO 2040
  1980   Q =  SQR (DS(I) ^ 2 + DS(3) ^ 2 - 2 * DS(I) * DS(3) *  COS (Z))
  1990   P = (DS(I) + DS(3) + Q) / 2
  2000   AC =  SQR (((P * (P - DS(I))) / (DS(3) * Q)))
  2010   IF AC = 1 THEN AC = .999999
  2020   X = 2 *  FN ACO(AC)
  2030   REM  R.A.
  2040   IF Z < 0 THEN R =  FN DEG(A(3) + 3.14159 - X) / 15
  2050   IF Z > 0 THEN R =  FN DEG(A(3) + 3.14159 + X) / 15
  2060   IF R > 24 THEN R = R - 24: GOTO 2060
  2070   IF R < - 24 THEN R = R + 24: GOTO 2070
  2080   IF R < 0 THEN R = R + 24: GOTO 2080
  2090   REM  DECLINATION
  2100   IF Z < 0 THEN V =  SIN (A(3) + 3.14159 - X) * 23.44194 +  FN DEG(L(I))
  2110   IF Z > 0 THEN V =  SIN (A(3) + 3.14159 + X) * 23.44194 +  FN DEG(L(I))
  2120   X =  FN DEG(X)
  2130   R(3) =  FN DEG(A(3) + 3.14159) / 15
  2140   IF R(3) > 24 THEN R(3) = R(3) - 24: GOTO 2140
  2150   IF R(3) < - 24 THEN R(3) = R(3) + 24: GOTO 2150
```

───────────────────── **RISES** *(continued)* ─────────────────────

```
2160  IF R(3) < 0 THEN R(3) = R(3) + 24
2170  V(3) =  SIN (A(3) + 3.141159) * 23.44194
2180  RETURN
2190  REM  PICK PLANET, SUN, OR MOON
2200  HOME : PRINT : PRINT
2210  PRINT : PRINT : PRINT
2220  PRINT "SELECT PLANET, SUN, OR MOON BY NUMBER"
2230  PRINT
2240  PRINT   TAB( 8)"MERCURY.......1"
2250  PRINT   TAB( 8)"VENUS.........2"
2260  PRINT   TAB( 8)"            3....SUN"
2270  PRINT   TAB( 8)"MARS..........4"
2280  PRINT   TAB( 8)"JUPITER.......5"
2290  PRINT   TAB( 8)"SATURN........6"
2300  PRINT   TAB( 8)"URANUS........7"
2310  PRINT   TAB( 8)"NEPTUNE.......8"
2320  PRINT   TAB( 8)"PLUTO.........9"
2330  PRINT   TAB( 8)"           10....MOON"
2340  PRINT
2350  INPUT "PLEASE SELECT NUMBER .. ";SS$
2360  SS =  VAL (SS$)
2370  IF SS < 1 OR SS > 10 THEN  PRINT "INVALID RESPONSE": PRINT : GOTO 2190
2380  CI = SS
2390  RETURN
2400  REM  CALC SIDEREAL TIME T2
2410  GC = 11.927485
2420  TC = .065711
2430  T2 = TC * (NI - 7020) + GC
2440  IF T2 > 24 THEN T2 = T2 - 24: GOTO 2440
2450  IF T2 <  - 24 THEN T2 = T2 + 24: GOTO 2450
2460  IF T2 < 0 THEN T2 = T2 + 24
2470  RETURN
2480  REM  CALC RISE, TRANSIT, AND SET TIMES
2490  TA =  TAN ( FN RAD(LA)) *  TAN ( FN RAD(V(I)))
2500  IF TA < 0 THEN TA =  ABS (TA): GOTO 2520
2510  IF TA > 0 THEN TA =  - TA
2520  TA =  FN ACO(TA)
2530  TA =  FN DEG(TA)
2540  B =  - TA / 15
2550  TZ = 0
2560  TR = (B + R(I) + TZ - T2) * .99727
2570  TR = TR - 1 / 60
2580  IF TR < 0 THEN TR = TR + 24
2590  IF TR > 24 THEN TR = TR - 24
2600  C = TA / 15
2610  TS = (C + R(I) + TZ - T2) * .99727
2620  TS = TS + 3 / 60
2630  IF TS < 0 THEN TS = TS + 24
2640  IF TS > 24 THEN TS = TS - 24
2650  TT = R(I) - T2 + 1 / 60
2660  IF TT < 0 THEN TT = TT + 24
2670  IF TT > 24 THEN TT = TT - 24
2680  RETURN
2690  REM  CALC RISE, TRANSIT, SET FOR MOON
2700  HOME : PRINT
```

RISES (continued)

```
2710  REM  RA AND DEC OF MOON
2720 LZ = 311.1687
2730 LE = 178.699
2740 LP = 255.7433
2750 PG = .111404 * NI + LP
2760  IF PG <  - 360 THEN PG = PG + 360: GOTO 2760
2770  IF PG < 0 THEN PG = PG + 360
2780  IF PG > 360 THEN PG = PG - 360: GOTO 2780
2790 LMD = LZ + 360 * NI / 27.32158
2800 PG = LMD - PG
2810 DR = 6.2886 *  SIN ( FN RAD(PG))
2820 LMD = LMD + DR
2830  IF LMD <  - 360 THEN LMD = LMD + 3600: GOTO 2830
2840  IF LMD <  - 360 THEN LMD = LMD + 360: GOTO 2840
2850  IF LMD < 0 THEN LMD = LMD + 360: GOTO 2850
2860  IF LMD > 3600 THEN LMD = LMD - 3600: GOTO 2860
2870  IF LMD > 360 THEN LMD = LMD - 360: GOTO 2870
2880 RM = LMD / 15
2890  IF RM > 24 THEN RM = RM - 24: GOTO 2890
2900  IF RM < 0 THEN RM = RM + 24
2910 AL = LE - NI * .052954
2920  IF AL <  - 3600 THEN AL = AL + 3600: GOTO 2920
2930  IF AL <  - 360 THEN AL = AL + 360: GOTO 2920
2940  IF AL < 0 THEN AL = AL + 360: GOTO 2940
2950  IF AL > 3600 THEN AL = AL - 3600: GOTO 2950
2960  IF AL > 360 THEN AL = AL - 360: GOTO 2960
2970 AL = LMD - AL
2980  IF AL < 0 THEN AL = AL + 360
2990  IF AL > 360 THEN AL = AL - 360
3000 HE = 5.1454 *  SIN (AL * 3.14159 / 180)
3010 DM = HE + 23.1444 *  SIN (LMD * 3.14159 / 180)
3020  PRINT : PRINT
3030 RA$ =  STR$ (RM):DE$ =  STR$ (DM)
3040 RA$ =  LEFT$ (RA$,5):DE$ =  LEFT$ (DE$,5)
3050  IF  VAL (RA$) < 10 THEN RA$ =  LEFT$ (RA$,4)
3060  IF  VAL (DE$) <  - 9 THEN DE$ =  LEFT$ (DE$,5): GOTO 3080
3070  IF  VAL (DE$) < 10 AND  VAL (DE$) >  - 10 THEN DE$ =  LEFT$ (DE$,4)
3080  PRINT "MOON DATA REQUESTED FOR ..."
3090  PRINT  TAB( 5)"YEAR ";YX;" MONTH ";MX;" DAY ";DX
3100  PRINT
3110  PRINT "AT NOON"
3120  PRINT "R.A. OF MOON IS ";RA$;" HRS"
3130  PRINT "DECLINATION IS..";DE$;" DEG"
3140  PRINT "-----------------------------"
3150  REM  CALC RISE, TRANSIT, SET TIMES
3160 R(I) = RM:V(I) = DM
3170 P$(I) = "MOON"
3180  GOSUB 2480
3190  GOTO 1560
3200  END
```

The bizarre system of Saturn is shown in this photomosaic. In the foreground is the satellite Dione, shown only as a spot on the earlier picture. Tethys and Mimas are to the right of the planet, Enceladus and Rhea to the left, and Titan is at the top right in the far distance.

Program 16: SKYSET/ SKYPLT

Horizon Plots of Visible Planets, Sun, Moon, and Stars for Any Date, Time, and Location_____

Before home computers became available to generate displays, the configuration of the stars had to be shown on planispheres or on monthly star charts. But the planets, Sun, and Moon had to be inserted on such charts by hand. In 1700 mechanical devices called orreries were devised to duplicate the orbital motions of the celestial bodies and show their locations. These devices required great skills in craftsmanship and engraving and were very expensive. In the current century the planetarium was developed by Zeis in Germany, and many planetaria associated with colleges and museums were established after World War II. In a planetarium an optical system projects stars and planets onto a domed ceiling in a theaterlike room. Today the personal computer can bring the planetarium into your living room. The monitor screen becomes the domed ceiling, and on it you can see the sky for any date and time, at any location on our planet.

For the Apple computer this program has two parts, SKYSET and SKYPLT. It requires use of the CHAIN program from the system master diskette. SKYSET is loaded first and calls SKYPLT when it is run. The program has all the instructions needed to run it. It is also adequately supplied with remark (REM) statements in case you wish to modify it in any way for your use or special needs. When you run the program there is some delay as the various arrays are loaded, but you will see the PLEASE WAIT, LOADING DATA message before the program begins.

Then you will be asked for date and other input information, as well as whether you want to change the location parameters of time zone, latitude, and longitude. Next you are asked to select a horizon of 180 degrees centered on east, south, west, or north. You can then select whether you want the planets, Sun, and Moon displayed without showing any stars (which gives a quick display), or with stars included (which takes somewhat longer to complete all the coordinate conversions). As the program begins its calculations the screen displays the message COMPUTING .. PLEASE WAIT. During this period it calls CHAIN and the subprogram SKYPLT.

Next the horizon chart is generated. For the particular horizon you have requested, the azimuth is shown below the display and the elevation is shown at the left. Date and time information are displayed below the chart. If you select planets only, the Sun, Moon, and planets are plotted and the stars are omitted. If you select stars as well, the stars are plotted first (this takes about five minutes). Next the program plots the Sun and any planets above that horizon at the time and date selected. Finally it plots the Moon and shows it as) before full, @ when close to full, and (after full. Since the projection is Mercator, constellations toward the zenith are somewhat distorted by being stretched out horizontally, but they are still recognizable.

Variables have been set initially for your latitude, longitude, and time zone. While running the program you can change the variables to other latitudes, longitudes, and time zones. The program should not be expected to run accurately at latitudes exceeding 85 degrees north and south.

Planets are identified by numbers; 1, Mercury; 2, Venus; 4, Mars; 5, Jupiter; 6, Saturn; 7, Uranus; 8, Neptune; and 0, Pluto. The shape table can be changed if you wish to have symbols, letters, or a different set of numbers identify the planets, Sun, and Moon. Letters are used in the alternative Program 16A (see Appendix). Note that if planets are within one pixel of each other, the outermost planet will overprint the innermost and only the symbol for the outer planet will be displayed.

To see how the program works, you might choose to display the sky on 26 November 1981 at 11:30 A.M. for the south horizon (Figure 16.1). This shows all the planets, the Sun, and the Moon above the horizon at the same time. If you then ask for the same date and time in the Southern Hemisphere (for example, for −40 degrees latitude) and request the north horizon, you will see these same planets and constellations inverted (Figure 16.2). Another interesting plot (Figure 16.3) is the south horizon at 7:30 A.M. on 5 February 1982, when all the planets are again displayed (but not the Sun and Moon). This display also demonstrates how the program can more quickly display the planets without the stars.

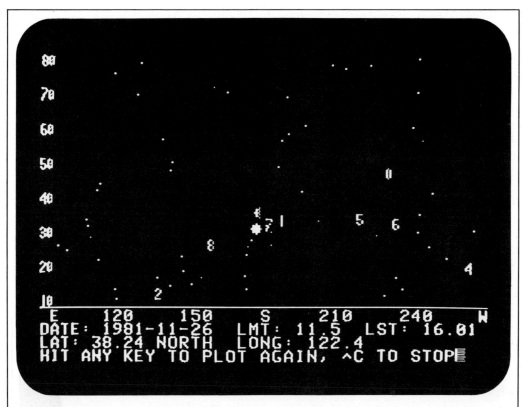

The SKYSET/SKYPLT program generates this display of all the planets, Sun, and Moon above the horizon on 26 November 1981 at 11:30 A.M. local time. The constellation Scorpio is directly south, Virgo is to the southwest, and Sagittarius is to the southeast.

Figure 16.1

Be extremely careful when keying in the star coordinates. If at any place you use a comma instead of a period, or vice versa, the positions of all the stars following the error will be incorrect and constellations will be unrecognizable. You can check yourself by seeing if your program produces the same displays as those shown in the Figures for the same time, date, and location.

Since it is extremely difficult to adapt this Apple-oriented program to other computers, another version (developed for an Exidy Sorcerer) is given in the Appendix as SKYPLA, Program 16A. This alternative program is more easily adaptable to computers with more characters and lines on the monitor screen than can be displayed by a typical Apple II computer installation. It does not require the chaining and subprogram required in the Apple program, and it can provide improved horizon graphics, depending on your computer.

   To run the program you load SKYSET. You must make sure that your disk contains the CHAIN program from the Apple DOS software. When you run SKYSET it calls SKYPLT. To end the program you give the command Control C when asked. You also have the option at this point of rerunning the program for a different date, time, or location. You press any key and after a short delay SKYSET is reloaded and you are asked if you want to change any variables.

   The listings of the SKYSET and SKYPLT programs follow.

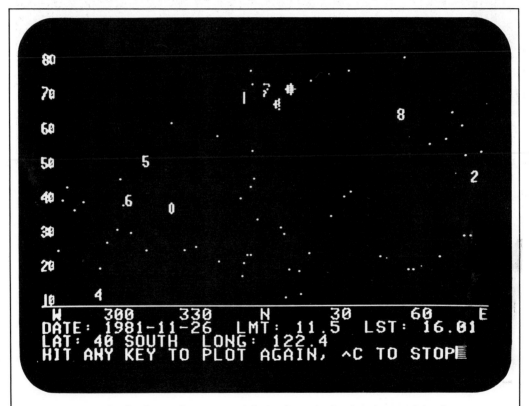

By requesting a display for 40 degrees south latitude and selecting the north horizon, you can see the planetary configuration of November 1981 as it appeared from Australia. The constellation Serpens is directly north and Scorpio is near the zenith in the northern sky. Sagittarius is to the northeast, and Virgo is to the northwest.

Figure 16.2

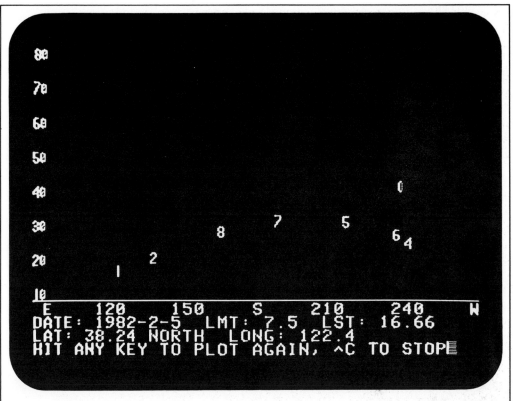

When you choose the option to display only the planets, the SKYSET/SKYPLT program generates this display for 5 February 1982 at 7:30 A.M.

Figure 16.3

┌─ **SKYSET** ───

```
10    HOME
12    IF DF = 1 THEN  GOTO 100
15    PRINT : PRINT
20    PRINT : PRINT : PRINT
25    PRINT   TAB( 8)"ASTRONOMY PROGRAM"
30    PRINT
35    PRINT   TAB( 10)"--------------"
40    PRINT   TAB( 10)"I    SKYSET    I"
45    PRINT   TAB( 10)"--------------"
50    PRINT : PRINT
55    PRINT   TAB( 5)"BY ERIC BURGESS F.R.A.S."
57    PRINT : PRINT : PRINT
60    PRINT   TAB( 5)"ALL RIGHTS RESERVED BY"
65    PRINT   TAB( 5)"S & T SOFTWARE SERVICE"
70    PRINT
75    PRINT   TAB( 10)"(VERSION 4/82)"
80    FOR K = 1 TO 2000: NEXT K
100   REM --- SET FUNCTIONS ---
110   DEF  FN ASN(X) =  ATN (X / SQR ( - X * X + 1))
120   DEF  FN ACO(X) =  - ATN (X / SQR ( - X * X + 1)) + 3.141593 / 2
130   DEF  FN RAD(X) = (X) * 3.14159 / 180
140   DEF  FN DEG(X) = (X) * 180 / 3.14159
150   IF DF = 1 GOTO 1320
160   REM --- INITIALIZE DATA VARIABLES ---
170   DIM PD(8,8),ST(236,1)
180   RESTORE
190   HOME : PRINT "PLEASE WAIT, LOADING DATA"
200   FOR I = 0 TO 8
210   FOR J = 0 TO 8
220   READ PD(I,J)
230   NEXT J,I
240   FOR I = 0 TO 236
250   FOR J = 0 TO 1
260   READ ST(I,J)
270   NEXT J,I
280 PI = 3.141593
290   REM --- SET HGR SHAPE TABLE ---
300   HIMEM: 38143
310   POKE 232,0: POKE 233,149
320 BA = 38144
330   READ N: POKE BA,N:BA = BA + 2:BD = BA + 2 * N
340   FOR X = 1 TO N
350 BR = BD - 38144
360   POKE BA,(BR -  INT (BR / 256) * 256)
370   POKE BA + 1, INT (BR / 256)
380 BA = BA + 2
390   READ Y: POKE BD,Y
400 BD = BD + 1
410   IF Y < > 0 GOTO 390
420   NEXT X
1000  REM --- INTRO PRINT ROUTINE ---
1010  HOME
1020  PRINT : PRINT : PRINT
```

───

SKYSET *(continued)*

```
1040  PRINT  TAB( 6);"ASTRONOMY PROGRAM FOR APPLE II"
1050 ZN = 8:LA =  FN RAD(38.24):L$ = "SEBASTOPOL,CAL.":LO = 122.49
1060  PRINT : PRINT
1070  PRINT "INITIAL CONDITIONS ARE SET"
1075  PRINT "FOR ... "L$: PRINT
1080  PRINT
1090  PRINT  TAB( 10);"LATITUDE "; FN DEG(LA)
1100  PRINT  TAB( 10);"LONGITUDE ";LO
1110  PRINT  TAB( 10);"TIME ZONE ";ZN: PRINT : PRINT
1120  PRINT "YOU CAN CHANGE THESE LATER IF YOU WISH"
1130 DF = 0
1140  FOR I = 1 TO 4400: NEXT I
1150  HOME : PRINT : PRINT : PRINT : PRINT
1160  PRINT : PRINT
1180  PRINT : PRINT
1190  PRINT : PRINT  TAB( 7)"PLOTS THE STARS FOR A REQUESTED"
1200  PRINT  TAB( 8)"TIME AND DATE, AND SHOWS THE"
1210  PRINT  TAB( 7)"PLANETS, SUN AND MOON ABOVE THE"
1220  PRINT  TAB( 16)"HORIZON"
1230  PRINT : PRINT  TAB( 4)"CLOSE TO FULL THE MOON IS SHOWN AS @"
1240  PRINT  TAB( 7)"BEFORE FULL IT IS SHOWN AS ) AND"
1250  PRINT  TAB( 8)"AFTER FULL IT IS SHOWN AS (": PRINT : PRINT
1260  PRINT : PRINT
1280  PRINT : PRINT
1290  FOR I = 1 TO 6500: NEXT I: HOME : PRINT
1300  REM  *** BEGIN MAIN PROGRAM ***
1310  IF DF = 0 THEN  PRINT : PRINT : PRINT : GOTO 1460
1320 NC = 0
1330  PRINT : INPUT "DO YOU WANT TO SEE THE VARIABLES? (Y/N) ";YN$
1340  IF YN$ = "N" THEN  GOTO 1380
1350  PRINT : PRINT "DATE.. ";YR;"-";MO;"-";DA;"     LMT.. ";T1;" HRS"
1355  PRINT
1360  PRINT "     TIME ZONE.. ";ZN
1365  PRINT
1370  PRINT "LAT.. "; FN DEG(LA);"    LONG.. ";LO: PRINT
1380  PRINT
1390  PRINT "DO YOU WANT TO ENTER OR CHANGE ANY": INPUT "VARIABLES?
      (Y)(N) ";YN$
1400  IF YN$ = "N" THEN NC = 1: GOTO 1770
1405  HOME : PRINT : PRINT : PRINT
1410  PRINT : PRINT
1420  PRINT "DO YOU WANT TO CHANGE:": PRINT "     THE DATE OF: "YR;"-";MO;
      "-";DA
1430  PRINT
1440  INPUT "(Y)(N)? ";YN$
1450  IF YN$ = "N" GOTO 1550
1460  PRINT : PRINT "WHAT IS THE DATE TO BE DISPLAYED?"
1465  PRINT
1470  INPUT "...THE YEAR: ";Y
1475  PRINT
1480  INPUT "...THE MONTH: ";M
1485  PRINT
1490  INPUT "...THE DAY: ";D
1500 MO = M:DA = D:YR = Y
1510  IF Y > 1800 GOTO 1550
```

SKYSET *(continued)*

```
1520   PRINT : PRINT "IS ";Y;: INPUT " THE YEAR YOU WANT(Y)(N)? ";YN$
1530   IF YN$ = "N" THEN  PRINT : GOTO 1470
1540   IF YN$ <  > "Y" GOTO 1520
1550   PRINT : PRINT
1560   IF DF = 0 THEN  PRINT "WHAT IS THE LMT TO BE DISPLAYED?": GOTO 1590
1570   PRINT "     THE LMT OF:";T1;: INPUT " (Y)(N)? ";YN$
1580   IF YN$ = "N" GOTO 1640
1590   GOSUB 3000
1600   IF DF <  > 0 THEN 1640
1605   HOME : PRINT : PRINT : PRINT
1610   PRINT : PRINT "IF YOU WANT TO CHANGE VARIABLES"
1620   PRINT "ANSWER 'Y' WHEN ASKED, OTHERWISE 'N'"
1625   PRINT
1630   PRINT "DO YOU WANT TO CHANGE:"
1640   PRINT
1650   PRINT "     THE TIME ZONE OF:";ZN;: INPUT " (Y)(N)? ";YN$
1660   IF YN$ = "N" GOTO 1680
1670   PRINT : INPUT "WHAT IS THE TIME ZONE(0-23)? ";ZN
1680   PRINT
1690   PRINT "     THE LONGITUDE OF: ";LO;: INPUT " (Y)(N)? ";YN$
1700   IF YN$ = "N" GOTO 1720
1710   PRINT : INPUT "WHAT IS THE LONGITUDE? ";LO: GOSUB 3300
1720   PRINT
1730   PRINT "     THE LATITUDE OF: "; FN DEG(LA);: INPUT " (Y)(N)? ";YN$
1740   IF YN$ = "N" GOTO 1770
1750   PRINT : INPUT "WHAT IS THE LATITUDE? ";LA
1760 LA =  FN RAD(LA)
1770   HOME : PRINT : PRINT : PRINT : PRINT
1780   PRINT "WHICH HORIZON DO YOU WANT TO SEE?"
1790   PRINT
1800   PRINT "     1 EAST HORIZON ... 0 TO 180 DEG"
1810   PRINT "     2 WEST HORIZON ... 180 TO 360 DEG"
1820   PRINT "     3 SOUTH HORIZON .. 90 TO 270 DEG"
1830   PRINT "     4 NORTH HORIZON .. 270 TO 90 DEG"
1840   PRINT : INPUT "SELECT (1-4): ";HZ
1850   IF (HZ = 1 OR HZ = 2 OR HZ = 3 OR HZ = 4) THEN 2000
1860   PRINT "INVALID ENTRY": GOTO 1780
2000   REM  *** BEGIN CALCS ***
2001   PRINT : PRINT
2002 SP = 0
2003   PRINT "DO YOU WANT PLANETS ONLY SHOWN"
2004   PRINT "WITHOUT SHOWING STARS AS WELL?"
2006   PRINT : PRINT "PLANETS IDENTIFIED AS.. "
2007   PRINT "MERCURY   1      VENUS 2"
2008   PRINT "MARS      4      JUPITER  5"
2009   PRINT "SATURN    6      URANUS   7"
2010   PRINT "NEPTUNE   8      PLUTO 0"
2011   PRINT : INPUT "(Y/N)";A$
2012   IF A$ = "Y" THEN SP = 1
2015   IF NC = 1 GOTO 2090
2020   HOME : PRINT "COMPUTING...PLEASE WAIT": PRINT
2030   REM *** SET PLANETS ***
2040   GOSUB 4000
2050   REM --- POSITION SUN ---
2060   GOSUB 5000
```

SKYSET *(continued)*

```
2070  REM --- POSITION MOON ---
2080  GOSUB 7000
2090  PRINT  CHR$ (4);"BLOAD CHAIN, A520"
2100  CALL 520"SKYPLT"
2110  REM *=*=*=*=*=*=*=*=*=*=*=*=*=*
3000  REM  *** TIME INPUT ***
3010  PRINT "DO YOU WANT INPUT IN DEC.HRS (D)"
3020  INPUT "     OR IN HR, MI, SE (H)? ";YN$
3030  IF YN$ = "D" THEN 3060
3040  IF YN$ = "H" THEN 3080
3050  PRINT "INVALID REPLY": PRINT : GOTO 3010
3060  PRINT : INPUT "WHAT IS THE LMT(HH.XXX)? ";T1
3070  PRINT : GOTO 3100
3080  PRINT : INPUT "WHAT IS THE LMT(HR,MI,SE)? ";HR,MI,SE
3090  T1 = HR + MI / 60 + SE / 3600: PRINT
3100  RETURN
3300  REM  *** CORRECT ZN FOR LONGITUDE ***
3320  LGC = (ZN * 15) - LO
3330  IF LGC < 0 THEN 3350
3340  ZN = ZN +  ABS (LGC / 15): GOTO 3360
3350  ZN = ZN + LGC / 15
3360  RETURN
4000  REM  *** ENTER VAR.FOR CALC LST ***
4010  GOSUB 4400
4020  GOSUB 4300:ND = G - 715875 + T1 / 24
4030  REM --- POSITION PLANETS ---
4040  FOR J = 0 TO 8
4050  GOSUB 4500
4060  A(J + 1) = A:D(J + 1) = D:L(J + 1) = L
4070  NEXT J
4080  FOR I = 1 TO 9
4090  IF I = 3 GOTO 4120
4100  GOSUB 4700
4110  AL(I) = AL:AZ(I) = AZ
4120  NEXT I
4130  RETURN
4300  REM  *** DAYS FROM EPOCH ***
4310  G = 365 * YR + DA + ((MO - 1) * 31)
4320  IF MO >  = 3 GOTO 4340
4330  G = G +  INT ((YR - 1) / 4) -  INT (.75 *  INT ((YR - 1) / 100 + 1)):
      GOTO 4350
4340  G = G -  INT (2.3 + MO * .4) +  INT (YR / 4)
      -  INT (.75 *  INT ((YR / 100) + 1))
4350  RETURN
4400  REM  *** CALCUL OF LST ***
4410  GOSUB 4300:NS = G - 722895
4420  SG = .065711
4430  T2 = SG * NS + 12.064707 + (((ZN + T1) / 24) * SG) + T1
4440  IF T2 > 24 THEN T2 = T2 - 24: GOTO 4440
4450  IF T2 < 0 THEN T2 = T2 + 24: GOTO 4450
4460  RETURN
4500  REM  *** SUB FOR A,D,L ***
4510  A = ND * PD(J,0) + PD(J,1)
4520  IF A > PI * 2 THEN A = (A / (PI * 2) -  INT (A / (PI * 2))) * PI * 2
4530  IF A < 0 THEN A = A + PI * 2: GOTO 4530
```

SKYSET *(continued)*

```
4540 C = PD(J,2) *  SIN (A - PD(J,3))
4550 A = A + C
4560  IF A > PI * 2 THEN A = A - PI * 2: GOTO 4560
4570  IF A < 0 THEN A = A + PI * 2: GOTO 4570
4580 D = PD(J,4) + PD(J,5) *  SIN (A - PD(J,6))
4590 L = PD(J,7) *  SIN (A - PD(J,8))
4600  RETURN
4700  REM --- ELEV & AZ OF PLANETS ---
4710 Z = A(3) - A(I)
4720  IF  ABS (Z) > PI AND Z < 0 THEN Z = Z + (PI * 2)
4730  IF  ABS (Z) > PI AND Z > 0 THEN Z = Z - (PI * 2)
4740 Q =  SQR (D(I) ^ 2 + D(3) ^ 2 - 2 * D(I) * D(3) *  COS (Z))
4750 P = (D(I) + D(3) + Q) / 2
4760 X = 2 *  FN ACO( SQR (((P * (P - D(I))) / (D(3) * Q))))
4770 T = X * (12 / PI)
4780  IF Z < 0 THEN R =  FN DEC(A(3) + PI - X) / 15
4790  IF Z > 0 THEN R =  FN DEC(A(3) + PI + X) / 15
4800  IF R > 24 THEN R = R - 24: GOTO 4800
4810  IF R < 0 THEN R = R + 24: GOTO 4810
4820  IF Z < 0 THEN V =  SIN (A(3) + PI - X) * 23.44194 +  FN DEG(L(I))
4830  IF Z > 0 THEN V =  SIN (A(3) + PI + X) * 23.44194 +  FN DEG(L(I))
4840 HA = T2 - R
4850  IF HA <  - 12 THEN HA = HA + 24
4860  IF HA > 12 THEN HA = HA - 24
4870 HA =  FN RAD(HA * 15):V =  FN RAD(V)
4880 AL =  FN ASN( SIN (V) *  SIN (LA) +  COS (V) *  COS (LA) *  COS (HA))
4890 AZ =  FN ACO(( SIN (V) -  SIN (LA) *  SIN (AL)) / ( COS (LA)
     *  COS (AL)))
4900  IF HA > 0 THEN AZ = PI * 2 - AZ
4910 AL =  FN DEG(AL):AZ =  FN DEG(AZ)
4920  RETURN
5000  REM --- POSITION SUN ---
5010 RS =  FN DEG(A(3) + PI) / 15
5020  IF RS > 24 THEN RS = RS - 24: GOTO 5020
5030  IF RS < 0 THEN RS = RS + 24: GOTO 5030
5040 VS =  SIN (A(3) + PI) * 23.44194
5050 HS = T2 - RS
5060  IF HS <  - 12 THEN HS = HS + 24
5070  IF HS > 12 THEN HS = HS - 24
5080 HS =  FN RAD(HS * 15):VS =  FN RAD(VS)
5090 AS =  FN ASN( SIN (VS) *  SIN (LA) +  COS (VS) *  COS (LA) *  COS (HS))
5100 ZS =  FN ACO(( SIN (VS) -  SIN (LA) *  SIN (AS)) / ( COS (LA)
     *  COS (AS)))
5110  IF HS > 0 THEN ZS = PI * 2 - ZS
5120 AS =  FN DEG(AS):ZS =  FN DEG(ZS)
5130  RETURN
7000  REM --- POSITION MOON ---
7010 ND = ND - .5
7020 LP = 255.7433
7030 LZ = 311.1687:LE = 178.699
7035 LM = LZ + 360 * ND / 27.32158
7040 MD = LM: GOSUB 7700:LM = MD
7070 PG = .111404 * ND + LP
7080 MD = PG: GOSUB 7700:PG = MD
7090 PG = LM - PG
```

─── **SKYSET** *(continued)* ───

```
7095   REM  CORRECT FOR ELLIPTICAL ORBIT
7100   DR = 6.2886 *  SIN ( FN RAD(PG))
7110   LM = LM + DR
7115   IF LM > 360 THEN LM = LM - 360: GOTO 7120
7117   IF LM < 0 THEN LM = LM + 360
7120   RQ = LM:RM = LM / 15
7130   IF RM > 24 THEN RM = RM - 24: GOTO 7130
7140   IF RM < 0 THEN RM = RM + 24
7150   AL = LE - ND * .052954
7160   ND = ND + .5
7170   MD = AL: GOSUB 7700:AL = MD
7180   AL = RQ - AL
7190   IF AL < 0 THEN AL = AL + 360
7200   IF AL > 360 THEN AL = AL - 360
7210   HE = 5.1454 *  SIN ( FN RAD(AL))
7220   DM = HE + 23.1444 *  SIN ( FN RAD(RQ))
7230   HD = T2 - RM
7240   IF HD <  - 12 THEN HD = HD + 24
7250   IF HD > 12 THEN HD = HD - 24
7260   IF (HD > 12 OR HD < - 12) THEN RETURN
7270   HA =  FN RAD(HD * 15):DM =  FN RAD(DM)
7280   ML =  FN ASN( SIN (DM) *  SIN (LA) +  COS (DM) *  COS (LA) *  COS (HA))
7290   MZ =  FN ACO(( SIN (DM) -  SIN (LA) *  SIN (ML)) / ( COS (LA)
       *  COS (ML)))
7300   IF HA > 0 THEN MZ = PI * 2 - MZ
7310   MZ =  FN DEG(MZ)
7320   ML =  FN DEG(ML)
7330   GOSUB 7600
7340   RETURN
7600   PM = RS + 12 - RQ / 15
7620   IF PM > 12 THEN PM = PM - 24
7630   IF (PM > = - 2 AND PM < = 2) THEN MS = 11: RETURN
7640   IF LA < 0 AND PM < - 2 THEN MS = 12
7650   IF PM < - 2 THEN MS = 13
7660   IF LA < 0 AND PM > 2 THEN MS = 13: RETURN
7670   IF PM > 2 THEN MS = 12
7680   RETURN
7700   IF MD < - 3600 THEN MD = MD + 3600: GOTO 7700
7710   IF MD < 0 THEN MD = MD + 360: GOTO 7710
7730   IF MD > 3600 THEN MD = MD - 3600: GOTO 7730
7740   IF MD > 360 THEN MD = MD - 360: GOTO 7740
7750   RETURN
9000   REM  *** ORBITAL ELEMENTS OF PLANETS, MERCURY-PLUTO ***
9010   DATA .071425,3.8494,.388301,1.34041,.3871,.07974,2.73514,.122173,
       .836013
9020   DATA .027962,3.02812,.013195,2.28638,.7233,.00506,3.85017,.059341,
       1.33168
9030   DATA .017202,1.74022,.032044,1.78547,1,.017,3.33926,0,0
9040   DATA .009146,4.51234,.175301,5.85209,1.5237,.141704,1.04656,.03142,
       .858702
9050   DATA .001451,4.53364,.090478,.23911,5.2028,.249374,1.76188,.01972,
       1.74533
9060   DATA .000584,4.89884,.105558,1.61094,9.5385,.534156,3.1257,.043633,
       1.977458
9070   DATA .000205,2.46615,.088593,2.96706,19.182,.901554,4.49084,.01396,
       1.28805
```

SKYSET *(continued)*

```
9080   DATA .000104,3.78556,.016965,.773181,30.06,.27054,2.33498,.031416,
       2.29162
9090   DATA .000069,3.16948,.471239,3.91303,39.44,9.86,5.23114,.300197,1.91812
9500   REM  *** DATA ON RA AND DEC OF STARS ***
9510   REM  URSA MINOR
9520   DATA 2,89,18,86,17,82,16,78,15,75,15.4,72,16.3,76
9530   REM  CEPHEUS
9540   DATA 20.8,61,21.5,70
9550   REM  CASSIOPEIA
9560   DATA 1.9,63,1.4,60,0.9,60,0.6,56,0.1,59
9570   REM  PERSEUS
9580   DATA 3.3,50,3.0,53,3.7,48,3.1,41,3.9,40,3.9,32
9590   REM  URSA MAJOR
9600   DATA 11,57,11,63,11.9,54,12.2,58,12.9,57,13.4,55,13.7,50
9610   REM  DRACO
9620   DATA 16,59,16.4,62,17.1,66,17.5,52,17.9,51,18.3,73,19.2,68
9650   REM  CEPHEUS
9640   DATA 23.8,78,21.3,62,22.1,58,22.8,67
9650   REM  ANDROMEDA
9660   DATA 2,42,1.1,35,.6,31
9670   REM  TRIANGULUM
9680   DATA 2.1,35,1.8,29,2.2,34
9690   REM  PEGASUS
9700   DATA  22.7,30,0.1,29,0.2,14,21.7,10,22.2,6,22.7,10,23,4,23,28
9710   REM  AURIGA
9720   DATA 5.2,46,5.9,45,5.9,37,4.9,33,5,41
9730   REM  BOOTES
9740   DATA 14.5,39,15,40,15.3,33,14.2,20,13.9,19,14.7,27,15.5,27,15.4,29
9750   REM  CORONA
9760   DATA 15.6,27
9770   REM  HERCULES
9780   DATA 16.7,39,16.7,31,17,31,17.2,37,17.2,25,16.5,21,16.4,19
9790   REM  LYRA
9800   DATA 18.7,39,18.8,33,19,32
9810   REM  CYGNUS
9820   DATA 20.7,45,20.3,40,19.8,45,20.8,34,19.5,28
9830   REM  TAURUS
9840   DATA 3.6,24,3,4,2.7,3,4.5,17,5.4,29,5.6,21,3.7,24,4.3,15,4.45,19
9850   REM  ARIES
9860   DATA 2.1,23,1.8,21,1.8,19
9870   REM  ERIDANUS
9880   DATA 3.9,-13,3.3,-20
9890   REM  PISCES
9900   DATA 1.5,-9,1.2,-10
9910   REM  CETUS
9920   DATA .7,-18,1.1,-10,1.3,-9,2,2
9930   REM  ORION
9940   DATA  5.8,8,5.4,8,5.75,-2,5.6,-1,5.45,0,5.8,-10,5.6,-6,5.6,10,5.5,-21,
       5.2,-9
9943   REM  CANIS MAJOR
9946   DATA 6.7,-17,6.3,-18,6.9,-29,7.2,-27,7.4,-29
9950   REM  CANIS MINOR
9960   DATA 7.6,7,7.4,9
9970   REM  GEMINI
```

———— **SKYSET** *(continued)* ————

```
9980    DATA 7.6,32,7.7,28,7.3,22,6.7,25,6.6,16,6.4,22,6.3,22
9990    REM  LEO
10000   DATA 10.1,12,10.1,17,10.3,20,10.3,24,11.2,20,11.2,16,11.8,15,9.8,
        28,9.7,26
10010   REM  CANCER
10020   DATA 8.7,29,8.6,21
10030   REM  HYDRA
10040   DATA 9.5,-9,8.7,7,8.9,7,9.2,2,10.4,-17
10050   REM  VIRGO
10060   DATA 11.8,2,13.4,-11,13,11,12.9,3,12.7,-1,12.3,-1,13.1,-5
10070   REM  CRATER
10080   DATA 10.8,-16,10.9,-18,11.3,-15,11.4,-18
10090   REM  CORVUS
10100   DATA 12.5,-16,12.2,-17,12.5,-23,12.2,-22
10110   REM  SERPENS
10120   DATA 15.8,17,15.5,10,15.7,7,15.8,5,15.8,-3
10130   REM  LIBRA
10140   DATA 15.3,-9,14.8,-16
10150   REM  OPHIUCHUS
10160   DATA 17.5,12,17.2,25,17.6,5,17.7,3
10170   REM  SAGITTARIUS
10180   DATA 18.3,-30,18,-30,18.4,-25,18.9,-26,19,-30,19.1,-21,18.3,-21
10190   REM  SCORPIO
10200   DATA 16.5,-26,16.6,-28,16.4,-24,16,-20,15.9,-22,15.9,-26,18.6,-43,
        16.7,-34,18.5,-37,18.7,-40,16.7,-38,22.9,-30
10210   REM  CAPRICORNUS
10220   DATA 21.7,-18,21.6,-18,21.4,-22,20.8,-28,20.7,-26,20.3,-14,20.2,-12,
        22.9,-30
10230   REM  DELPHINUS
10240   DATA 20.5,11,20.6,15,20.7,15,20.6,16,20.8,16
10250   REM  AQUARIUS
10260   DATA 22.6,0,22.5,0,22.4,1,22.3,-2,22,0,21.5,-6
10270   REM  AQUILA
10280   DATA 19.8,9,19.7,10.5,19.9,6,19.1,13,18.95,14,20.1,-1
10290   REM  SOUTHPOLAR REGION
10300   DATA 12.2,-59,12.1,-50,12.4,-57,12.7,-59,12.3,-63
10310   DATA 14,-60,14.7,-60,14.7,-65,15.9,-63,15.1,-69,16.9,-69
10320   DATA 20.3,-57,1.7,-57,2,-62,0.4,-63,6.3,-52,6.8,-51
10330   DATA 8.8,-55,9.3,-55,9.2,-59,8.3,-60,9.1,-70,9.8,-65
10340   DATA 3.9,-75,12.5,-69,12.6,-68
11000   REM *** SHAPE TABLE DATA ***
11005   DATA 13
11010   DATA 36,12,21,54,54,30,7,32,4,0
11020   DATA 32,12,173,182,246,63,32,12,45,0
11030   DATA 146,100,100,12,32,63,63,0
11040   DATA 54,14,45,32,28,63,32,12,45,5,0
11050   DATA   22,21,45,32,28,63,32,44,45,05,00
11060   DATA 12,12,12,54,54,54,196,63,4,0
11070   DATA   41,12,228,191,150,114,45,32,4,0
11080   DATA 64,99,173,246,30,30,46,45,5,0
11090   DATA 146,9,36,36,36,4,0
11100   DATA 36,188,119,247,45,62,46,53,36,53,37,44,63,44,60,7,0
11110   DATA 36,188,55,21,63,14,46,14,12,37,12,39,60,22,214,7,0
11120   DATA 182,12,37,12,63,12,60,28,54,6,0
11130   DATA 182,28,39,28,45,28,44,12,54,6,0
```

SKYPLT

```
100   REM --- SKYPLT PROGRAM ---
1000  REM --- RESET FUNCTIONS ---
1010  DEF  FN ASN(X) =  ATN (X / SQR ( - X * X + 1))
1020  DEF  FN ACO(X) =  -  ATN (X /  SQR ( - X * X + 1)) + 3.14159 / 2
1030  DEF  FN RAD(X) = X * 3.14159 / 180
1040  DEF  FN DEG(X) = X * 180 / 3.14159
1050  REM --- SET COORDINATES AND DATA ---
1060  HOME
1070  HGR
1080  HCOLOR= 7: SCALE= 1: ROT= 0
1090 X =   FRE (0)
1100  FOR X = 1 TO 8
1110  DRAW X + 1 AT 0,X * 21 - 13
1120  DRAW 1 AT 5,X * 21 - 13
1130  NEXT X
1140  HPLOT 0,159 TO 279,159
1150  VTAB (21)
1160  ON HZ GOTO 1170,1180,1190,1200
1170  PRINT " N    30    60    E   120    150    S";: GOTO 1210
1180  PRINT " S   210   240    W   300    330    N";: GOTO 1210
1190  PRINT " E   120   150    S   210    240    W";: GOTO 1210
1200  PRINT " W   300   330    N    30     60    E";
1210  PRINT "DATE: "YR"-"MO"-"DA"  LMT: " LEFT$ ( STR$ (T1),5)" LST:"
      LEFT$ ( STR$ (T2),5)
1220  PRINT "LAT: "; LEFT$ ( STR$ ( ABS ( FN DEG(LA))),5);
1230  IF LA > 0 THEN  PRINT " NORTH";
1240  IF LA < 0 THEN  PRINT " SOUTH";
1250  PRINT "  LONG: " LEFT$ ( STR$ (LO),5)
1260  REM --- GET AZ AND EL FOR EACH STAR AND POKE IT ON CHART ---
1265  IF SP = 1 THEN  GOTO 1490
1270  FOR K = 0 TO 236
1280 SR = ST(K,0):SD = ST(K,1)
1290 HD = T2 - SR
1300  IF HD < - 12 THEN HD = HD + 24
1310  IF HD > 12 THEN HD = HD - 24
1320 HA = HD * 15
1330 HA =  FN RAD(HA):SD =  FN RAD(SD)
1340 SL =  FN ASN( SIN (SD) *  SIN (LA) +  COS (SD) *  COS (LA) *  COS (HA))
1350 SZ = ( SIN (SD) -  SIN (LA) * SIN (SL)) / ( COS (LA) *  COS (SL))
1360  IF SZ > = 1 THEN SZ = 0: GOTO 1400
1370  IF SZ < = - 1 THEN SZ = PI: GOTO 1400
1380 SZ =  FN ACO(SZ)
1390  IF HA > 0 THEN SZ = PI * 2 - SZ
1400 SZ =  FN DEG(SZ)
1410  IF SZ > 360 THEN SZ = SZ - 360
1420  IF SZ < 0 THEN SZ = SZ + 360
1430 SL =  FN DEG(SL)
1440 X1 = SZ:Y1 = SL
1450  GOSUB 2000
1460  IF X2 < 0 GOTO 1480
1470  HPLOT X2,Y2
1480  NEXT K
1490  REM --- PLOT SUN ---
```

— **SKYPLT** *(continued)* —

```
1500  X1 = ZS:Y1 = AS
1510  GOSUB 2000
1520  IF X2 < 0 GOTO 1540
1530  DRAW 10 AT X2,Y2
1540  REM --- PLOT PLANETS ---
1550  FOR X = 1 TO 9
1560  IF X = 3 THEN 1610
1570  X1 = AZ(X):Y1 = AL(X)
1580  GOSUB 2000
1590  IF X2 < 0 GOTO 1610
1600  DRAW 10 - X AT X2,Y2
1610  NEXT X
1620  REM --- PLOT MOON ---
1630  X1 = MZ:Y1 = ML
1640  GOSUB 2000
1650  IF X2 < 0 GOTO 1670
1660  DRAW MS AT X2,Y2
1670  REM --- END ---
1680  PRINT "HIT ANY KEY TO PLOT AGAIN, ^C TO STOP";
1690  GET YN$
1700  TEXT : HOME
1710  IF YN$ = CHR$ (3) THEN  END
1720 DF = 1
1730  PRINT : PRINT  CHR$ (4);"BLOAD CHAIN, A520"
1740  CALL 520"SKYSET"
1750  REM =*=*=*=*=*=*=*=*=*=*=*=*
2000  REM --- BRACKET VALUES ---
2010  IF Y1 < 10 OR Y1 > 81.4 GOTO 2150
2020  ON HZ GOTO 2030,2050,2070,2090
2030  IF X1 > 0 AND X1 < 180 THEN GOTO 2120
2040  GOTO 2150
2050  IF X1 > 180 AND X1 < 360 THEN X1 = X1 - 180: GOTO 2120
2060  GOTO 2150
2070  IF X1 > 90 AND X1 < 270 THEN X1 = X1 - 90: GOTO 2120
2080  GOTO 2150
2090  IF X1 >  = 0 AND X1 < 90 THEN X1 = X1 + 90: GOTO 2120
2100  IF X1 > 270 AND X1 <  = 360 THEN X1 = X1 - 270: GOTO 2120
2110  GOTO 2150
2120  X2 = (265 / 180) * X1 + 11
2130  Y2 = (150 / 71.4) * (81.4 - Y1) + 5
2140  RETURN
2150  X2 =  - 1
2160  RETURN
```

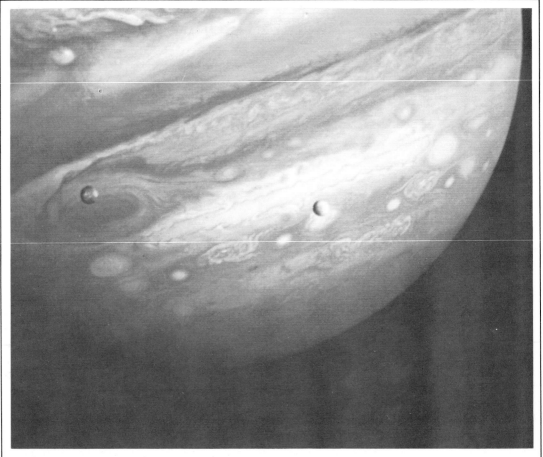

Photo Credit: NASA/Jet Propulsion Laboratory

Jupiter, the giant planet of our solar system, was explored for the first time by a Pioneer spacecraft in 1973. It was later revealed as a planet of incredibly complex detail by the Voyager flybys. Even in a small telescope Jupiter presents an interesting pattern of light and dark bands with a large red spot. In close-up pictures like the one shown here, the light and dark bands are resolved into intricate details of swirling storms. This Voyager picture shows the satellites Io and Europa in transit across the face of Jupiter.

Program 17: PLNTF

Finds and Plots Planets, Sun, and Moon in Constellations for Any Date and Time_____

How do you find out where a planet, the Sun, or the Moon is on any date? Some astronomical books give this type of information, but it is usually out of date when you most want to use it. The astronomical magazines provide the information on a monthly basis. But if you want the information for a few months ahead you must refer to a yearly almanac. If you want the information a year or years ahead you have to make laborious calculations, perhaps helped by a handheld calculator. However, with a personal computer and this program you can quickly and easily locate a planet, the Moon, or the Sun at *any* date.

When you select a date and a planet, the Sun, or the Moon, this program calculates where the object is located among the "fixed" stars of the celestial sphere. It selects a suitable zodiacal star chart (2 hours of right ascension and 30 degrees of declination), displays its name, and shows the object among the stars. The dotted line across the middle of each chart is the ecliptic. Because of the limitations of resolution of any monitor screen, the selected object's accurate right ascension and declination is displayed before the object is shown on the star chart.

The program then offers the option of having other planets, the Sun and the Moon displayed on the chart if they are located on the same chart region on the date requested (see Figure 17.1).

When you have obtained the information, the computer asks you if you want another planet on the same date. If you answer 'Y', it will then select an appropriate chart to display the new selection and offer again the ability to chart other planets in that new star chart.

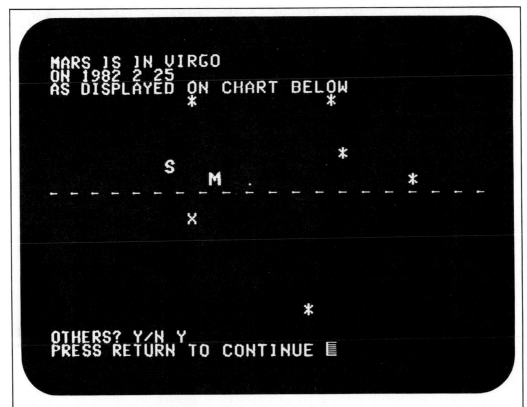

This display results if you ask the PLNTF program to find Mars on 25 February 1982. The planet (M) is plotted near to the star Spica in the constellation of Virgo. The program then asks if you want to see other planets in the same region of the zodiac. An affirmative answer of 'Y' places Saturn (S) on the display, close to Mars and Spica at the date selected.

── Figure 17.1 ──

Other alternatives available are to ask for another date for the same planet, or another date and another planet.

You can also select, when determining the date, whether you require a series of plots. If you do not, insert 1's in answer to the interval in days and the number of plots required. If you require a series, you must input the number of plots and the time interval (in days) between each plot. If your series of plots runs off the screen, the monitor will display the name of the planet and the message OFF CHART for every plot that is off the chart. Figure 17.2 illustrates the option of displaying a series of plots.

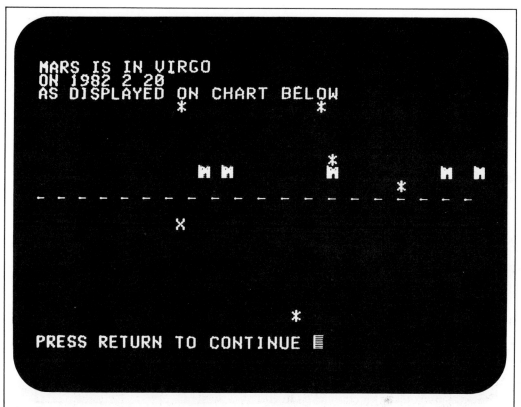

```
MARS IS IN VIRGO
ON 1982 2 20
AS DISPLAYED ON CHART BELOW
              *                    *


          H H            H          H H
                                  *
   - - - - - - - - - - - - - - - - - - - - -
         X



                    *
PRESS RETURN TO CONTINUE ▤
```

An alternative use of the PLNTF program is to plot the positions of a planet for a series of dates. This display shows the motion of Mars at 20-day intervals through the constellation of Virgo starting on 20 February 1982.

Figure 17.2

The program operates by first loading planetary data, as in other programs, and then computing the right ascension and declination of the planet you wish to find for the date you have requested. It displays this information on the monitor screen. Next the program determines (in instructions 2490 through 2590) which zodiacal region must be displayed. It then jumps to the part of the program (beginning at instruction 3660) containing the screen coordinates for the stars in that region of the sky, and displays the appropriate stars on the monitor screen. Converting the right ascension and declination to screen coordinates, the program places the planet at its correct location in the constellation by poking its ASCII character at the screen coordinate. Use of POKE statements enables the celestial objects to be displayed on the monitor screen without scrolling or erasing the other characters on the display.

When you request the display of other planets in the same star chart region, the program checks the right ascension of each of the other planets, the Sun, and the Moon. If any of these fall within the range of right ascensions covered by the chart displayed on the monitor, the program displays them at the correct location in the zodiacal constellation.

If you have requested a series of displays, the program loops until the number of displays matches the number requested.

The program will need modification to suit each individual computer's display format. The parts of the program requiring such modification are those that display the constellations and POKE the planets, Sun, and Moon in the star chart. Refer to the alternative program for guidance in adapting this program to your computer. You can also modify this program to use the high resolution graphics of the Apple by changing the VTAB and TAB instructions to high resolution graphic coordinates. If you do this, you must incorporate a short routine to allow the main program to continue beyond the region of memory reserved for the high resolution graphics.

For computers with screens having 64 characters and 30 lines, the alternative PLNTA, Program 17A in the Appendix, identifies and displays nearby constellations over 4 hours of right ascension and 60 degrees of declination, provides a grid of right ascension and declination, and names the bright stars. You can elect to have the grid lines deleted.

The basic calculations of planetary positions and the subroutines and flags used to produce the different types of displays and select the chart zones will operate in connection with any new display formats that you may wish to develop. Both programs have been sprinkled with remark (REM) statements to help you modify them to suit your system.

The listing of the PLNTF program follows.

┌─── **PLNTF** ───┐

```
10   CLEAR
20   HOME
30   DIM PD(9,9)
40   PRINT : PRINT
50   DEF   FN ACO(X) =  - ATN (X / SQR ( - X * X + 1)) + 1.5707963
60   DEF   FN RAD(X) = .01745328 * (X)
70   DEF   FN DEG(X) = 57.29578 * (X)
80   PRINT : PRINT : PRINT
90   PRINT  TAB( 8)"ASTRONOMY PROGRAM"
100  PRINT
110  PRINT  TAB( 5)"----------------------"
120  PRINT  TAB( 5)"I    PLANET FINDER    I"
130  PRINT  TAB( 5)"----------------------"
140  PRINT
150  PRINT  TAB( 5)"BY ERIC BURGESS F.R.A.S."
160  PRINT : PRINT
170  PRINT  TAB( 5)"ALL RIGHTS RESERVED BY"
180  PRINT  TAB( 5)"S & T SOFTWARE SERVICE"
190  PRINT
200  PRINT  TAB( 8)"(VERSION 4.82)"
210  FOR K = 1 TO 3000: NEXT K
220  HOME : PRINT : PRINT
230  INPUT "DO YOU WANT INSTRUCTIONS? Y/N ";A$
240  IF A$ = "N" THEN 480
250  IF A$ < > "Y" THEN  PRINT : PRINT "INVALID RESPONSE": PRINT : GOTO 220
260  HOME : PRINT : PRINT
270  PRINT "THIS PROGRAM PLACES A PLANET, OR THE"
280  PRINT "SUN, OR THE MOON AMONG THE ZODIACAL"
290  PRINT "CONSTELLATIONS FOR ANY DATE."
300  PRINT
310  PRINT "ANSWER THE PROMPT 'OTHERS?' WITH 'Y'"
320  PRINT "AND OTHER PLANETS, SUN, OR MOON"
330  PRINT "IN THE SAME CHART WILL BE DISPLAYED"
340  PRINT
350  PRINT "YOU CAN ALSO PLOT THE POSITIONS OF"
360  PRINT "A PLANET FOR A SERIES OF INTERVALS"
370  PRINT
380  PRINT "BECAUSE OF THE LIMITATIONS OF SCREEN"
390  PRINT "RESOLUTION, THE RIGHT ASCENSION AND"
400  PRINT "DECLINATION OF THE CHOSEN PLANET, SUN,"
410  PRINT "OR MOON IS GIVEN BEFORE THE MONITOR"
420  PRINT "DISPLAYS THE ZODIACAL STAR CHART"
430  PRINT
440  PRINT "THE DOTTED LINE ACROSS THE MIDDLE"
450  PRINT "OF EACH CHART IS THE ECLIPTIC"
460  PRINT : PRINT : PRINT
470  INPUT "PRESS RETURN TO CONTINUE ";A$
480  HOME : PRINT : PRINT
490  PRINT "ENTER THE DATE": PRINT
500  INPUT "  THE YEAR ";YD$:Y = VAL (YD$)
510  IF Y = 0 THEN  PRINT "INVALID RESPONSE": PRINT : GOTO 500
520  IF Y > 1800 THEN  GOTO 590
530  PRINT "IS  ";Y;" THE CORRECT YEAR? ";
```

└──┘

PLNTF *(continued)*

```
540   INPUT Y$
550   IF PS = 10 OR F8 = 1: THEN  GOSUB 7210
560   IF Y$ = "Y" THEN  GOTO 590
570   IF Y$ < > "N" THEN  PRINT "INVALID RESPONSE": PRINT : GOTO 530
580  ·IF Y$ = "N" THEN  PRINT : GOTO 500
590   PRINT : INPUT "  THE MONTH ";MD$:M =  VAL (MD$)
600   IF M = 0 OR M > 12 THEN  PRINT "INVALID RESPONSE": PRINT : GOTO 590
610   PRINT : PRINT "  THE DAY "
620   INPUT "(USE DECIMAL FOR HRS) ";DD$
630   D =  VAL (DD$)
640   REM  STORES INITIAL DATE IN D2,M2,Y2
650   IF F9 = 1 THEN  GOTO 670
660   D2 =  VAL (DD$):Y2 = Y:M2 = M
670   IF D = 0 OR D > 31 THEN  PRINT "INVALID RESPONSE": PRINT : GOTO 610
680   IF M = 2 AND D > 29 THEN  PRINT "INVALID RESPONSE": PRINT : GOTO 610
690   PRINT : PRINT
700   PRINT "SELECT INTERVALS AND HOW MANY"
710   PRINT  TAB( 5)"ENTER 1'S IF YOU NEED ONE PLOT ONLY"
720   PRINT : PRINT
730   INPUT "WHAT IS THE INTERVAL (DAYS) ";TI$: PRINT
740  TI =  VAL (TI$): IF TI = 0 THEN PRINT "INVALID RESPONSE": PRINT :
      GOTO 730
750   INPUT "HOW MANY INTERVALS ";IN$: PRINT
760  IN =  VAL (IN$): IF IN = 0 THEN PRINT "INVALID RESPONSE": PRINT :
      GOTO 879
770   REM  SETS INTERVAL COUNT AT 1
780  NC = 1
790   GOSUB 800: GOTO 920
800   REM  CALC DAYS FROM 1960,1,1 EPOCH TO DATE
810   HOME : PRINT : PRINT "PLEASE WAIT"
820  DC = 365 * Y + D + ((M - 1) * 31)
830   IF M > = 3 GOTO 870
840   REM  CALC FOR JAN OR FEB
850  DC = DC +  INT ((Y - 1) / 4) -  INT ((.75) *  INT ((Y - 1) / 100 + 1))
860   GOTO 890
870   REM  CALC FOR MAR THRU DEC
880  DC = DC -  INT (M * .4 + 2.3) +  INT (Y / 4)
      -  INT ((.75) *  INT ((Y / 100) + 1))
890  NI = DC - 715875
900   IF F9 = 1 THEN 930
910   RETURN
920   REM  JUMPS PLANETARY INPUTS IF NEW INTERVAL
930   IF F9 = 1 THEN  GOTO 1490
940   REM  INPUT PLANETARY DATA, ORBITAL PARAMETERS
950   REM  JUMPS PLANETARY INPUT IF NEW DATE
960   IF FL = 1 THEN  GOTO 1230
970   IF F8 = 1 THEN  GOTO 1250
980   IF F6 = 1 THEN  GOTO 1250
990   RESTORE
1000   FOR YP = 0 TO 8: FOR XP = 0 TO 8
1010   READ PD(YP,XP)
1020   NEXT XP,YP
1030   REM  MERCURY
1040   DATA .071422,3.8484,.388301,1.34041,.3871,.07974,2.73514,.122173,
       .836013
```

─────────── **PLNTF** *(continued)* ───────────

```
1050   REM   VENUS
1060   DATA  .027962,3.02812,.013195,2.28638,.7233,.00506,3.85017,.059341,
       1.33168
1070   REM   EARTH (FOR SUN)
1080   DATA  .017202,1.74022,.032044,1.78547,1,.017,3.33926,0,0
1090   REM   MARS
1100   DATA  .009146,4.51234,.175301,5.85209,1.5237,.141704,1.04656,.03142,
       .858702
1110   REM   JUPITER
1120   DATA   .001450,4.53364, .090478,.23911,5.2028,.249374,1.76188,.01972,
       1.74533
1130   REM   SATURN
1140   DATA   .000584,4.89884,.105558,1.61094,9.5385,.534156,3.1257,.043633,
       1.977458
1150   REM   URANUS
1160   DATA  .000205,2.46615,.088593,2.96706,19.182,.901554,4.49084,.01396,
       1.28805
1170   REM   NEPTUNE
1180   DATA  .000104,3.78556,.016965,.773181,30.06,.27054,2.33498,.031416,
       2.29162
1190   REM   PLUTO
1200   DATA  .000069,3.16948,.471239,3.91303,39.44,9.86,5.23114,.300197,
       1.91812
1210   FOR I9 = 1 TO 9: READ P$(I9): NEXT I9
1220   DATA      MERCURY,VENUS,SUN,MARS,JUPITER,SATURN,URANUS,NEPTUNE,PLUTO
1230   F = 0
1240   IF F9 = 2 THEN  GOTO 1260
1250   FL = 0
1260   REM   CALC DATA FOR PLANETS
1270   IF F9 = 2 THEN  GOTO 1490
1280   IF F8 = 1 THEN F8 = 0: GOTO 1490
1290   HOME
1300   PRINT : PRINT : PRINT
1310   PRINT "WHICH PLANET, SUN, OR MOON"
1320   PRINT "    DO YOU WANT TO FIND?"
1330   PRINT
1340   PRINT   TAB( 5)"MERCURY (+)...1"
1350   PRINT   TAB( 5)"VENUS (V).....2"
1360   PRINT   TAB( 5)"            3...SUN (0)"
1370   PRINT   TAB( 5)"MARS (M)......4"
1380   PRINT   TAB( 5)"JUPITER (J)...5"
1390   PRINT   TAB( 5)"SATURN (S)....6"
1400   PRINT   TAB( 5)"URANUS (U)....7"
1410   PRINT   TAB( 5)"NEPTUNE (N)...8"
1420   PRINT   TAB( 5)"PLUTO (P).....9"
1430   PRINT   TAB( 5)"           10...MOON ()
1440   PRINT
1450   PRINT   TAB( 10): INPUT "SELECT 1 THRU 10 ";PS$
1460   PS =  VAL (PS$)
1470   IF PS = 0 OR PS > 10 THEN  PRINT "INVALID CHOICE": PRINT : GOTO 1450
1480   REM   STORES SELECTION IN P2
1490   P2 = PS
1500   IF PS = 10 THEN  GOSUB 7210
1510   I = 1
1520   REM   CALC. PLANETARY DATA AT DATE
```

┌─**PLNTF** *(continued)*──┐

```
1530  REM   AND STORE IN ARRAYS
1540  FOR J = 0 TO 8: GOSUB 1710
1550 A(I) = A:D(I) = DS:L(I) = L
1560 I = I + 1: NEXT J
1570  FOR I = 1 TO 9
1580  REM  SKIP EARTH
1590  IF I = 3 THEN  NEXT I
1600  GOSUB 1860
1610 Q(I) = Q:X(I) = X:R(I) = R:V(I) = V
1620  NEXT I
1630  FOR I = 1 TO 9:A(I) =  FN DEG(A(I))
1640  IF I = 3 THEN  NEXT I
1650  NEXT
1660 I = PS
1670 R(3) = (A(3) - 180) / 15
1680  IF R(3) < 0 THEN R(3) = R(3) + 24
1690 V(3) = ( SIN ( FN RAD(A(3) - 180))) * 23.44194
1700  GOTO 2070
1710  REM  CALC A, DS, AND L
1720  REM  AT DATE
1730  REM  HELIOCENTRIC LONGITUDE,A
1740  A = NI * PD(J,0) + PD(J,1)
1750  IF A > 6.28318 THEN A = ((A / 6.28318) -  INT (A / 6.28318)) * 6.28318
1760  IF A < 0 THEN A = A + 6.28318: GOTO 1760
1770  C = PD(J,2) *  SIN (A - PD(J,3))
1780  A = A + C
1790  IF A > 6.28318 THEN A = A - 6.28318
1800  IF A < 0 THEN A = A + 6.28318: GOTO 1800
1810  REM  CALC DIST OF PLANET FROM SUN DS
1820 DS = PD(J,4) + PD(J,5) *  SIN (A - PD(J,6))
1830  REM  CALC DISTANCE FROM ECLIPTIC,L
1840 L = PD(J,7) *  SIN (A - PD(J,8))
1850  RETURN
1860  REM   CALC Z,Q,X,R,V
1870  REM  CALC ANG. DIST. FROM SUN Z
1880 Z = A(3) - A(I)
1890  IF  ABS (Z) > 3.14159 AND Z < 0 THEN Z = Z + 6.28318
1900  IF  ABS (Z) > 3.14159 AND Z > 0 THEN Z = Z - 6.28318
1910  REM  CALC DISTANCE OF PLANET FROM EARTH Q
1920 Q =  SQR (D(I) ^ 2 + D(3) ^ 2 - 2 * D(I) * D(3) *  COS (Z))
1930  REM  CALC ANG. DIST. FROM SUN, X
1940 PP = (D(I) + D(3) + Q) / 2
1950 X = 2 *  FN ACO( SQR (((PP * (PP - D(I))) / (D(3) * Q))))
1960  REM  CALC RIGHT ASCENSION, R
1970  IF Z < 0 THEN R =  FN DEG(A(3) + 3.14159 - X) / 15
1980  IF Z > 0 THEN R =  FN DEG(A(3) + 3.14159 + X) / 15
1990  IF R > 24 THEN R = R - 24: GOTO 1990
2000  IF R <  - 24 THEN R = R + 24: GOTO 2000
2010  IF R < 0 THEN R = R + 24: GOTO 2010
2020  REM  CALC DECLINATION, V
2030  IF Z < 0 THEN V =  SIN (A(3) + 3.14159 - X) * 23.44194 + FN DEG(L(I))
2040  IF Z > 0 THEN V =  SIN (A(3) + 3.14159 + X) * 23.44194 + FN DEG(L(I))
2050 X =  FN DEG(X)
2060  RETURN
2070 RA = R(PS)
```

└──┘

```
2080  DE = V(PS)
2090  REM   JUMPS PRINTING RA AND DEC IF NEW INTERVAL
2100  IF F9 = 1 THEN 2210
2110  IF F9 = 2 THEN 2210
2120  PRINT : PRINT
2130  RA$ =  STR$ (RA):DE$ =  STR$ (DE)
2140  RA$ =  LEFT$ (RA$,5):DE$ =  LEFT$ (DE$,5)
2150  PRINT "R.A. OF ";P$(PS);" IS ";RA$
2160  PRINT "DECLINATION IS ";DE$
2170  PRINT
2180  PRINT "PRESS RETURN TO DISPLAY PLANET,"
2190  INPUT "SUN, OR MOON ON STAR CHART ";A$
2200  REM  STORES RA FOR SELECTED PLANET
2210  R3 = RA:DC = DE
2220  GOSUB 2250: GOTO 2360
2230  REM  ASCII CODES FOR POKING PLANETS
2240  REM  AS FLASHING SYMBOLS
2250  IF PS = 1 THEN P$ = "MERCURY":P = 107: GOTO 2350
2260  IF PS = 2 THEN P$ = "VENUS" :P = 86: GOTO 2350
2270  IF PS = 3 THEN P$ = "SUN":P = 79: GOTO 2350
2280  IF PS = 4 THEN P$ = "MARS":P = 77: GOTO 2350
2290  IF PS = 5 THEN P$ = "JUPITER":P = 74: GOTO 2350
2300  IF PS = 6 THEN P$ = "SATURN":P = 83: GOTO 2350
2310  IF PS = 7 THEN P$ = "URANUS":P = 85: GOTO 2350
2320  IF PS = 8 THEN P$ = "NEPTUNE":P = 78: GOTO 2350
2330  IF PS = 9 THEN P$ = "PLUTO":P = 80: GOTO 2350
2340  IF PS = 10 THEN P$ = "MOON":P = 105: GOTO 2350
2350  RETURN
2360  REM
2370  IF F9 = 1 OR F5 = 1 GOTO 2390
2380  PN = P
2390  REM  SELECT STAR CHART FOR DISPLAY
2400  GOSUB 2440
2410  IF F9 = 1 THEN 6580
2420  IF (F7 = 1 AND F5 = 1) THEN RETURN
2430  GOTO 2720
2440  REM  SELECT CHART AND ADJUST
2450  REM  HORIZONTAL PLOT
2460  PDE = DE
2470  REM  JUMPS CHART SELECTION IF SHOWING OTHER PLANETS
2480  IF F9 = 1 OR F5 = 1 THEN  GOTO 2610
2490  IF RA > 22 AND RA < 23.99999 THEN RA = RA - 22:CH = 12: GOTO 2640
2500  IF RA > 20 AND RA < 21.99999 THEN RA = RA - 20:CH = 11: GOTO 2640
2510  IF RA > 18 AND RA < 19.99999 THEN RA = RA - 18:CH = 10: GOTO 2640
2520  IF RA > 16 AND RA < 17.99999 THEN RA = RA - 16:CH = 9: GOTO 2640
2530  IF RA > 14 AND RA < 15.99999 THEN RA = RA - 14:CH = 8: GOTO 2640
2540  IF RA > 12 AND RA < 13.99999 THEN RA = RA - 12:CH = 7: GOTO 2640
2550  IF RA > 10 AND RA < 11.99999 THEN RA = RA - 10:CH = 6: GOTO 2640
2560  IF RA > 8 AND RA < 9.99999 THEN RA = RA - 8:CH = 5: GOTO 2640
2570  IF RA > 6 AND RA < 7.99999 THEN RA = RA - 6:CH = 4: GOTO 2640
2580  IF RA > 4 AND RA < 5.99999 THEN RA = RA - 4:CH = 3: GOTO 2640
2590  IF RA > 2 AND RA < 3.99999 THEN RA = RA - 2:CH = 2: GOTO 2640
2600  CH = 1
2610  IF F9 = 1 THEN  GOSUB 6790
2620  IF F5 = 1 THEN  GOSUB 6790
```

PLNTF *(continued)*

```
2630  IF F7 = 1 THEN  RETURN
2640  IF F9 = 1 THEN  GOSUB 6670
2650  IF F5 = 1 THEN  GOSUB 6670
2660 RA = RA * 15
2670  IF (RA -  INT (RA)) > .49 THEN RA = 1 +  INT (RA)
2680 PL =  INT (1.3 * RA)
2690  IF F5 = 1 THEN  GOTO 2710
2700 PX = PL
2710  RETURN
2720  REM  IN EACH CHART POKE PL,P
2730  REM  WHERE P IS NAME OF PLANET
2740  REM    SUCH AS +,V,M,J,S,U,N,P,O,)
2750  HOME
2760 S = 1487
2770  GOSUB 2790
2780  GOTO 2930
2790  REM  SUB FOR ABOVE AND BELOW ECLIPTIC PLANE
2800  IF PS = 10 THEN CF = HE: GOTO 2820
2810 CF =  FN DEG(L(PS))
2820  IF CF -  INT (CF) > .49 THEN CF = CF + 1
2830 CF =  INT (CF)
2840  IF CF <  - 9 OR CF > 8 THEN S = 0:OC$ = "OFF CHART": GOTO 2910
2850  IF CF > 0 AND CF < 4 THEN CF = 128 * CF: GOTO 2890
2860  IF CF < 9 AND CF > 3 THEN CF = (CF * 128) - 984: GOTO 2890
2870  IF CF < 0 AND CF >  - 5 THEN CF = 128 * CF: GOTO 2890
2880  IF CF <  - 4 AND CF >  - 10 THEN CF = 984 + 128 * CF: GOTO 2890
2890 S = S - CF
2900  RETURN
2910  REM  PRINT ECLIPTIC
2920  RETURN
2930  FOR J = 1 TO 39 STEP 2
2940  VTAB 12: PRINT  TAB( J)"-";
2950  NEXT J
2960  PRINT
2970  GOTO 3290
2980  HOME : PRINT : PRINT : PRINT
2990  PRINT "WANT ANOTHER PLANET, SAME DATE? Y/N "
3000  PRINT : PRINT : INPUT ".......  ";A$
3010  IF A$ = "N" THEN  GOTO 3100
3020  IF A$ < > "Y" THEN  PRINT "INVALID RESPONSE": PRINT : GOTO 2990
3030  REM  RESET Y,M,D TO ORIGINAL SELECTION
3040 RA = 0:DE = 0:RA$ = "":DE$ = ""
3050 Y = Y2:M = M2:D = D2
3060 F9 = 0:F8 = 0:F6 = 0
3070 FL = 0:F1 = 0:F2 = 0:F3 = 0:F4 = 0:F5 = 0:F7 = 0:FX = 0
3080  HOME : PRINT : PRINT : PRINT
3090  GOTO 770
3100  HOME
3110  PRINT : PRINT : PRINT : PRINT
3120  PRINT "WANT ANOTHER DATE, SAME PLANET? Y/N "
3130  PRINT : PRINT : INPUT "........ ";A$
3140  IF A$ = "N" THEN  HOME : PRINT : PRINT : GOTO 3200
3150  IF A$ < > "Y" THEN  PRINT "INVALID RESPONSE": PRINT : GOTO 3120
3160 PL = P2:F9 = 0:FL = 0:F1 = 0:F2 = 0:F3 = 0:F4 = 0:F5 = 0:F6 = 0:F7
      = 0:F8 = 0:FX = 0
```

```
3170 F9 = 2
3180  GOTO 480
3190  PRINT
3200  PRINT "WANT ANOTHER DATE, ANOTHER PLANET? Y/N "
3210  PRINT : PRINT
3220  INPUT "........ ";A$
3230  IF A$ = "N" THEN  HOME : GOTO 3280
3240  IF A$ < > "Y" THEN  PRINT "INVALID RESPONSE": PRINT : GOTO 3200
3250  HOME
3260 FL = 0:F1 = 0:F2 = 0:F3 = 0:F4 = 0:F5 = 0:F6 = 0:F7 = 0:F9 = 0:F8 = 0
3270 FX = 0: GOTO 480
3280  END
3290  REM   SELECT ZODIACAL CHART REGION
3300  IF CH = 1 GOTO 3660
3310  IF CH = 2 GOTO 3910
3320  IF CH = 3 GOTO 4160
3330  IF CH = 4 GOTO 4330
3340  IF CH = 5 GOTO 4580
3350  IF CH = 6 GOTO 4810
3360  IF CH = 7 GOTO 5020
3370  IF CH = 8 GOTO 5170
3380  IF CH = 9 GOTO 5280
3390  IF CH = 10 GOTO 5510
3400  IF CH = 11 GOTO 5720
3410  IF CH = 12 GOTO 5930
3420  REM  PLOT OTHER PLANETS
3430  IF IN > 1 THEN F6 = 1: GOTO 3520
3440  INPUT "OTHERS? Y/N ";A$
3450  IF A$ = "N" THEN F6 = 1: GOTO 3520
3460  IF A$ < > "Y" THEN  GOTO 3440
3470  IF A$ = "Y" THEN F5 = 1
3480 F6 = 0
3490  REM  SUB TO PLOT OTHER PLANETS
3500 CF = 0
3510  GOSUB 6120
3520  REM  CHECKS IF MORE CYCLES
3530  IF NC < IN GOTO 3560
3540  IF NC > = IN GOTO 3590
3550  REM  SUB TO INCREASE DATE BY INTERVAL AND REPEAT PROGRAM
3560  GOSUB 6480
3570  IF F9 = 1 THEN  GOTO 3530
3580  REM  RESETS F9 WHEN ALL CYCLES COMPLETED
3590  REM  RESETS NC,D,M AND Y TO INITIAL VALUES
3600 NC = 1:D = D2:M = M2:Y = Y2
3610 PS = P2:RA = R3:DE = DC
3620 P = PN:PL = PX
3630  INPUT "PRESS RETURN TO CONTINUE ";A$
3640  :
3650  GOTO 2980
3660  REM   CHART 1 PISCES
3670 C$ = "PISCES"
3680  GOSUB 3700
3690  GOTO 3810
3700  IF  FN DEG(L(PS)) < - 10 THEN W$ = "OFF CHART BOTTOM"
```

PLNTF *(continued)*

```
3710    IF  FN DEG(L(PS)) > 10 THEN W$ = "OFF CHART TOP"
3720    VTAB 2
3730    PRINT P$;" IS IN ";C$;" ";W$
3740    VTAB 3
3750    PRINT "ON ";Y2;" ";M2;" ";D2
3760    VTAB 4
3770    IF  LEFT$ (W$,3) = "OFF" THEN GOTO 3790
3780    PRINT "AS DISPLAYED ON CHART BELOW"
3790    W$ = ""
3800    RETURN
3810    VTAB 7: PRINT  TAB( 9)"."
3820    VTAB 7: PRINT  TAB( 37)"*"
3830    VTAB 10: PRINT  TAB( 23)"."
3840    VTAB 11: PRINT  TAB( 19)"."
3850    VTAB 13: PRINT  TAB( 5)"."
3860    VTAB 15: PRINT  TAB( 9)"."
3870    VTAB 19: PRINT  TAB( 5)"*"
3880    VTAB 20: PRINT  TAB( 38)"."
3890    POKE S - PL,P
3900    GOTO 7190
3910    REM   CHART 2 ARIES
3920    C$ = "ARIES"
3930    GOSUB 3700
3940    VTAB 5
3950    PRINT  TAB( 33)"X"
3960    VTAB 6
3970    PRINT  TAB( 37)"*"
3980    VTAB 7
3990    PRINT  TAB( 37)"."
4000    VTAB 10
4010    PRINT  TAB( 4)".::"
4020    VTAB 10
4030    PRINT  TAB( 15)"."
4040    VTAB 15
4050    PRINT  TAB( 36)"."
4060    VTAB 16
4070    PRINT  TAB( 27)"."
4080    VTAB 19
4090    PRINT  TAB( 33)"."
4100    VTAB 18
4110    PRINT  TAB( 5)"*"; TAB( 16)". ."
4120    VTAB 21
4130    PRINT  TAB( 24)"*"; TAB( 30)"*"
4140    POKE S - PL,P
4150    GOTO 7190
4160    REM   CHART 3 TAURUS
4170    C$ = "TAURUS"
4180    GOSUB 3700
4190    VTAB 6
4200    PRINT  TAB( 11)"*"
4210    VTAB 13
4220    PRINT  TAB( 9)"*"; TAB( 19)"."; TAB( 29)"."
4230    VTAB 14
4240    PRINT  TAB( 30)"."
4250    VTAB 16
```

PLNTF *(continued)*

```
4260    PRINT  TAB( 27)"X"; TAB( 29)":"; TAB( 32)"."
4270    VTAB 21
4280    PRINT  TAB( 21)"."
4290    VTAB 22
4300    PRINT  TAB( 10)"."; TAB( 22)"."
4310    POKE S - PL,P
4320    GOTO 7190
4330    REM   CHART 4 GEMINI
4340 C$ = "GEMINI"
4350    GOSUB 3700
4360    VTAB 6
4370    PRINT  TAB( 6)"*"
4380    VTAB 7
4390    PRINT  TAB( 9)"."
4400    VTAB 8
4410    PRINT  TAB( 6)"."
4420    VTAB 9
4430    PRINT  TAB( 23)"*"
4440    VTAB 11
4450    PRINT  TAB( 15)"*"
4460    VTAB 13
4470    PRINT  TAB( 19)"."; TAB( 31)"*"; TAB( 33)"*"
4480    VTAB 14
4490    PRINT  TAB( 29)"."
4500    VTAB 16
4510    PRINT  TAB( 26)"*"
4520    VTAB 19
4530    PRINT  TAB( 23)"."
4540    VTAB 22
4550    PRINT  TAB( 11)"*"
4560    POKE S - PL,P
4570    GOTO 7190
4580    REM   CHART 5 CANCER
4590 C$ = "CANCER"
4600    GOSUB 3700
4610    VTAB 9
4620    PRINT  TAB( 26)"."
4630    VTAB 10
4640    PRINT  TAB( 25)"."
4650    VTAB 14
4660    PRINT  TAB( 6)"*"
4670    VTAB 15
4680    PRINT  TAB( 19)"."
4690    VTAB 18
4700    PRINT  TAB( 18)"*"; TAB( 21)"*"
4710    VTAB 19
4720    PRINT  TAB( 26)"."; TAB( 29)"*"
4730    VTAB 20
4740    PRINT  TAB( 12)"."; TAB( 23)"."
4750    VTAB 21
4760    PRINT  TAB( 20)"."
4770    VTAB 22
4780    PRINT  TAB( 22)"."
4790    POKE S - PL,P
4800    GOTO 7190
```

PLNTF *(continued)*

```
4810   REM   CHART 6 LEO
4820  C$ = "LEO"
4830   GOSUB 3700
4840   VTAB 5
4850   PRINT   TAB( 20)"*"; TAB( 36)"*"
4860   VTAB 6
4870   PRINT   TAB( 15)"."
4880   VTAB 7
4890   PRINT   TAB( 3)"."; TAB( 37)"."
4900   VTAB 8
4910   PRINT   TAB( 7)"."
4920   VTAB (10)
4930   PRINT   TAB( 14)"."; TAB( 19)"."
4940   VTAB 11
4950   PRINT   TAB( 3)"*"; TAB( 28)"."; TAB( 36)"X"
4960   VTAB 14
4970   PRINT   TAB( 7)"."
4980   VTAB 18
4990   PRINT   TAB( 12)"."
5000   POKE S - PL,P
5010   GOTO 7190
5020   REM   CHART 7 VIRGO
5030  C$ = "VIRGO"
5040   GOSUB 3700
5050   VTAB 5
5060   PRINT   TAB( 14)"*"; TAB( 26)"*"
5070   VTAB 9
5080   PRINT   TAB( 27)"*"
5090   VTAB 11
5100   PRINT   TAB( 19)"."; TAB( 33)"*"
5110   VTAB 14
5120   PRINT   TAB( 14)"X"
5130   VTAB 21
5140   PRINT   TAB( 24)"*"
5150   POKE S - PL,P
5160   GOTO 7190
5170   REM   CHART 8 LIBRA
5180  C$ = "LIBRA"
5190   GOSUB 3700
5200   VTAB 6
5210   PRINT   TAB( 18)"*"
5220   VTAB 11
5230   PRINT   TAB( 22)"*"
5240   VTAB 19
5250   PRINT   TAB( 16)"."
5260   POKE S - PL,P
5270   GOTO 7190
5280   REM   CHART 9 SCORPIO
5290  C$ = "SCORPIO"
5300   GOSUB 3700
5310   VTAB 5
5320   PRINT   TAB( 8)"."; TAB( 13)"."
5330   VTAB 7
5340   PRINT   TAB( 10)"*"; TAB( 18)"*"
5350   VTAB 10
```

```
5360    PRINT   TAB( 36)"*"
5370    VTAB 13
5380    PRINT   TAB( 37)"*"
5390    VTAB 14
5400    PRINT   TAB( 29)"*"
5410    VTAB 15
5420    PRINT   TAB( 27)"X"
5430    VTAB 16
5440    PRINT   TAB( 37)
5450    VTAB 17
5460    PRINT   TAB( 25)"*"
5470    VTAB 21
5480    PRINT   TAB( 21)"*"
5490    POKE S - PL,P
5500    GOTO 7190
5510    REM   CHART 10 SAGITTARIUS
5520    C$ = "SAGITTARIUS"
5530    GOSUB 3700
5540    VTAB 10
5550    PRINT   TAB( 29)"."
5560    VTAB 11
5570    PRINT   TAB( 18)"."; TAB( 21)"."
5580    VTAB 14
5590    PRINT   TAB( 30)"*"
5600    VTAB 15
5610    PRINT   TAB( 22)"*"
5620    VTAB 16
5630    PRINT   TAB( 24)"."
5640    VTAB 17
5650    PRINT   TAB( 24)"*"; TAB( 36)"*"
5660    VTAB 18
5670    PRINT   TAB( 20)"*"
5680    VTAB 22
5690    PRINT   TAB( 31)"*"
5700    POKE S - PL,P
5710    GOTO 7190
5720    REM   CHART 11 CAPRICORNUS
5730    C$ = "CAPRICORNUS"
5740    GOSUB 3700
5750    VTAB 5
5760    PRINT   TAB( 6)"*"
5770    VTAB 6
5780    PRINT   TAB( 20)"."; TAB( 30)"*"
5790    VTAB 8
5800    PRINT   TAB( 30)"*"
5810    VTAB 13
5820    PRINT   TAB( 13)"."; TAB( 18)"*"
5830    VTAB 14
5840    PRINT   TAB( 6)"*"; TAB( 8)"*"
5850    VTAB 17
5860    PRINT   TAB( 13)".."
5870    VTAB 18
5880    PRINT   TAB( 26)"."
5890    VTAB 19
5900    PRINT   TAB( 20)"."; TAB( 25)"."
```

┌─ **PLNTF** *(continued)* ─────────────────────────────────────┐

```
5910   POKE S - PL,P
5920   GOTO 7190
5930   REM  CHART 12 AQUARIUS
5940 C$ = "AQUARIUS"
5950   GOSUB 3700
5960   VTAB 5
5970   PRINT  TAB( 7)"."; TAB( 14)"."; TAB( 33)"*"
5980   VTAB 6
5990   PRINT  TAB( 5)"."; TAB( 11)"."; TAB( 25)"*"; TAB( 29)"*"
6000   VTAB 13
6010   PRINT  TAB( 23)"."
6020   VTAB 16
6030   PRINT  TAB( 18)".."
6040   VTAB 17
6050   PRINT  TAB( 26)"."
6060   VTAB 19
6070   PRINT  TAB( 26)"*"
6080   VTAB 20
6090   PRINT  TAB( 2)"*"
6100   POKE S - PL,P
6110   GOTO 7190
6120   REM  SUB TO PLOT NEARBY PLANETS
6130 PN = P
6140   FOR I = 1 TO 10
6150   PS = I:F5 = 1
6160   REM  JUMPS R RANGE IF BEEN THROUGH TABLE ONCE
6170   IF F4 = 1 THEN 6300
6180   IF CH = 1 THEN R2 = 0: GOTO 6300
6190   IF CH = 2 THEN R2 = 2: GOTO 6300
6200   IF CH = 3 THEN R2 = 4: GOTO 6300
6210   IF CH = 4 THEN R2 = 6: GOTO 6300
6220   IF CH = 5 THEN R2 = 8: GOTO 6300
6230   IF CH = 6 THEN R2 = 10: GOTO 6300
6240   IF CH = 7 THEN R2 = 12: GOTO 6300
6250   IF CH = 8 THEN R2 = 14: GOTO 6300
6260   IF CH = 9 THEN R2 = 16: GOTO 6300
6270   IF CH = 10 THEN R2 = 18: GOTO 6300
6280   IF CH = 11 THEN R2 = 20: GOTO 6300
6290   IF CH = 12 THEN R2 = 22
6300   IF R(I) < R2 OR R(I) > (R2 + 1.99999) THEN  GOTO 6410
6310   REM  GET PLANET SYMBOL
6320   GOSUB 2250
6330 RA = R(I):DE = V(I)
6340   REM  GET POKE FOR PLANET
6350   GOSUB 2440
6360   IF F7 = 1 THEN F7 = 0: GOTO 6400
6370 CF = 0:S = 1487
6380   GOSUB 2790
6390   POKE S - PL,P
6400   F4 = 1
6410   NEXT I
6420 PS = P2
6430 P = PN
6440   GOSUB 7210
6450 RA = R3:DE = DC
```

└──┘

```
6460 F4 = 0:F5 = 0:F = 1: RETURN
6470   REM   INCREASE COUNTER BY ONE
6480 NC = NC + 1
6490 PS = P2
6500 F9 = 1
6510 D = D + TI
6520   REM   INCREMENTS DATE
6530   IF D > 30 THEN D = D - 30:M = M + 1: GOTO 6530
6540   IF M > 12 THEN M = M - 12:Y = Y + 1
6550   REM   GOES THROUGH PROG. AGAIN FOR NEW DATE
6560   GOSUB 820
6570   REM   PLOT PLANET'S NEW POSITION
6580   IF F7 = 1 THEN F7 = 0: GOTO 6600
6590   POKE S - PL,P
6600   REM   POKE OTHER PLANETS ON NEW DATE
6610   IF F6 = 1 THEN 6630
6620   F9 = 0: GOSUB 6120
6630   REM   REPEATS IF MORE INTERVALS
6640   F9 = 1
6650   RETURN
6660   REM   ADJUST RA PLOT TO CHART
6670   IF RA > 22 AND RA < 23.99999 THEN RA = RA - 22: RETURN
6680   IF RA > 20 AND RA < 21.99999 THEN RA = RA - 20: RETURN
6690   IF RA > 18 AND RA < 19.99999 THEN RA = RA - 18: RETURN
6700   IF RA > 16 AND RA < 17.99999 THEN RA = RA - 16: RETURN
6710   IF RA > 14 AND RA < 15.99999 THEN RA = RA - 14: RETURN
6720   IF RA > 12 AND RA < 13.99999 THEN RA = RA - 12: RETURN
6730   IF RA > 10 AND RA < 11.99999 THEN RA = RA - 10: RETURN
6740   IF RA > 8 AND RA < 9.99999 THEN RA = RA - 8: RETURN
6750   IF RA > 6 AND RA < 7.99999 THEN RA = RA - 6: RETURN
6760   IF RA > 4 AND RA < 5.99999 THEN RA = RA - 4: RETURN
6770   IF RA > 2 AND RA < 3.99999 THEN RA = RA - 2: RETURN
6780   RETURN
6790   IF CH = 1 THEN   GOTO 6920
6800   IF CH = 2 THEN   GOTO 6940
6810   IF CH = 3 THEN   GOTO 6960
6820   IF CH = 4 THEN   GOTO 6980
6830   IF CH = 5 THEN   GOTO 7000
6840   IF CH = 6 THEN   GOTO 7020
6850   IF CH = 7 THEN   GOTO 7040
6860   IF CH = 8 THEN   GOTO 7060
6870   IF CH = 9 THEN   GOTO 7080
6880   IF CH = 10 THEN   GOTO 7100
6890   IF CH = 11 THEN   GOTO 7120
6900   GOTO 7140
6910   REM   REJECT OBJECTS NOT ON CHART
6920   IF RA > 1.99999 THEN   GOTO 7160
6930   RETURN
6940   IF RA < 2 OR RA > 3.9999 THEN GOTO 7160
6950   RETURN
6960   IF RA < 4 OR RA > 5.99999 THEN GOTO 7160
6970   RETURN
6980   IF RA < 6 OR RA > 7.99999 THEN GOTO 7160
6990   RETURN
7000   IF RA < 8 OR RA > 9.99999 THEN GOTO 7160
```

PLNTF *(continued)*

```
7010   RETURN
7020   IF RA < 10 OR RA > 11.99999 THEN   GOTO 7160
7030   RETURN
7040   IF RA < 12 OR RA > 13.99999 THEN   GOTO 7160
7050   RETURN
7060   IF RA < 14 OR RA > 15.99999 THEN   GOTO 7160
7070   RETURN
7080   IF RA < 16 OR RA > 17.99999 THEN   GOTO 7160
7090   RETURN
7100   IF RA < 18 OR RA > 19.99999 THEN   GOTO 7160
7110   RETURN
7120   IF RA < 20 OR RA > 21.99999 THEN   GOTO 7160
7130   RETURN
7140   IF RA < 22 OR RA > 23.99999 THEN   GOTO 7160
7150   RETURN
7160 F7 = 1
7170   PRINT P$(PS);" OFF CHART";
7180   RETURN
7190   VTAB 23
7200   GOTO 3420
7210   REM  POKE MOON
7220   REM   LONG OF MOON
7230 LZ = 311.1687:LP = 255.7433:LE = 178.699
7240 NM = NI - .5
7250 PG = .111404 * NM + LP
7260   IF PG <  - 3600 THEN PG = PG + 3600: GOTO 7260
7270   IF PG <  - 360 THEN PG = PG + 360: GOTO 7270
7280   IF PG < 0 THEN PG = PG + 360
7290   IF PG > 3600 THEN PG = PG - 3600: GOTO 7290
7300   IF PG > 360 THEN PG = PG - 360: GOTO 7300
7310 LMD = LZ + 360 * NM / 27.32158
7320 PG = LMD - PG
7330 DR = 6.2886 * SIN ( FN RAD(PG))
7340 LMD = LMD + DR
7350   IF LMD <  - 3600 THEN LMD = LMD + 3600: GOTO 7350
7360   IF LMD <  - 360 THEN LMD = LMD + 360: GOTO 7360
7370   IF LMD < 0 THEN LMD = LMD + 360: GOTO 7370
7380   IF LMD > 3600 THEN LMD = LMD - 3600: GOTO 7380
7390   IF LMD > 360 THEN LMD = LMD - 360: GOTO 7390
7400 RM = LMD / 15
7410 RQ = RM
7420   IF RM > 24 THEN RM = RM - 24: GOTO 7420
7430   IF RM < 0 THEN RM = RM + 24
7440 AL = LE - NM * .052954
7450   IF AL <  - 3600 THEN AL = AL + 3600: GOTO 7450
7460   IF AL <  - 360 THEN AL = AL + 360: GOTO 7460
7470   IF AL < 0 THEN AL = AL + 360: GOTO 7470
7480   IF AL > 3600 THEN AL = AL - 3600: GOTO 7480
7490   IF AL > 360 THEN AL = AL - 360: GOTO 7490
7500 AL = LMD - AL
7510   IF AL < 0 THEN AL = AL + 360
7520   IF AL > 360 THEN AL = AL - 360
7530 HE = 5.1333 * SIN (AL * 3.14159 / 180)
7540 DM = HE + 23.1444 * SIN (LMD * 3.14159 / 180)
7550 R(10) = RM:V(10) = DM
```

PLNTF *(continued)*

```
7560 P$(10) = "MOON"
7570 RETURN
7580 GOSUB 2090: GOSUB 2470
7590 IF RA < 0 OR RA > 1.99999 THEN 7610
7600 GOTO 2090
7610 PRINT "MOON OFF CHART"
7620 RETURN
```

Io is one of the four Galilean satellites of Jupiter. All four are strange, planet-sized worlds, ranging from Io with its active volcanoes to the frigid outer worlds of Ganymede and Callisto. These large satellites are shown in this montage of pictures returned by the Voyager spacecraft. Io is bottom right; Europa, bottom left; Ganymede, top right; and Callisto, top left. Their sizes are shown proportionately; Ganymede is slightly larger than the planet Mercury.

Program 18: JSATS

Positions of Galilean Satellites of Jupiter for Any Date and Time_____

In 1610 Galileo in Padua startled his contemporaries with his discovery that Jupiter has four large satellites. He thereby provided evidence that ultimately helped to shatter the dogma of an Earth-centered universe and led to the Copernican revolution of human thought. In Paris 65 years later, Roemer used the timing of the motion of these satellites to measure the velocity of light across Earth's orbit. His observations showed that light travels at 186,000 miles per second.

Even the smallest telescope reveals the four Galilean satellites of Jupiter moving over the course of days from side to side of the planet. The close-up photographs returned by the Voyager spacecraft revealed much greater detail. Io, the innermost of the large moons, is a volcanic world of brimstone and sulfur with volcanic fountains shooting material hundreds of miles into space. It is a world torn by the gravities of mighty Jupiter and the other satellites, and bombarded by energetic particles trapped in the magnetic field of the giant planet. Europa is a very different but equally strange world. At first it appeared to be as smooth as a bowling ball, but it was later revealed as a world whose icy surface is marked by an intricate pattern of eggshell-like cracks. Mighty Ganymede, larger than the planet Mercury, displays several different types of terrain, very different from the surfaces of Io and Europa. There are great plains patterned by strange areas of grooves and ghostlike rings of ruined craters. Unlike the inner satellites, Ganymede has regions of heavily cratered terrain. The outermost of the big satellites, Callisto, is more like our Moon. Its surface is pockmarked with innumerable craters.

This program allows you to compute the positions of the four satellites relative to Jupiter as seen from Earth, and it identifies the satellites.

With the program you will never be in doubt as to the identity of each of the four starlike objects. The program can be used worldwide; it asks for your time zone and local time. (Times are displayed in both universal and local time.) It will display several time intervals on the screen. More can be requested, but the display will scroll on the screen after these have been displayed. If you habitually need more time intervals displayed together on the monitor screen, you can tighten the display by eliminating instructions 1410 and 1500.

The display can be oriented with either north or south at the top to suit observers with different types of telescopes and for Northern or Southern Hemisphere viewing.

When you run this program, the computer first displays the distances of the satellites in terms of Jupiter radii for exact observations (see Figure 18.1). When you are ready to continue it displays the satellites and the planet, identified as follows: J, Jupiter; I, Io; E, Europa; G, Ganymede; and C, Callisto (see Figure 18.2).

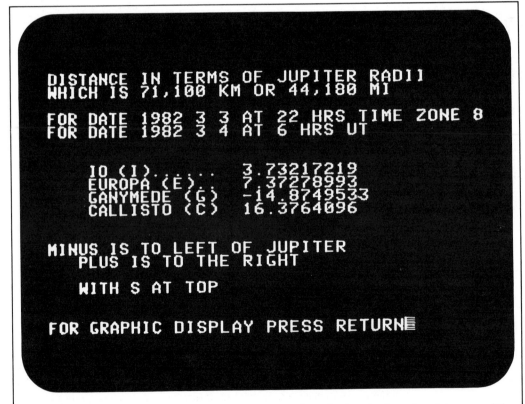

The JSATS program provides this information in its first display. It gives the distances of the Galilean satellites from Jupiter in terms of Jovian radii (44,180 miles).

Figure 18.1

```
JUPITER..J                    S                    IO     I
EUROPA...E                    ^                  GANYMEDE G
CALLISTO C                    |
- - - - - - - - - - - - - - - | - - - - - - - - - - - - - - - -
1982 3 3 AT 22 HR   ZONE 8
 .    .    .    .    .    . UT 6 HRS 3 4
          G                  J  I   E      C
- - - - - - - - - - - - - - - - - - - - - - - - - - - - - - -
1982 3 4 AT 22 HR   ZONE 8
 .    .    .    .    .    . UT 6 HRS 3 5
          G      I  J  E      C
- - - - - - - - - - - - - - - - - - - - - - - - - - - - - - -
1982 3 5 AT 22 HR   ZONE 8
 .    .    .    .    .    . UT 6 HRS 3 6
          E         CIJ   G

DO YOU WANT ANOTHER DATE Y/N? ▮
```

Next the JSATS program generates a graphic display showing the positions of the satellites relative to Jupiter. For this display three configurations were selected, 24 hours apart. The movements of the satellites relative to the planet are clearly shown.

—Figure 18.2—

Jupiter is always centrally located on the display line. It also shows which format, north or south at top, is being used. The program does not show positions out of the horizontal plane, so sometimes if two of the satellites are visible, one slightly above the other at the same angular distance from Jupiter, the display shows them side by side. You can quickly ascertain which is which by asking for displays a few hours on either side. The display also shows the satellites alongside Jupiter when in occultation or transit—they are not erased from the screen. Again, by asking for other displays at earlier times you can identify transits and occultations. With north at top, satellites moving toward Jupiter from the right of the planet will go into occultation behind Jupiter; those moving in the opposite direction will go into transit across Jupiter. With south at top, satellite movement from the right results in transit and from the left it results in occultation.

You can add to the program and simplify the sorting out of occultations and transits by showing an orbit projection, using the high resolution graphics technique given in the MARSP program.

The program first calculates the number of days from the epoch of 1900, adjusting for time and time zone. It uses this calculation to determine the position of each satellite in its orbit around Jupiter at the requested date (instructions 920 through 1040). Next it determines the relative positions of Earth and Jupiter at the requested date (instructions 1050 through 1120) and corrects for the viewing angle from Earth. Then it calculates the radial distance of each satellite from Jupiter (in terms of the radius of Jupiter) as viewed from Earth's position, and it displays this information on the monitor screen. When you are ready to continue, the program uses a simple sort subroutine (starting at instruction 2100) to arrange the satellites in the correct order. They can then be displayed together with Jupiter on one line across the screen, at proportional distances from Jupiter.

Next the program increments the time and date as requested and develops a second and third display. Housekeeping routines adjust for month and year ends in the display of dates for each configuration of the satellites.

As with other programs, you may have to change HOME statements (instructions 30, 200, 370, 660, 1380, 1570, 2060, and 2080) and change RETURN to ENTER in instruction 1560.

The listing of the JSATS program follows.

── **JSATS** ────

```
10    CLEAR
20    REM   SATELLITES OF JUPITER
30    HOME : PRINT : PRINT : PRINT : PRINT
40    DIM Z$(4),B$(4),Y$(4),M$(2),DA$(2),LT$(5)
50    DIM TZ$(2),ND$(1),IN$(2)
60    PRINT   TAB( 8)"AN ASTRONOMY PROGRAM"
70    PRINT
80    PRINT   TAB( 6)"-------------------------"
90    PRINT   TAB( 6)"I  GALILEAN SATELLITES   I"
100   PRINT   TAB( 6)"-------------------------"
110   PRINT : PRINT
120   PRINT   TAB( 7)"BY ERIC BURGESS F.R.A.S."
130   PRINT
140   PRINT   TAB( 12)"VERSION APR.82"
150   PRINT
160   PRINT   TAB( 8)"ALL RIGHTS RESERVED BY"
170   PRINT   TAB( 8)"S & T SOFTWARE SERVICE"
180   PRINT : PRINT
190   FOR K = 3000 TO 1 STEP  - 1: NEXT
200   HOME : PRINT : PRINT : PRINT : PRINT
210   PRINT   TAB( 5)"THIS PROGRAM SHOWS THE GALILEAN"
220   PRINT "SATELLITES POSITIONED RELATIVE"
230   PRINT   TAB( 5)"TO JUPITER AT THEIR APPROXIMATE"
240   PRINT   TAB( 11)"RELATIVE POSITIONS"
250   PRINT : PRINT
260   PRINT "YOU CAN CHOOSE TO HAVE THEM SHOWN WITH"
270   PRINT "NORTH AT THE TOP, OR WITH SOUTH AT THE"
280   PRINT "TOP AS SEEN IN A TELESCOPE, AND YOU"
290   PRINT "CAN SELECT UP TO FIVE DISPLAYS OF THEIR"
300   PRINT "CONFIGURATIONS AT UP TO 24 HRS APART"
310   PRINT : PRINT : PRINT .
320   INPUT "DO YOU WANT SOUTH AT TOP (Y/N)? ";A$
330   IF A$ = "Y" THEN FL = 1: GOTO 360
340   IF A$ < > "N" THEN  PRINT "INVALID RESPONSE": PRINT : GOTO 320
350   FL = 0
360   DEF  FN RAD(X) = (X) * 3.14159 / 180
370   HOME : PRINT : PRINT : PRINT
380   REM  ENTER DATE
390   LY = 0: PRINT "ENTER THE DATE"
400   PRINT : PRINT
410   INPUT "YEAR ";Y$:Y =  VAL (Y$)
420   IF Y = 0 THEN  PRINT "INVALID RESPONSE": PRINT : GOTO 410
430   IF Y / 4 -  INT (Y / 4) = 0 AND Y / 100 -  INT (Y / 100) < > 0 THEN
      LY = 1
440   PRINT
450   INPUT "MONTH ";M$
460   M =  VAL (M$)
470   IF M = 0 OR M > 12 THEN  PRINT "INVALID RESPONSE": PRINT : GOTO 450
480   PRINT
490   INPUT "DAY ";D$:DA =  VAL (D$)
500   IF DA = 0 OR DA > 31 THEN  PRINT "INVALID RESPONSE": PRINT : GOTO 490
510   IF LY = 1 AND M = 2 AND DA < 30 THEN  GOTO 550
520   IF LY = 0 AND M = 2 AND DA < 29 THEN  GOTO 550
```

JSATS (continued)

```
530   IF M < > 2 THEN  GOTO 550
540   PRINT "INVALID RESPONSE": PRINT : GOTO 490
550   PRINT : INPUT "TIME ZONE ";TZ$:TZ = VAL (TZ$)
560   IF TZ < - 12 OR TZ > 12 THEN PRINT "INVALID RESPONSE": PRINT :
      GOTO 550
570   PRINT : INPUT "LOCAL TIME (DEC.HRS.) ";LT$
580 LT = VAL (LT$)
590   IF LT < 0 OR LT > 24 THEN  PRINT "INVALID RESPONSE": PRINT : GOTO 570
600   PRINT
610   INPUT "HOW MANY DISPLAYS (1 TO 5)? ";ND$:ND = VAL (ND$)
620   IF ND < 1 OR ND > 5 THEN  PRINT "INVALID RESPONSE": PRINT : GOTO 610
630   PRINT
640   INPUT "WHAT IS INTERVAL (UP TO 24 HR)? ";IN
650   IF IN < 0 OR IN > 24 THEN  PRINT "INVALID RESPONSE": PRINT : GOTO 640
660   HOME : PRINT : PRINT : PRINT
670   PRINT "PROGRAM RUNNING...PLEASE WAIT"
680 L = 1
690   YP = Y:MP = M:DP = DA
700   IF L > 1 GOTO 720
710 UT = LT + TZ
720   REM  CORRECT FOR DAY MONTH AND YR ENDS
730   IF UT > 24 THEN UT = UT - 24:DA = DA + 1
740   IF (LY = 1 AND M = 2 AND DA = 30) THEN M = 3:DA = 1: GOTO 790
750   IF (LY = 0 AND M = 2 AND DA = 29) THEN M = 3:DA = 1: GOTO 790
760   IF DA < 31 GOTO 790
770   IF (M = 4 OR M = 6 OR M = 9 OR M = 11) THEN M = M + 1:DA = 1: GOTO 790
780   IF DA > 31 THEN M = M + 1:DA = 1
790   IF M = 13 THEN M = 1:Y = Y + 1
800   IF FL = 1 THEN 820
810   YP = Y:MP = M:DP = DA
820 T = UT / 24
830   YC = Y:MC = M:DC = DA + T
840   IF MC > 2 THEN 880
850   YC = Y - 1
860   MC = M + 12
870 TD = INT (365.25 * YC) + INT ((MC + 1) * 30.6001) + DC
880 :
890 YZ = INT (YC / 100)
900   REM  CALC DAYS SINCE 1900 EPOCH
910 I = TD - YZ + 2 + INT (YZ / 4) - 694025.5
920   REM  DEG/DAY*NUMBR DAYS + POSTN AT EPOCH
930 A = I * 203.40586 + 84.55061
940   ZZ = A: GOSUB 2290
950 A = ZZ
960 B = I * 101.2916323 + 41.50155
970   ZZ = B: GOSUB 2290
980 B = ZZ
990 C = I * 50.23451687 + 109.97702
1000  ZZ = C: GOSUB 2290
1010 C = ZZ
1020 D = I * 21.48798021 + 176.35864
1030  ZZ = D: GOSUB 2290
1040 D = ZZ
1050  REM  ORBITAL MOTION IN PERIOD
1060  REM  EARTH AND JUPITER
```

```
1070 P1 = I * .9856003 + 358.476
1080 ZZ = P1: GOSUB 2290
1090 P1 = ZZ
1100 I2 = I * .0830853 + 225.328
1110 ZZ = I2: GOSUB 2290:I2 = ZZ
1120 P2 = I * .9025179 + 221.647
1130 ZZ = P2: GOSUB 2290
1140 P2 = ZZ
1150 P2 = P2 + ( SIN ( FN RAD(2 * P1)) / 50) + ( SIN ( FN RAD(P1)) / .521)
1160 P3 = ( SIN ( FN RAD(2 * I2)) * .1673) + ( SIN ( FN RAD(I2)) * 5.5372)
1170 P2 = P2 - P3
1180 P1 =  SQR (28.07 - ( COS ( FN RAD(P2)) * 10.406))
1190 J = ( SIN ( FN RAD(P2)) / P1)
1200 I3 = 57.2958 * ( ATN (( SIN ( FN RAD(J))) /
      SQR (1 - ( SIN ( FN RAD(J))) ^ 2)))
1210 I3 = I3 * 57.2958
1220 P1 = P1 / 173
1230 REM   CALC RADIAL DISTANCE FROM JUPITER
1240 S1 = 5.906 *  SIN ( FN RAD(A + I3 - P3 - (P1 * 203.405863)))
1250 S2 = 9.397 *  SIN ( FN RAD(B + I3 - P3 - (P1 * 101.2916323)))
1260 S3 = 14.989 *  SIN ( FN RAD(C + I3 - P3 - (P1 * 50.23451687)))
1270 S4 = 26.364 *  SIN ( FN RAD(D + I3 - P3 - (P1 * 21.48798021)))
1280 IF FL < > 1 THEN  GOTO 1370
1290 IF S1 < 0 THEN S1 = 0 +  ABS (S1): GOTO 1310
1300 IF S1 > 0 THEN S1 = 0 -  ABS (S1)
1310 IF S2 < 0 THEN S2 = 0 +  ABS (S2): GOTO 1330
1320 IF S2 > 0 THEN S2 = 0 -  ABS (S2)
1330 IF S3 < 0 THEN S3 = 0 +  ABS (S3): GOTO 1350
1340 IF S3 > 0 THEN S3 = 0 -  ABS (S3)
1350 IF S4 < 0 THEN S4 = 0 +  ABS (S4): GOTO 1370
1360 IF S4 > 0 THEN S4 = 0 -  ABS (S4)
1370 IF L > 1 THEN  GOTO 1580
1380 HOME : PRINT : PRINT : PRINT
1390 PRINT "DISTANCE IN TERMS OF JUPITER RADII"
1400 PRINT "WHICH IS 71,100 KM OR 44,180 MI"
1410 PRINT
1420 PRINT "FOR DATE ";YP;" ";MP;" ";DP;" AT ";LT;" HRS TIME ZONE ";TZ
1430 PRINT "FOR DATE ";Y;" ";M;" ";DA;" AT ";UT;" HRS UT"
1440 EM = M
1450 PRINT : PRINT
1460 PRINT  TAB( 5)"IO (I)......  ";S1
1470 PRINT  TAB( 5)"EUROPA (E)..  ";S2
1480 PRINT  TAB( 5)"GANYMEDE (G)  ";S3
1490 PRINT  TAB( 5)"CALLISTO (C)  ";S4
1500 PRINT : PRINT
1510 PRINT "MINUS IS TO LEFT OF JUPITER"
1520 PRINT "   PLUS IS TO THE RIGHT": PRINT
1530 IF FL = 1 THEN  PRINT  TAB( 4)"WITH S AT TOP ": GOTO 1550
1540 PRINT  TAB( 4)"WITH N AT TOP"
1550 PRINT : PRINT
1560 INPUT "FOR GRAPHIC DISPLAY PRESS RETURN";A$
1570 HOME
1580 GOSUB 2100
1590 REM  REGENERATE LMT
1600 LT = UT - TZ
```

---- **JSATS** *(continued)* ----

```
1610  IF LT > 0 THEN  GOTO 1690
1620  DP = DA - 1:LT = LT + 24: IF DP > 0 THEN 1690
1630  IF DP < 1 THEN MP = EM - 1
1640  IF (LY = 1 AND MP = 2) THEN DP = 29: GOTO 1690
1650  IF (LY = 0 AND MP = 2) THEN DP = 28: GOTO 1690
1660  IF (MP = 4 OR MP = 6 OR MP = 9 OR MP = 11) THEN DP = 30: GOTO 1690
1670  DP = 31
1680  PRINT
1690  IF MP = 0 THEN MP = 12:YP = Y - 1
1700  IF L > 1 THEN  GOTO 1760
1710  IF FL = 1 THEN  PRINT  TAB( 20);"S": GOTO 1730
1720  PRINT  TAB( 20);"N"
1730  PRINT "JUPITER..J"; TAB( 20)"^"; TAB( 30);"IO ..... I"
1740  PRINT "EUROPA...E"; TAB( 20)"!"; TAB( 30);"GANYMEDE G"
1750  PRINT "CALLISTO C"; TAB( 20)"!"
1760  PRINT "-------------------------------------"
1770  PRINT YP;" ";MP;" ";DP;" AT ";LT;" HR  ZONE ";TZ
1780  PRINT " . . . . . . UT ";UT;" HRS ";EM;" ";DA
1790  PRINT
1800  M1 = 20 +  INT (.7 *  VAL (B$(0)))
1810  PRINT  TAB( M1); RIGHT$ (B$(0),1);
1820  M2 = 20 +  INT (.7 *  VAL (B$(1)))
1830  PRINT  TAB( M2); RIGHT$ (B$(1),1);
1840  M3 = 20 +  INT (.7 *  VAL (B$(2)))
1850  PRINT  TAB( M3); RIGHT$ (B$(2),1);
1860  M4 = 20 +  INT (.7 *  VAL (B$(3)))
1870  PRINT  TAB( M4); RIGHT$ (B$(3),1);
1880  M5 = 20 +  INT (.7 *  VAL (B$(4)))
1890  PRINT  TAB( M5); RIGHT$ (B$(4),1)
1900  IF L = ND THEN  GOTO 2020
1910  REM  INCREMENT FOR INTERVAL
1920  REM  AND ADJUST MONTH ENDS
1930  UT = UT + IN
1940  IF UT > 24 THEN UT = UT - 24:DA = DA + 1
1950  IF (LY = 1 AND M = 2 AND DA = 30) THEN M = 3:DA = 1: GOTO 2000
1960  IF (LY = 0 AND M = 2 AND DA = 29) THEN M = 3:DA = 1: GOTO 2000
1970  IF DA < 31 GOTO 2000
1980  IF (M = 4 OR M = 6 OR M = 9 OR M = 11) THEN M = M + 1:DA = 1:
      GOTO 2000
1990  IF DA > 31 THEN M = M + 1:DA = 1
2000  IF M = 13 THEN M = 1:Y = Y + 1
2010  L = L + 1: GOTO 690
2020  PRINT : PRINT
2030  INPUT "DO YOU WANT ANOTHER DATE Y/N? ";A$
2040  IF A$ = "N" THEN  GOTO 2080
2050  IF A$ < > "Y" THEN  PRINT "INVALID RESPONSE": PRINT : GOTO 2030
2060  HOME
2070  FL = 0: GOTO 320
2080  HOME
2090  END
2100  REM  SUB TO DISPLAY SATELLITES
2110  Z$(0) =  STR$ (S1) + "I":Z$(1) =  STR$ (S2) + "E"
2120  Z$(2) =  STR$ (S3) + "G":Z$(3) =  STR$ (S4) + "C"
2130  Z$(4) =  STR$ (0) + "J"
2140  FOR I = 0 TO 4
```

—— **JSATS** *(continued)* ——

```
2150 B$(I) = "****"
2160  NEXT I
2170 K = 0
2180  FOR I = 0 TO 4
2190  IF Z$(I) = "****" GOTO 2270
2200  FOR J = I TO 4
2210  IF Z$(J) = "****" GOTO 2230
2220  IF  VAL (Z$(J)) <  VAL (Z$(I)) THEN I = J: GOTO 2230
2230  NEXT J
2240 B$(K) = Z$(I):Z$(I) = "****"
2250 K = K + 1
2260 I = - 1
2270  NEXT I
2280  RETURN
2290  REM  SUB FOR REDUCING ANGLES
2300 ZZ = (ZZ / 360 -  INT (ZZ / 360)) * 360
2310  IF ZZ < 0 THEN ZZ = ZZ + 360
2320  RETURN
```

PART
4

GENERAL AND TUTORIAL

THERE IS A BASIC HUMAN fascination in the night sky, which is especially captivating when our view is unhindered by the glaring lights of the cities. The stars beckon—bright diamonds sparkling in the blackness, some glaring to our dark-adjusted night vision, others elusively winking in and out at the limit of visibility. At first the stars and planets seem distant and aloof, but once you learn to recognize them, they represent both a comforting presence and a new horizon.

This group of programs helps you find and recognize constellations and bright stars, know when there are meteor showers, and have data about the solar system at your fingertips. One of the programs gives helpful information for photographing planets, and another provides useful astronomical conversions.

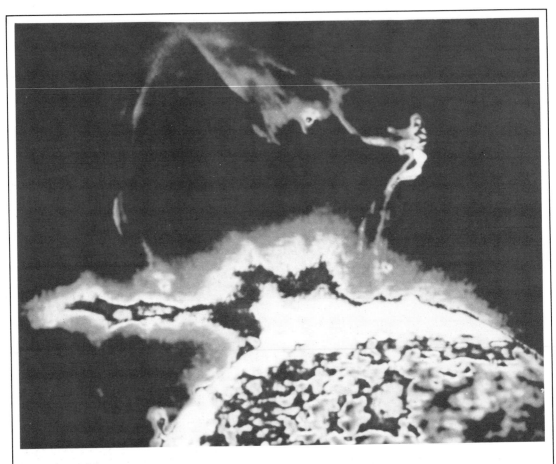

The Sun is our nearest star. The central body of the solar system, it is an enormous nuclear power station converting the nuclear energy of fusion into life-giving heat and light for our planet. This photograph, taken from the telescope carried by the Skylab space-craft while it was still in orbit, shows a solar eruption rising over half a million miles from the solar surface.

Program 19: ACONV

Useful Astronomical Conversions_____

This program is a straightforward conversion routine useful when astronomical measurements need to be expressed in other units. It includes light years, astronomical units, parsecs, telescope resolving power, and metric conversions. It allows conversions either way. If you desire other conversions, these can be added easily, following the pattern shown.

The listing of the ACONV program follows.

```
        ASTRONOMY CONVERSION PROGRAM

  YOU CAN HAVE....

    KM TO MI....(1)  OR  MI TO KM....(2)
    LY TO MI....(3)  OR  MI TO LY....(4)
    PARS. TO LY.(5)  OR  LY TO PARS..(6)
    'C TO 'F....(7)  OR  'F TO 'C....(8)
    A.U. TO MI..(9)  OR  MI TO A.U..(10)
    A.U. TO KM..(9)  OR  KM TO A.U..(10)
    DEG.MIN.SEC. TO DECIMAL DEG....(11)
    DECIMAL DEG. TO DEG.MIN.SEC...(12)
    HRS.MIN.SEC. TO DECIMAL HRS...(13)
    DECIMAL HRS. TO HRS.MIN.SEC...(14)
    RES. PWR. TO INS. OF APERTURE (15)
    INS. OF APERTURE TO RES. PWR. (16)

  TYPE 999 TO END OR TO CHANGE
  TO ANOTHER CONVERSION

  SELECT CONVERSION REQUIRED 1-16 ▦
```

The menu of astronomical conversions available on the ACONV program

— Figure 19.1 —

```
┌─ ACONV ──────────────────────────────────────────────────────────┐
│                                                                    │
│                                                                    │
│   10   HOME : PRINT : PRINT : PRINT : PRINT : PRINT                │
│   20   PRINT  TAB( 7)"---------------------------"                 │
│   30   PRINT  TAB( 7)"I  C O N V E R S I O N S  I"                 │
│   40   PRINT  TAB( 7)"---------------------------"                 │
│   50   PRINT : PRINT                                               │
│   60   PRINT  TAB( 11)"AN ASTRONOMY PROGRAM": PRINT : PRINT        │
│   70   PRINT  TAB( 10)"BY ERIC BURGESS F.R.A.S.": PRINT : PRINT    │
│   80   PRINT  TAB( 11)"ALL RIGHTS RESERVED BY"                     │
│   90   PRINT  TAB( 11)"S & T SOFTWARE SERVICE": PRINT : PRINT      │
│  100   FOR K = 2000 TO 1 STEP  - 1: NEXT K                         │
│  110   HOME : PRINT                                                │
│  120   PRINT : PRINT  TAB( 10)"ASTRONOMY CONVERSION PROGRAM"       │
│  130   PRINT : PRINT : PRINT "YOU CAN HAVE....": PRINT             │
│  140   PRINT  TAB( 5);"KM TO MI....(1) OR MI TO KM....(2)"         │
│  150   PRINT  TAB( 5);"LY TO MI....(3) OR MI TO LY....(4)"         │
│  160   PRINT  TAB( 5);"PARS. TO LY.(5) OR LY TO PARS..(6)"         │
│  170   PRINT  TAB( 5);"'C TO 'F....(7) OR 'F TO 'C....(8)"         │
│  180   PRINT  TAB( 5);"A.U. TO MI..(9) OR MI TO A.U. (10)"         │
│  190   PRINT  TAB( 5);"A.U. TO KM..(9) OR KM TO A.U. (10)"         │
│  200   PRINT  TAB( 5);"DEG.MIN.SEC. TO DECIMAL DEG.. (11)"         │
│  210   PRINT  TAB( 5)"DECIMAL DEG. TO DEG.MIN.SEC.. (12)"          │
│  220   PRINT  TAB( 5);"HRS.MIN.SEC. TO DECIMAL HRS.. (13)"         │
│  230   PRINT  TAB( 5);"DECIMAL HRS. TO HRS.MIN.SEC...(14)"         │
│  240   PRINT  TAB( 5);"RES. PWR. TO INS. OF APERTURE (15)"         │
│  250   PRINT  TAB( 5);"INS. OF APERTURE TO RES. PWR. (16)"         │
│  260   PRINT : PRINT "TYPE 999 TO END OR TO CHANGE"                │
│  270   PRINT "TO ANOTHER CONVERSION"                               │
│  280   PRINT : INPUT "SELECT CONVERSION REQUIRED 1-16 ";D          │
│  290   IF D = 999 THEN  GOTO 1680                                  │
│  300   ON D GOTO 310,420,510,580,660,740,820,890,970,1060,1220,1290,1380,1450, │
│        1540,1610                                                   │
│  310   HOME                                                        │
│  320   PRINT                                                       │
│  330   PRINT "KM TO MI CONVERSION"                                 │
│  340   PRINT "TO STOP ENTER 999"                                   │
│  350   PRINT : INPUT "KM ";X                                       │
│  360   IF X = 999 THEN  GOTO 110                                   │
│  370   Y = X * (1 / 1.60934)                                       │
│  380   PRINT "MI = ";Y                                             │
│  390   PRINT : GOTO 350                                            │
│  400   GOTO 130                                                    │
│  410   IF X = 999 THEN  GOTO 110                                   │
│  420   HOME : PRINT                                                │
│  430   PRINT "MI TO KM CONVERSION"                                 │
│  440   PRINT "TO STOP ENTER 999"                                   │
│  450   PRINT : INPUT "MI ";X                                       │
│  460   IF X = 999 THEN  GOTO 110                                   │
│  470   Y = X * 1.609344                                            │
│  480   PRINT "KM = ";Y                                             │
│  490   GOTO 450                                                    │
│  500   GOTO 10                                                     │
│  510   HOME : PRINT                                                │
│  520   PRINT "LY TO TRILLION MI CONVERSION"                        │
│                                                                    │
└────────────────────────────────────────────────────────────────────┘
```

─── **ACONV** (continued) ───

```
530   PRINT : INPUT "LY ";X
540   IF X = 999 THEN  GOTO 110
550  Y = X * 5.88
560   PRINT "TRILLION MI = "; LEFT$ ( STR$ (Y),5)
570   GOTO 530
580   HOME : PRINT
590   PRINT "TRILLION MI TO LY CONVERSION"
600   PRINT "TO STOP ENTER 999"
610   PRINT : INPUT "TRILLION MI ";X
620   IF X = 999 THEN  GOTO 110
630  Y = X * 1 / 5.88
640   PRINT "LY = "; LEFT$ ( STR$ (Y),5)
650   PRINT : GOTO 610
660   HOME : PRINT
670   PRINT "PARSECS TO LY CONVERSION"
680   PRINT "TO STOP ENTER 999"
690   PRINT : INPUT "PARSECS ";X
700   IF X = 999 THEN  GOTO 110
710  Y = X * 3.26
720   PRINT "LY = "; LEFT$ ( STR$ (Y),5)
730   PRINT : GOTO 690
740   HOME : PRINT
750   PRINT "LY TO PARSECS CONVERSION"
760   PRINT "TO STOP ENTER 999"
770   PRINT : INPUT "LY ";X
780   IF X = 999 THEN  GOTO 110
790  Y = X * 1 / 3.26
800   PRINT "PARSECS = "; LEFT$ ( STR$ (Y),5)
810   GOTO 770
820   HOME : PRINT : PRINT "'C TO 'F CONVERSION"
830   PRINT "TO STOP ENTER 999"
840   PRINT : INPUT "'C ";X
850   IF X = 999 GOTO 110
860  Y = X * 1.8 + 32
870   PRINT "'F = "; LEFT$ ( STR$ (Y),5)
880   GOTO 840
890   HOME : PRINT : PRINT "'F TO 'C CONVERSION"
900   PRINT "TO STOP ENTER 999"
910   PRINT : INPUT "'F ";X
920   IF X = 999 GOTO 110
930  Y = (X - 32) * .555555556
940   PRINT "'C = "; LEFT$ ( STR$ (Y),5);
950   PRINT " = "; LEFT$ ( STR$ (Y + 273),5);"K"
960   GOTO 910
970   HOME : PRINT : PRINT "A.U. TO MILES/KILOMETERS CONVERSION"
980   PRINT "TO STOP ENTER 999"
990   PRINT : INPUT "A.U.";X
1000   IF X = 999 GOTO 110
1010  Y = X * 92.95721: REM   MILES
1020  Z = X * 149.598: REM  KILOMETERS
1030   PRINT "MILES (MILLIONS) = "; LEFT$ ( STR$ (Y),5)
1040   PRINT "KILOMETERS (MILLIONS) = "; LEFT$ ( STR$ (Z),5)
1050   GOTO 990
1060   HOME : PRINT : PRINT "MILES/KMS TO A.U. CONVERSION"
1070   PRINT "TO STOP TYPE 999"
```

ACONV (continued)

```
1080   PRINT : INPUT "TYPE KMS OR MI ";Q$
1090   IF Q$ = "KMS" GOTO 1120
1100   IF Q$ = "MI" GOTO 1170
1110   PRINT "PLEASE TYPE KMS OR MI ": GOTO 1080
1120   PRINT : INPUT "KILOMETERS (MILLIONS) ";X
1130   IF X = 999 GOTO 110
1140 Y = X * (1 / 149.598)
1150   PRINT "A.U. = "; LEFT$ ( STR$ (Y),5)
1160   GOTO 1120
1170   PRINT : INPUT "MILES (MILLIONS) ";X
1180   IF X = 999 GOTO 110
1190 Y = X * (1 / 92.15721)
1200   PRINT "A.U = "; LEFT$ ( STR$ (Y),5)
1210   GOTO 1170
1220   HOME : PRINT : PRINT "DEG.MIN.SEC. TO DECIMAL DEG."
1230   PRINT "TO STOP ENTER 999,0,0"
1240   PRINT : INPUT "DEG,MIN,SEC ";X,X1,X2
1250   IF X = 999 GOTO 110
1260 Y = X + X1 / 60 + X2 / 3600
1270   PRINT "DECIMAL DEG.= "; LEFT$ ( STR$ (Y),5)
1280   GOTO 1240
1290   HOME : PRINT : PRINT "DECIMAL DEGREES TO DEG.MIN.SEC."
1300   PRINT "TO STOP GO TO 999"
1310   PRINT : INPUT "DECIMAL DEGREES";X
1320   IF X = 999 GOTO 110
1330 Y =  INT (X)
1340 Y1 = 60 * (X -  INT (X))
1350 Y2 = 60 * (Y1 -  INT (Y1))
1360   PRINT "DEG.MIN.SEC.="Y; INT (Y1);Y2
1370   GOTO 1310
1380   HOME : PRINT "HRS.MIN.SEC. TO DECIMAL HRS "
1390   PRINT "TO STOP ENTER 999,0,0"
1400   PRINT : INPUT "HRS,MIN,SEC ";X,X1,X2
1410   IF X = 999 GOTO 110
1420 Y = X + X1 / 60 + X2 / 3600
1430   PRINT "DECIMAL HRS - ";Y
1440   GOTO 1400
1450   HOME : PRINT "DECIMAL HRS TO HRS.MIN.SEC. "
1460   PRINT "TO  STOP ENTER 999"
1470   PRINT : INPUT "DEC.HRS. ";X
1480   IF X = 999 GOTO 110
1490 Y =  INT (X)
1500 Y1 = 60 * (X -  INT (X))
1510 Y2 = 60 * (Y1 -  INT (Y1))
1520   PRINT "HRS.MIN.SEC.= ";Y;" "; INT (Y1);" "; INT (Y2)
1530   GOTO 1470
1540   HOME : PRINT "RESOLVING POWER TO APERTURE"
1550   PRINT "TO  STOP ENTER 999"
1560   PRINT : INPUT "RESOLVING POWER SEC. OF ARC ";X
1570   IF X = 999 GOTO 110
1580 Y = 4.56 / X
1590   PRINT "APERTURE IN INCHES MUST BE ";Y
1600   GOTO 1560
1610   HOME : PRINT "APERTURE TO RESOLVING POWER"
1620   PRINT "TO STOP ENTER 999"
```

ACONV *(continued)*

```
1630   PRINT : INPUT "APERTURE IN INCHES ";X
1640   IF X = 999 GOTO 110
1650   Y = 4.56 / X
1660   PRINT "RESOLVING POWER IN ARC SECS. IS ";Y
1670   GOTO 1630
1680   HOME
1690   END
```

Photo Credit: NASA/Ames

Earth and the other worlds of our solar system have been bombarded for billions of years by projectiles from space, believed to be the debris from the formation of the system. Today some of these projectiles still enter Earth's atmosphere and produce glowing trails of "shooting stars" as they burn up in the upper atmosphere. Spacecraft such as Pioneer Venus have been sent into the atmospheres of other worlds, and they also glow brightly like meteors as they plunge into those atmospheres.

Program 20: SSTAR

Dates and Radiants of Annual Meteor Showers_____

Meteors, often referred to as "shooting stars" or "falling stars," result from small particles of matter, called meteorites, that encounter Earth at high speed and burn up in its atmosphere. Many meteors are associated with the orbits of comets; an annual meteor shower results when Earth crosses the cometary orbit. Meteors are usually more numerous in the early morning hours because at that time the observer is on the face of Earth moving into the meteors, that is, the leading hemisphere of the planet in its orbital motion around the Sun.

The program asks you to select any month and then provides the details of any annual meteor shower expected during that month. If you use PHASE with this program you will be able to find out whether observations of the meteor shower will be affected by a bright Moon (between full and last quarter). The program graphically displays the radiant for the shower among the stars of the constellations for any selected month. It provides details of each meteor shower and names the surrounding constellations. Typical displays of the program are shown in Figures 20.1 and 20.2.

For other computers you can use PRINT @, CHR$(XX), and TAB statements to position the stars, instead of the VTAB and TAB instructions used for the Apple. For a more dynamic display, POKE statements can be used to randomly display meteors from the radiant. POKE statements allow ASCII characters to be displayed on the screen at a specified screen coordinate without scrolling the screen display. A routine to do this for the Exidy Sorcerer's 64 by 29 character display is shown in the Appendix and can be adapted to other computers. For the Apple, POKE instructions for screen visuals are more complex (see PLNTF, Program 17). To use high resolution graphics to plot stars and names in a long routine like this, the program must contain a routine to jump over the area of memory reserved for pages 1 and 2 of the high resolution graphics (see your Apple manual).

The listing of the SSTAR program follows.

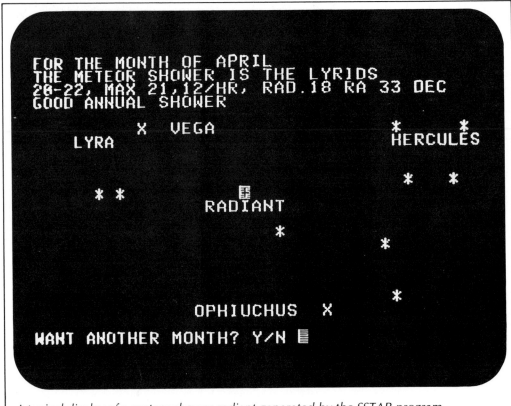

A typical display of a meteor shower radiant generated by the SSTAR program

Figure 20.1

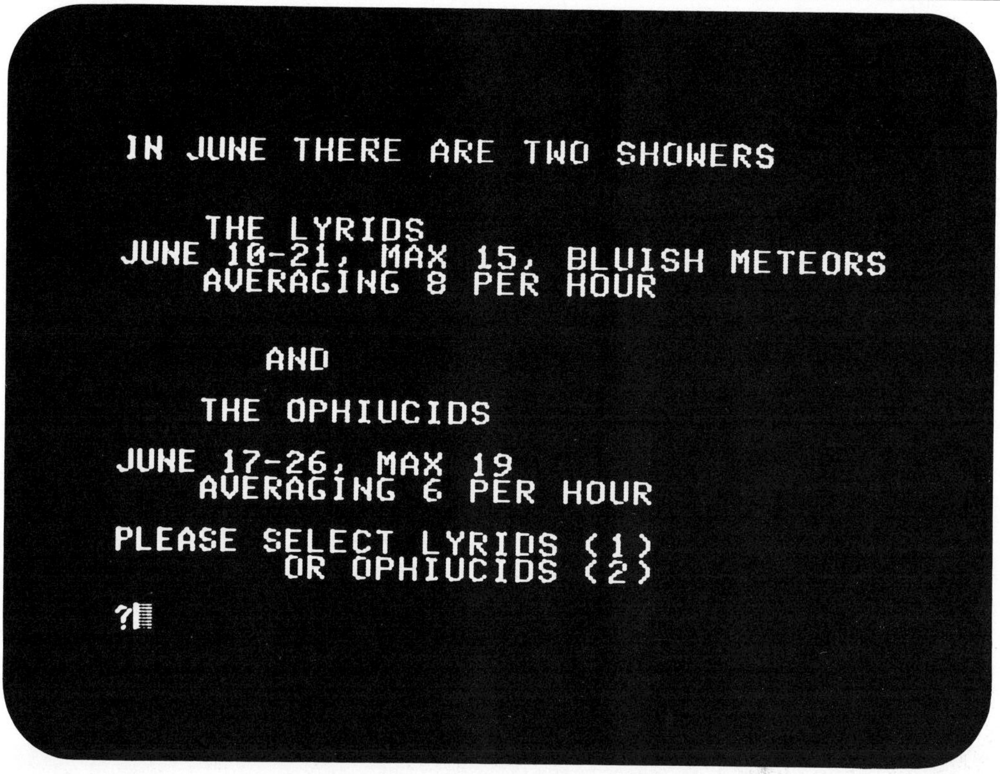

If there are two meteor showers in a requested month, the program develops the display shown here for you to select the shower you wish to be displayed.

Figure 20.2

```
┌─ SSTAR ─────────────────────────────────────────────────┐
│                                                          │
│ 10    HOME                                               │
│ 20    PRINT : PRINT : PRINT                              │
│ 30    PRINT  TAB( 6)"--------------------------"         │
│ 40    PRINT  TAB( 6)"I     M E T E O R S      I"         │
│ 50    PRINT  TAB( 6)"--------------------------"         │
│ 60    PRINT : PRINT : PRINT                              │
│ 70    PRINT  TAB( 10)"AN ASTRONOMY PROGRAM"              │
│ 80    PRINT : PRINT  TAB( 9)"BY ERIC BURGESS F.R.A.S."   │
│ 90    PRINT : PRINT : PRINT                              │
│ 100   PRINT  TAB( 9)"ALL RIGHTS RESERVED BY"             │
│ 110   PRINT  TAB( 9)"S & T SOFTWARE SERVICE"             │
│ 120   PRINT : PRINT                                      │
│ 130   PRINT  TAB( 3)"THIS PROGRAM PROVIDES PARTICULARS OF" │
│ 140   PRINT  TAB( 3)"ANNUAL METEOR SHOWERS FOR ANY MONTH" │
│ 150   FOR K = 1 TO 3000: NEXT K                          │
│ 160   HOME : PRINT : PRINT                               │
│ 170   INPUT "SELECT MONTH (1 THRU 12) ";M                │
│ 180   PRINT                                              │
│ 190   IF M < 1 OR M > 12 THEN  PRINT "INVALID RESPONSE": PRINT : GOTO 170 │
│ 200   HOME : PRINT : PRINT                               │
│ 210 M2 = M                                               │
│ 220   IF M2 = 1 THEN M$ = "JANUARY": GOTO 340            │
│ 230   IF M2 = 2 THEN M$ = "FEBRUARY": GOTO 470           │
│ 240   IF M2 = 3 THEN M$ = "MARCH": GOTO 520              │
│ 250   IF M2 = 4 THEN M$ = "APRIL": GOTO 570              │
│ 260   IF M2 = 5 THEN M$ = "MAY": GOTO 750                │
│ 270   IF M2 = 6 THEN M$ = "JUNE": GOTO 900               │
│ 280   IF M2 = 7 THEN M$ = "JULY": GOTO 1140              │
│ 290   IF M2 = 8 THEN M$ = "AUGUST": GOTO 1340            │
│ 300   IF M2 = 9 THEN M$ = "SEPTEMBER": GOTO 1300         │
│ 310   IF M2 = 10 THEN M$ = "OCTOBER": GOTO 1490          │
│ 320   IF M2 = 11 THEN M$ = "NOVEMBER": GOTO 1680         │
│ 330   IF M2 = 12 THEN M$ = "DECEMBER": GOTO 2280         │
│ 340   PRINT "FOR THE MONTH OF ";M$                       │
│ 350   PRINT "THE METEOR SHOWER IS QUANDRANTIDS"          │
│ 360   PRINT "4 JAN MAX 24 HRS, RAD 15.5 RA 50 DEC"       │
│ 370   PRINT "GOOD ANNUAL SHOWER 40 PER HOUR"             │
│ 380   PRINT : GOSUB 3330: PRINT : PRINT                  │
│ 390   PRINT  TAB( 19)"*"                                 │
│ 400   PRINT  TAB( 26)"X": PRINT                          │
│ 410   PRINT  TAB( 15)"*"; TAB( 28)"BOOTES"               │
│ 420   PRINT  TAB( 30)"."                                 │
│ 430   PRINT  TAB( 13)"."; TAB( 29)".": PRINT  TAB( 2)"CORONA" │
│ 440   PRINT  TAB( 12)"*"; TAB( 24)"X": PRINT             │
│ 450   PRINT  TAB( 30)"X  ARCTURUS"                       │
│ 460   GOTO 3270                                          │
│ 470   PRINT "DURING THE MONTH OF ";M$                    │
│ 480   PRINT                                              │
│ 490   PRINT "THERE IS NO MAJOR ANNUAL METEOR SHOWER"     │
│ 500   PRINT : PRINT : PRINT                              │
│ 510   GOTO 3280                                          │
│ 520   PRINT "DURING THE MONTH OF ";M$                    │
│ 530   PRINT                                              │
│ 540   PRINT "THERE IS NO MAJOR ANNUAL METEOR SHOWER"     │
│                                                          │
└──────────────────────────────────────────────────────────┘
```

SSTAR *(continued)*

```
550    PRINT : PRINT : PRINT
560    GOTO 3280
570    PRINT "FOR THE MONTH OF ";M$
580    PRINT "THE METEOR SHOWER IS THE LYRIDS"
590    PRINT "20-22, MAX 21,12/HR, RAD.18 RA 33 DEC"
600    PRINT "GOOD ANNUAL SHOWER": PRINT
610    PRINT   TAB( 11)"X  VEGA"; TAB( 33)"*"; TAB( 39)"*"
620    PRINT   TAB( 5)"LYRA"; TAB( 33)"HERCULES"
630    PRINT
640    PRINT   TAB( 34)"*"; TAB( 38)"*"
650    PRINT   TAB( 7)"* *";: GOSUB 3330
660    PRINT : PRINT
670    PRINT   TAB( 23)"*"
680    PRINT   TAB( 32)"*"
690    PRINT
700    PRINT : PRINT
710    PRINT   TAB( 33)"*"
720    PRINT   TAB( 16)"OPHIUCHUS"; TAB( 27)"X"
730    PRINT
740    GOTO 3270
750    PRINT "FOR THE MONTH OF ";M$
760    PRINT "THE METEOR SHOWER IS ETA AQUARIDS"
770    PRINT "4-13,MAX 6, RAD 22.5 RA, 0 DEC"
780    PRINT "WEAK DISPLAY"
790    PRINT   TAB( 15)"."; TAB( 30)"*"; TAB( 36)".."
800    PRINT   TAB( 24)"."; TAB( 35)".": PRINT : PRINT
810    PRINT   TAB( 17)"*"; TAB( 24)"*"
820    PRINT   TAB( 15)"*";: GOSUB 3330
830    PRINT   TAB( 30)"AQUARIUS": PRINT
840    PRINT   TAB( 33)"X": PRINT : PRINT
850    PRINT : PRINT
860    PRINT   TAB( 12)"."
870    PRINT   TAB( 11)"."; TAB( 29)"*"; TAB( 31)"."; TAB( 35)"."; TAB( 39)"."
880    PRINT   TAB( 19)"CAPRICORNUS"
890    GOTO 3270
900    PRINT : PRINT
910    PRINT "IN JUNE THERE ARE TWO SHOWERS"
920    PRINT : PRINT
930    PRINT   TAB( 5)"THE LYRIDS"
940    PRINT "JUNE 10-21, MAX 15, BLUISH METEORS"
950    PRINT   TAB( 5)"AVERAGING 8 PER HOUR"
960    PRINT : PRINT : PRINT   TAB( 8)"AND"
970    PRINT : PRINT   TAB( 5)"THE OPHIUCIDS"
980    PRINT : PRINT "JUNE 17-26, MAX 19"
990    PRINT   TAB( 5)"AVERAGING 6 PER HOUR"
1000   PRINT : PRINT "PLEASE SELECT LYRIDS (1)"
1010   PRINT "         OR OPHIUCIDS (2)"
1020   PRINT : INPUT A$:JS =  VAL (A$)
1030   IF JS < 1 OR JS > 2 THEN  PRINT "INVALID RESPONSE": PRINT : GOTO 1000
1040   IF JS = 1 THEN   GOSUB 2820: GOTO 1070
1050   IF JS = 2 THEN   GOSUB 3050
1060   IF A$ = "N" THEN  GOTO 3280
1070   FOR J = 1 TO 3000: NEXT J
1080   INPUT "DO YOU WANT TO SEE OTHER SHOWER Y/N ";A$
1090   HOME
1100   IF A$ = "Y" AND JS = 1 THEN GOSUB 3050: GOTO 1130
1110   IF A$ = "Y" AND JS = 2 THEN GOSUB 2820
```

```
┌─ SSTAR (continued) ──────────────────────────────────────┐
│                                                           │
│                                                           │
│   1120    IF A$ = "N" THEN   GOTO 3280                     │
│   1130    GOTO 3270                                        │
│   1140    PRINT "FOR THE MONTH OF ";M$                     │
│   1150    PRINT "THE METEOR SHOWER IS DELTA AQUARIDS"      │
│   1160    PRINT "7/25-8/4,MAX 28,RAD 22.8 RA -16 DEC"      │
│   1170    PRINT "DIFFUSE SHOWER"                           │
│   1180    PRINT   TAB( 17)"."                              │
│   1190    PRINT : PRINT                                    │
│   1200    PRINT   TAB( 32)"AQUARIUS"                       │
│   1210    PRINT   TAB( 28)"."                              │
│   1220    PRINT   TAB( 11)"DELTA *";: GOSUB 3330           │
│   1230    PRINT   TAB( 33)"*"; TAB( 37)"."                 │
│   1240    PRINT                                            │
│   1250    PRINT   TAB( 39)".": PRINT                       │
│   1260    PRINT : PRINT   TAB( 19)"X": PRINT               │
│   1270    PRINT   TAB( 11)"FOMALHAUT"                      │
│   1280    PRINT   TAB( 31)"*": PRINT : PRINT : PRINT       │
│   1290    GOTO 3270                                        │
│   1300    PRINT "DURING THE MONTH OF ";M$                  │
│   1310    PRINT                                            │
│   1320    PRINT "THERE IS NO MAJOR METEOR SHOWER"          │
│   1330    PRINT : PRINT : PRINT : GOTO 3280                │
│   1340    PRINT "FOR THE MONTH OF ";M$                     │
│   1350    PRINT "THE METEOR SHOWER IS THE PERSEIDS"        │
│   1360    PRINT "4-16,MAX 12, RAD 3 RA 58 DEC"             │
│   1370    PRINT "A SPECTACULAR SHOWER, 50 PER HR"          │
│   1380    PRINT   TAB( 6)"."; TAB( 39)"X"                  │
│   1390    PRINT : PRINT   TAB( 29)"CASSIOPEIA"             │
│   1400    PRINT : GOSUB 3330                               │
│   1410    PRINT   TAB( 23)"."                              │
│   1420    PRINT   TAB( 10)"PERSEUS";                       │
│   1430    PRINT   TAB( 20)"*": PRINT   TAB( 21)"."         │
│   1440    PRINT   TAB( 12)"*": PRINT : PRINT               │
│   1450    PRINT   TAB( 39)"X": PRINT   TAB( 8)"X"; TAB( 20)"X" │
│   1460    PRINT   TAB( 20)"."; TAB( 30)"ANDROMEDA"         │
│   1470    PRINT   TAB( 37)"*"                              │
│   1480    GOTO 3270                                        │
│   1490    PRINT "FOR THE MONTH OF ";M$                     │
│   1500    PRINT "THE METEOR SHOWER IS THE ORIONIDS"        │
│   1510    PRINT "16-26,MAX 21, RAD. 6.4 RA, 15 DEC"        │
│   1520    PRINT "SLOW PERSISTENT TRAIL METEORS, 30/HR"     │
│   1530    PRINT   TAB( 6)"."; TAB( 23)"*"; TAB( 26)"."     │
│   1540    PRINT   TAB( 11)"."; TAB( 22)"."; TAB( 34)"*"    │
│   1550    PRINT   TAB( 10)"GEMINI"                         │
│   1560    PRINT   TAB( 8)"."; TAB( 18)"*"                  │
│   1570    GOSUB 3330                                       │
│   1580    PRINT : PRINT   TAB( 16)"."                      │
│   1590    PRINT : PRINT   TAB( 8)"*"; TAB( 32)"."          │
│   1600    PRINT   TAB( 28)"X"; TAB( 36)"X"                 │
│   1610    PRINT   TAB( 4)"X"                               │
│   1620    PRINT "PROCYON"; TAB( 35)"ORION"                 │
│   1630    PRINT : PRINT   TAB( 35)"X"                      │
│   1640    PRINT   TAB( 33)"X": PRINT   TAB( 31)"X"; TAB( 38)"." │
│   1650    PRINT   TAB( 34)":": PRINT   TAB( 2)"."; TAB( 38)"X" │
│   1660    PRINT   TAB( 30)"*"                              │
│   1670    GOTO 3270                                        │
│                                                           │
└───────────────────────────────────────────────────────────┘
```

─────────── **SSTAR** *(continued)* ───────────

```
1680   PRINT : PRINT
1690   PRINT "IN ";M$;" THERE ARE TWO SHOWERS"
1700   PRINT : PRINT
1710   PRINT  TAB( 5)"THE TAURIDS"
1720   PRINT : PRINT "10/20-11/30,MAX 8/11,RAD 3.7 RA,14 DEC"
1730   PRINT "SLOW 12/HR, SCATTERED FIREBALLS"
1740   PRINT : PRINT  TAB( 8)"AND"
1750   PRINT : PRINT  TAB( 5)"THE LEONIDS"
1760   PRINT
1770   PRINT "15-19,MAX 17, RAD 10 RA, 22 DEC"
1780   PRINT "SWIFT METEORS, 10 PER HOUR"
1790   PRINT : PRINT
1800   PRINT "PLEASE SELECT TAURIDS (1)"
1810   PRINT "        OR LEONIDS (2)"
1820   PRINT : INPUT A$:JS =  VAL (A$)
1830   IF JS < 1 OR JS > 2 THEN  PRINT "INVALID RESPONSE": PRINT : GOTO 1790
1840   IF JS = 1 THEN  GOSUB 1930: GOTO 1860
1850   IF JS = 2 THEN  GOSUB 2120
1860   FOR J = 1 TO 3000: NEXT J
1870   INPUT "WANT TO SEE OTHER SHOWER Y/N ";A$
1880   HOME
1890   IF A$ = "Y" AND JS = 1 THEN GOSUB 2120: GOTO 1920
1900   IF A$ = "Y" AND JS = 2 THEN GOSUB 1930
1910   IF A$ = "N" THEN  GOTO 3280
1920   GOTO 3270
1930   REM  NOV TAURIDS
1940   HOME
1950   PRINT "FIRST ";M$;" SHOWER IS TAURIDS"
1960   PRINT "10/20-11/30,MAX 8, 12 PER HOUR"
1970   PRINT "WITH 2 RAD.,3.73 RA, 14 & 22 DEC"
1980   PRINT "SLOW METEORS, SCATTERED FIREBALLS"
1990   PRINT : PRINT  TAB( 16)"PLEIADES": PRINT
2000   PRINT  TAB( 19)".::"
2010   PRINT "TAURUS": GOSUB 3330
2020   PRINT  TAB( 39)"*"
2030   PRINT "ALDEBARAN"; TAB( 11)"."
2040   PRINT  TAB( 10)"X"; TAB( 11) "; TAB( 13)"."; TAB( 34)"ARIES"
2050   GOSUB 3330: PRINT ""
2060   PRINT
2070   PRINT
2080   PRINT
2090   PRINT  TAB( 30)"*"; TAB( 45)".": PRINT : PRINT
2100   PRINT
2110   RETURN
2120   REM  NOV LEONIDS
2130   HOME
2140   PRINT "SECOND ";M$;" SHOWER IS LEONIDS"
2150   PRINT "15-19,MAX 17, RAD 10.13 RA, 22 DEC"
2160   PRINT "SWIFT METEORS, 10 PER HOUR"
2170   PRINT  TAB( 3)"."; TAB( 10)".": PRINT  TAB( 7)"."
2180   PRINT : PRINT  TAB( 24)".": PRINT  TAB( 18)"."; TAB( 26)"*"
2190   GOSUB 3330
2200   PRINT : PRINT  TAB( 17)"*": PRINT  TAB( 3)"*"
2210   PRINT  TAB( 20)".": PRINT  TAB( 3)"X"; TAB( 8)"LEO"
2220   PRINT  TAB( 20)"X  REGULUS"
2230   PRINT "."; TAB( 28)"."
```

SSTAR *(continued)*

```
2240   PRINT  TAB( 37)"*   ."
2250   PRINT : PRINT : PRINT
2260   PRINT
2270   RETURN
2280   PRINT : PRINT : PRINT
2290   PRINT "IN ";M$;" THERE ARE TWO METEOR SHOWERS"
2300   PRINT
2310   PRINT : PRINT  TAB( 5)"THE GEMINIDS"
2320   PRINT "9-13, MAX 13, 60 PER HOUR"
2330   PRINT "SWIFT BRIGHT METEORS, SOME FIREBALLS"
2340   PRINT : PRINT : PRINT  TAB( 8)"AND"
2350   PRINT : PRINT : PRINT  TAB( 5)"THE URSIDS"
2360   PRINT "17-24, MAX 22, 5 PER HOUR"
2370   PRINT "WEAK DISPLAY"
2380   PRINT
2390   PRINT : PRINT "PLEASE SELECT GEMINIDS (1)"
2400   PRINT "          OR URSIDS (2)"
2410   PRINT : INPUT A$:JS =  VAL (A$)
2420   IF JS = 1 THEN  GOSUB 2510: GOTO 2440
2430   IF JS = 2 THEN  GOSUB 2660
2440   FOR J = 1 TO 3000: NEXT J
2450   INPUT "WANT TO SEE THE OTHER SHOWER Y/N ";A$
2460   HOME
2470   IF A$ = "Y" AND JS = 1 THEN GOSUB 2660: GOTO 2500
2480   IF A$ = "Y" AND JS = 2 THEN GOSUB 2510
2490   IF A$ = "N" THEN  GOTO 3280
2500   GOTO 3270
2510   REM   DEC. GEMINIDS
2520   HOME
2530   PRINT "FIRST ";M$;" SHOWER IS GEMINIDS"
2540   PRINT "9-13,MAX 13, RAD. 7.46 RA 32 DEC"
2550   PRINT "BRIGHT SWIFT METEORS 60 PER HOUR"
2560   PRINT  TAB( 40)"."
2570   PRINT : PRINT  TAB( 22)"*"
2580   GOSUB 3330
2590   PRINT ".": PRINT  TAB( 18)"*"; TAB( 33)"."
2600   PRINT  TAB( 18)".": PRINT " ."; TAB( 24)"*"; TAB( 38)"* *"
2610   PRINT  TAB( 28)".": PRINT  TAB( 25)"GEMINI"
2620   PRINT  TAB( 36)"*": PRINT
2630   PRINT  TAB( 35)".": PRINT
2640   PRINT  TAB( 23)"*"; TAB( 46)"X": PRINT  TAB( 19)"X  PROCYON"
2650   RETURN
2660   REM   DEC URSIDS
2670   HOME
2680   PRINT "SECOND ";M$;" SHOWER IS THE URSIDS"
2690   PRINT "17-24,MAX 22, RAD 14.6 RA, 78 DEC"
2700   PRINT "FAINT, MEDIUM SPEED, 5 PER HOUR"
2710   PRINT
2720   PRINT  TAB( 10)"POLARIS"; TAB( 20)"*"
2730   PRINT : PRINT  TAB( 16)".": PRINT : PRINT : PRINT
2740   PRINT  TAB( 12)".": PRINT
2750   PRINT : GOSUB 3330: PRINT "": PRINT  TAB( 13)"."
2760   PRINT "   URSA"; TAB( 39)"."
2770   PRINT "   MINOR"
2780   PRINT  TAB( 10)"."; TAB( 16)"*": PRINT
2790   PRINT  TAB( 14)"*"; TAB( 30)"DRACO"
2800   PRINT
```

SSTAR (continued)

```
2810    RETURN
2820    REM   JUNE LYRIDS
2830    HOME
2840    PRINT "FIRST ";M$;" SHOWER IS THE LYRIDS"
2850    PRINT "10-20,MAX 15, RAD 18.3 RA, 35 DEC"
2860    PRINT "BLUISH METEORS, 8 PER HOUR"
2870    PRINT
2880    PRINT  TAB( 40)"*"
2890    PRINT  TAB( 6)"VEGA"; TAB( 13)"X"; TAB( 30)"*";
2900    PRINT  TAB( 37)"*": GOSUB 3330
2910    PRINT ""
2920    PRINT  TAB( 9)"*  *"; TAB( 30)"  *    *"
2930    PRINT "    LYRA";
2940    PRINT  TAB( 32)"HERCULES"
2950    PRINT "  X"
2960    PRINT  TAB( 28)"*"
2970    PRINT  TAB( 39)"*": PRINT
2980    PRINT : PRINT : PRINT
2990    PRINT  TAB( 12)"."
3000    PRINT  TAB( 10)"*"; TAB( 38)"*"
3010    PRINT  TAB( 33)"*"
3020    PRINT "*"; TAB( 31)"OPHIUCHUS"
3030    PRINT
3040    RETURN
3050    REM   JUNE OPHIUCIDS
3060    HOME
3070    PRINT "SECOND ";M$;" SHOWER IS THE OPHIUCIDS"
3080    PRINT "17-26,MAX 19, RAD 17.3 RA, -20 DEC"
3090    PRINT "6 PER HOUR"
3100    PRINT  TAB( 18)"OPHIUCHUS";
3110    PRINT  TAB( 33)"."
3120    PRINT "."; TAB( 10)"."; TAB( 30)"*"
3130    PRINT : PRINT
3140    PRINT  TAB( 15)"."; TAB( 21)"*": PRINT
3150    GOSUB 3330
3160    PRINT  TAB( 38)"*": PRINT
3170    PRINT " *"; TAB( 39)"*"
3180    PRINT  TAB( 34)"*"
3190    PRINT  TAB( 22)"ANTARES";
3200    PRINT  TAB( 31)"X"; TAB( 39)"*"
3210    PRINT "   *"; TAB( 8)"*"; TAB( 30)"*"
3220    PRINT : PRINT "SAGITTARIUS"
3230    PRINT "  *"; TAB( 26)"*"
3240    PRINT  TAB( 5)"*"; TAB( 12)"."; TAB( 15)"**"
3250    PRINT  TAB( 12)"*"; TAB( 26)"*"; TAB( 31)"SCORPIO"
3260    RETURN
3270    FOR K = 1 TO 5000: NEXT K
3280    INPUT "WANT ANOTHER MONTH? Y/N ";A$
3290    IF A$ = "Y" THEN 160
3300    IF A$ < > "N" THEN  PRINT "INVALID RESPONSE": PRINT : GOTO 3280
3310    HOME
3320    END
3330    REM   SUB FOR METEOR RADIANT
3340    PRINT  TAB( 20);: FLASH : PRINT "+": NORMAL
3350    PRINT  TAB( 17)"RADIANT";
3360    RETURN
```

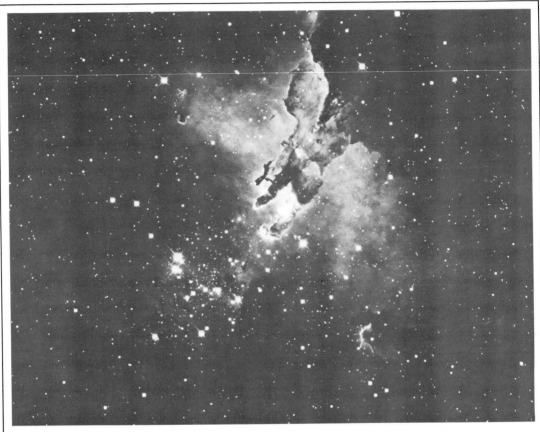

Photo: NASA/University of California

Beyond the solar system the stars beckon. Countless stars clump into our island universe, the Milky Way. Out in the farther depths of space we have discovered other island universes as numerous as the stars in our own Milky Way system. Between the stars are clouds of dust, some illuminated by nearby stars to form glowing areas of nebulosity. In this picture gas mingles with dust in the constellation of Serpens and shows us what may be the birth of new star systems.

Program 21: CONST and CONSH

Recognizing the Constellations_____

When the Sun has set and twilight deepens, the night sky begins to develop the pinprick points of starlight. If we watch these stars from night to night we soon see that they rise, move together in a procession across the sky, and set, making a complete circuit once each day and once each year. In the northern sky, however, some groups of stars do not rise and set, but daily and annually circle around the celestial north pole, approximately marked by the pole star, Polaris. Similarly, in the Southern Hemisphere other groups of stars are always visible around the celestial south pole. You can use Program 16 to show how these motions and the constellations appear in different parts of the world, at different times of the night and in different seasons. You will notice that the individual stars do not appear to move relative to one another.

The patterns of the stars can be imagined to be joined by lines to represent objects, animals, and people. Since ancient times people have grouped the stars into constellations in this way, forming imaginary figures of mythological heroes and heroines in the night sky. Orion and the Pleiades are mentioned in the Book of Job, and the division of the stars into constellations is ascribed by Josephus to Seth, a son of Adam. Often these constellations are in groups telling a story, such as that concerning Andromeda, Perseus, Cassiopeia, Pegasus, and their surrounding constellations. Legend has it that Queen Cassiopeia (Figure 21.1) of Ethiopia and her husband, Cepheus, consented to the sacrifice of their daughter Andromeda to placate the gods. Andromeda was chained to a rock to await the coming of a sea monster. But Perseus saw the maiden, fell in love with her, and came to her rescue on the winged horse Pegasus. He slew the monster as he had the Gorgon.

Because of the revolution of Earth around the Sun, the Sun appears to pass through a series of constellations, called the zodiacal constellations.

These zodiacal constellations figure prominently in astrology, and can be memorized by the old English rhyme:

> The Ram, the Bull, the Heavenly Twins,
> And next the Crab, the Lion, shines,
> The Virgin and the Scales;
> The Scorpion, Archer, and Sea-Goat,
> The Man that pours the water out,
> And Fish with glittering scales.

Many bright stars have also been given names, usually derived from the Arabic names given them during the Middle Ages. In modern astronomy the stars of the constellations are assigned Greek letters; alpha for the brightest, then beta, and so on through the Greek alphabet. When these letters have been used up, the stars are designated by numbers.

The constellations and the stars within them were aids to navigation by seafaring peoples, and some modern research suggests that birds migrate by reference to the stars. The latitude of any place on Earth in the Northern Hemisphere can be quickly determined to within a few degrees by measuring the elevation of Polaris (see Program 7). The longitude is somewhat more difficult to establish. However, traveling toward the east an observer finds that the stars cross the meridian earlier, one hour for each fifteen degrees of longitude. Toward the west, the meridian crossing occurs later. But to determine how much earlier or later, an observer must have a clock that tells him the time at his original location. This need to know time exactly was one reason for the establishment of Greenwich Observatory and the zero meridian of longitude. Without an accurate measurement of when stars crossed the meridian at a particular place, determined by a transit instrument, and accurate clocks to maintain this time on long voyages, exploration of our globe would have been much more difficult.

Once you recognize the constellations they become friends of the night sky, recognizable not only anywhere on our planet but also throughout our solar system. The groupings of stars that we see in the skies of Earth are the same as those in the skies of the Moon and of Mars, for example.

A well-known astronomer once wrote that the best way to learn the constellations is to get someone to teach them to you. An alternative is to refer to them in an astronomical textbook. But the lines shown on the star maps in such books are not available when you look at the night sky. This program is designed to help you learn the constellations by pattern recognition. It does not show the imaginary lines of the star charts. But when you remember the patterns of bright and faint stars you should have no problem in identifying the constellations in the night sky. It does not take long to become familiar with the constellations. Unfortunately, not all are visible at any one time or season. With this program you can learn to recognize 42 major constellations before going out to search for them in the night sky. A supplementary program (CONSH) is appended to identify the southern circumpolar constellations for readers in the Southern Hemisphere.

The program offers two alternatives. You can be shown a constellation and asked to name it, and then be asked to name the bright star within it. Or you can have questions mingling stars and constellations unrelated to each other.

The program randomly selects the constellations for display and gives you two tries at naming the chosen constellation correctly. Then it tells you the name, so you should remember to associate it with the star pattern the next time you see it. You can continue being tested as long as you

NAME THIS CONSTELLATION

Here the CONST program displays the constellation Cassiopeia. If you do not name it correctly after two tries, the name is displayed and you are asked if you would like another quiz.

Figure 21.1

wish, or you can quit at any time. Figure 21.1, which shows Cassiopeia, is typical of the displays generated by this program.

There are no complex algorithms in the program. It selects the constellations by use of the random function and jumps to the appropriate section of the program at instruction 450. Each section is a straightforward routine that prints stars on the monitor screen and then uses the subroutine (instruction 7450) to ask you to name the constellation. It checks if your answer is correct. If your answer is incorrect it allows you one more try before printing the name of the constellation. A similar subroutine (instruction 7540) is used to quiz you on the name of a bright star within the constellation.

For several of the constellations the name of a second star is given in the constellation routines, but it is not displayed with this version of the program. If you wish you can add a second subroutine to display these other stars for a quiz. You can also use the same techniques to display nebulae when you want a more advanced tutorial. These would be interesting projects for an astronomy class.

If you are in the Southern Hemisphere you will probably find it preferable to invert all the constellations when you key in the routines for each constellation by changing the VTAB and TAB instructions. To do this you should subtract each TAB instruction number from 40 and use the difference as the new TAB number; then subtract each VTAB instruction number from 24 and use the difference as the new VTAB number. This procedure will invert all the constellations when they are displayed on the monitor screen.

This program is easily adapted to other computers. The VTAB instructions have been purposely left in for all lines even though they often are not required. This has been done to provide line information so that VTAB instructions can be substituted to select a line position on the screen. For the TRS-80, for example, the VTAB/PRINTTAB(X) instructions can be changed to PRINT @ instructions. Other computers can use simple PRINT statements to count the number of lines in place of the VTAB statement.

The listings of the CONST and CONSH programs follow.

─── **CONST** ───

```
10    HOME
20    PRINT : PRINT : PRINT : PRINT
30    PRINT  TAB( 5)"AN ASTRONOMY PROGRAM"
40    PRINT  TAB( 5)"    TUTORIAL"
50    PRINT : PRINT
60    PRINT  TAB( 5)"--------------------"
70    PRINT  TAB( 5)"I  CONSTELLATIONS  I"
80    PRINT  TAB( 5)"--------------------"
90    PRINT : PRINT : PRINT
100   PRINT  TAB( 4)"BY ERIC BURGESS F.R.A.S."
110   PRINT
120   PRINT  TAB( 4)"ALL RIGHTS RESERVED BY"
130   PRINT  TAB( 4)"S & T SOFTWARE SERVICE"
140   FOR K = 1 TO 3000: NEXT K
150   PRINT : PRINT : PRINT
160   HOME
170   PRINT : PRINT : PRINT
180   INPUT "WANT INSTRUCTIONS Y/N ";A$
190   IF A$ = "N" THEN  GOTO 370
200   IF A$ < > "Y" THEN  PRINT "INVALID RESPONSE": PRINT : GOTO 180
210   HOME : PRINT : PRINT
220   PRINT "THIS PROGRAM DISPLAYS A CONSTELLATION"
230   PRINT "SELECTED AT RANDOM AND ASKS FOR ITS NAME"
240   PRINT "YOU HAVE TWO TRIES BEFORE THE NAME"
250   PRINT  TAB( 8)"IS REVEALED"
260   PRINT
270   PRINT "NEXT THE PROGRAM ASKS FOR THE NAME OF"
280   PRINT "A BRIGHT STAR IN THE CONSTELLATION"
290   PRINT "AGAIN YOU HAVE TWO CHANCES TO NAME IT"
300   PRINT "BEFORE BEING TOLD THE NAME"
310   PRINT : PRINT
320   PRINT : PRINT "THE PROGRAM OFFERS TWO ALTERNATIVES"
330   PRINT "CONSTELLATIONS FOLLOWED BY STARS"
340   PRINT "OR STARS AND CONSTELLATIONS MIXED"
350   PRINT : PRINT
360   INPUT "PRESS RETURN WHEN READY";A$
370   HOME : PRINT : PRINT : PRINT : PRINT
380   PRINT : PRINT "CHOOSE CONSTELLATIONS (1)"
390   PRINT "OR": INPUT "STARS AND CONSTELLATIONS (2)";AL$
400   IF AL$ = "2" THEN AL = 1
410   IF AL$ = "1" THEN AL = 0
420   Z = 1
430   C =  RND (Z) * 42
440   C =  INT (C) + 1
450   ON C GOSUB 460,590,790,990,1200,1320,1520,1710,1840,2050,2220,2360,2510,
      2740,2900,3090,3230,3420,3530,3640,3820,3960,4080,4280,4460,4710,4880,
      5040,5210,5430,5610,5690,5810,5950,6160,6330,6470,6660,6830,7010,7170,
      7310
460   C$ = "PISCES"
470   HOME
480   VTAB 5: PRINT  TAB( 10)"."
490   VTAB 9: PRINT  TAB( 8)"."
500   VTAB 12: PRINT  TAB( 12)"."; TAB( 16)"."
```

```
┌─── CONST (continued) ──────────────────────────────────────┐
│                                                             │
│                                                             │
│   510   VTAB 13: PRINT  TAB( 24)"."; TAB( 31)"."; TAB( 35)"."│
│   520   VTAB 14: PRINT  TAB( 4)"."; TAB( 28)"."             │
│   530   VTAB 15: PRINT  TAB( 33)"."                         │
│   540   VTAB 16: PRINT  TAB( 28)"."; TAB( 31)"."            │
│   550   VTAB 22                                             │
│   560   GOSUB 7450                                          │
│   570   INPUT "PRESS RETURN TO CONTINUE ";A$               │
│   580   GOTO 7610                                           │
│   590   C$ = "ARIES"                                        │
│   600   S$ = "HAMAL"                                        │
│   610   HOME                                                │
│   620   VTAB 6: PRINT  TAB( 23)"."                          │
│   630   VTAB 7: PRINT  TAB( 22)"."                          │
│   640   VTAB 9: PRINT  TAB( 26)"."                          │
│   650   VTAB 10: PRINT  TAB( 12)"."                         │
│   660   VTAB 11: PRINT  TAB( 13)"."                         │
│   670   VTAB 12: PRINT  TAB( 11)"."                         │
│   680   IF FL = 1 THEN  VTAB 15: PRINT TAB( 25)"";          │
│   690   IF FL = 1 THEN  FLASH : PRINT TAB( 25)"";           │
│   700   IF FL = 1 THEN  FLASH : PRINT "X": NORMAL : GOTO 720│
│   710   VTAB 15: PRINT  TAB( 25)"X"                         │
│   720   VTAB 16: PRINT  TAB( 9)"."; TAB( 28)"*"             │
│   730   VTAB 17: PRINT  TAB( 5)"."; TAB( 28)"."             │
│   740   VTAB 20                                             │
│   750   IF FL = 1 THEN FL = 0: GOTO 7540                    │
│   760   GOSUB 7450                                          │
│   770   INPUT "PRESS RETURN TO CONTINUE ";A$               │
│   780   GOTO 610                                            │
│   790   C$ = "TAURUS"                                       │
│   800   S$ = "ALDEBARAN"                                    │
│   810   S2$ = "PLEIADES":S3$ = "HYADES"                     │
│   820   HOME                                                │
│   830   VTAB 10: PRINT  TAB( 7)"*"                          │
│   840   VTAB 13: PRINT  TAB( 31)".::"                       │
│   850   VTAB 15: PRINT  TAB( 13)"."                         │
│   860   VTAB 16: PRINT  TAB( 3)"*"                          │
│   870   VTAB 17: PRINT  TAB( 20)"."                         │
│   880   VTAB 18: PRINT  TAB( 21)".."                        │
│   890   IF FL = 1 THEN  VTAB 19: PRINT TAB( 18)"";: FLASH   │
│   900   IF FL = 1 THEN  PRINT "X": NORMAL : GOTO 920        │
│   910   VTAB 19: PRINT  TAB( 18)"X"                         │
│   920   VTAB 21: PRINT  TAB( 32)"."; TAB( 37)"."            │
│   930   VTAB 23: PRINT  TAB( 36)"."                         │
│   940   VTAB 24: PRINT  TAB( 30)"."; TAB( 37)"."            │
│   950   IF FL = 1 THEN FL = 0: GOTO 7540                    │
│   960   GOSUB 7450                                          │
│   970   INPUT "PRESS RETURN TO CONTINUE ";A$               │
│   980   GOTO 820                                            │
│   990   C$ = "GEMINI"                                       │
│   1000  S$ = "CASTOR":S2$ = "POLLUX"                        │
│   1010  HOME                                                │
│   1020  VTAB 7: PRINT  TAB( 22)"."                          │
│   1030  IF FL = 1 THEN  VTAB 7: PRINT TAB( 9)"";            │
│   1040  IF FL = 1 THEN  FLASH : PRINT "X": NORMAL : GOTO 1060│
│   1050  VTAB 8: PRINT  TAB( 9)"X"                           │
│                                                             │
└─────────────────────────────────────────────────────────────┘
```

─── **CONST** *(continued)* ───

```
1060    VTAB 10: PRINT   TAB( 17)"."
1070    VTAB 12: PRINT   TAB( 6)"X"
1080    VTAB 13: PRINT   TAB( 10)"."
1090    VTAB 14: PRINT   TAB( 25)"*"
1100    VTAB 15: PRINT   TAB( 7)"."
1110    VTAB 16: PRINT   TAB( 15)"*"; TAB( 32)"*  *"
1120    VTAB 17: PRINT   TAB( 20)"."; TAB( 31)"."
1130    VTAB 19: PRINT   TAB( 15)"*"
1140    VTAB 20: PRINT   TAB( 29)"X"
1150    VTAB 22: PRINT   TAB( 27)"*"
1160    IF FL = 1 THEN FL = 0: GOTO 7540
1170    GOSUB 7450
1180    INPUT "PRESS RETURN TO CONTINUE ";A$
1190    GOTO 1010
1200 C$ = "CANCER"
1210    HOME
1220    VTAB 5: PRINT   TAB( 15)"."
1230    VTAB 11: PRINT   TAB( 15)"."
1240    VTAB 12: PRINT   TAB( 19)"."
1250    VTAB 15: PRINT   TAB( 15)"."; TAB( 26)"."
1260    VTAB 19: PRINT   TAB( 11)"."
1270    VTAB 21: PRINT   TAB( 25)"."
1280    PRINT
1290    GOSUB 7450
1300    INPUT "PRESS RETURN TO CONTINUE ";A$
1310    GOTO 7610
1320 C$ = "LEO"
1330 S$ = "REGULUS"
1340    HOME
1350    VTAB 7: PRINT   TAB( 34)"."
1360    VTAB 8: PRINT   TAB( 38)"*"
1370    VTAB 9: PRINT   TAB( 29)"."
1380    VTAB 11: PRINT   TAB( 14)"*"; TAB( 17)"."; TAB( 29)"*"
1390    VTAB 12: PRINT   TAB( 4)"."
1400    VTAB 14: PRINT   TAB( 31)".": VTAB 15: PRINT   TAB( 14)"*"
1410    VTAB 16: PRINT   TAB( 4)"*"
1420    IF FL = 1 THEN   VTAB 17: PRINT TAB( 30)"";
1430    IF FL = 1 THEN   FLASH : PRINT "X": NORMAL : GOTO 1450
1440    VTAB 17: PRINT   TAB( 30)"X"
1450    VTAB 19: PRINT   TAB( 12)"."; TAB( 25)"."; TAB( 36)"."
1460    VTAB 20: PRINT   TAB( 18)"."
1470    VTAB 22: PRINT   TAB( 14)".": PRINT
1480    IF FL = 1 THEN FL = 0: GOTO 7540
1490    GOSUB 7450
1500    INPUT "PRESS RETURN TO CONTINUE ";A$
1510    GOTO 1340
1520 C$ = "VIRGO"
1530 S$ = "SPICA"
1540    HOME
1550    VTAB 5: PRINT   TAB( 18)"*"
1560    VTAB 7: PRINT   TAB( 31)"."
1570    VTAB 8: PRINT   TAB( 33)"."; TAB( 37)"."
1580    VTAB 10: PRINT   TAB( 20)"*"
1590    VTAB 11: PRINT   TAB( 36)"*"
1600    VTAB 12: PRINT   TAB( 12)"."
```

CONST *(continued)*

```
1610   VTAB 13: PRINT  TAB( 22)"*"; TAB( 28)"*"
1620   VTAB 15: PRINT  TAB( 13)"."
1630   IF FL = 1 THEN  VTAB 19: PRINT TAB( 10)"";
1640   IF FL = 1 THEN  FLASH : PRINT "X": NORMAL : GOTO 1660
1650   VTAB 19: PRINT  TAB( 10)"X"
1660   PRINT
1670   IF FL = 1 THEN FL = 0: GOTO 7540
1680   GOSUB 7450
1690   INPUT "PRESS RETURN TO CONTINUE ";A$
1700   GOTO 1540
1710   C$ = "LIBRA"
1720   HOME
1730   VTAB 5: PRINT  TAB( 22)"."
1740   VTAB 10: PRINT  TAB( 18)"*"
1750   VTAB 12: PRINT  TAB( 7)"."
1760   VTAB 13: PRINT  TAB( 13)"."
1770   VTAB 14: PRINT  TAB( 8)"."
1780   VTAB 15: PRINT  TAB( 23)"*"
1790   VTAB 20: PRINT  TAB( 18)"."
1800   PRINT : PRINT
1810   GOSUB 7450
1820   INPUT "PRESS RETURN TO CONTINUE ";A$
1830   GOTO 7610
1840   C$ = "SCORPIO"
1850   S$ = "ANTARES"
1860   HOME
1870   VTAB 5: PRINT  TAB( 30)". *"
1880   VTAB 6: PRINT  TAB( 33)"."
1890   VTAB 7: PRINT  TAB( 34)"*"
1900   VTAB 10: PRINT  TAB( 27)"*"; TAB( 34)"*"
1910   IF FL = 1 THEN  VTAB 11: PRINT TAB( 25)"";
1920   IF FL = 1 THEN  FLASH : PRINT "X": NORMAL : GOTO 1940
1930   VTAB 11: PRINT  TAB( 25)"X"
1940   VTAB 12: PRINT  TAB( 24)"*"; TAB( 34)"."
1950   VTAB 16: PRINT  TAB( 8)"*"; TAB( 11)"**"; TAB( 21)"*"; TAB( 25)"."
1960   VTAB 17: PRINT  TAB( 24)"."
1970   VTAB 18: PRINT  TAB( 10)"*"
1980   VTAB 19: PRINT  TAB( 9)"*"; TAB( 21)"*"
1990   VTAB 22: PRINT  TAB( 12)"*"; TAB( 18)"*"
2000   PRINT
2010   IF FL = 1 THEN FL = 0: GOTO 7540
2020   GOSUB 7450
2030   INPUT "PRESS RETURN TO CONTINUE ";A$
2040   GOTO 1860
2050   C$ = "SAGITTARIUS"
2060   HOME
2070   VTAB 6: PRINT  TAB( 9)"."
2080   VTAB 7: PRINT  TAB( 9)"."
2090   VTAB 9: PRINT  TAB( 12)"*"; TAB( 15)"."
2100   VTAB 10: PRINT  TAB( 14)"."
2110   VTAB 12: PRINT  TAB( 15)"*"; TAB( 25)"*"
2120   VTAB 13: PRINT  TAB( 17)"*"; TAB( 27)"."
2130   VTAB 14: PRINT  TAB( 12)"*"; TAB( 29)"."
2140   VTAB 15: PRINT  TAB( 13)"*"; TAB( 26)"*"
2150   VTAB 16: PRINT  TAB( 29)"*"
```

─────── **CONST** *(continued)* ───────

```
2160    VTAB 19: PRINT   TAB( 24)"*"
2170    VTAB 21: PRINT   TAB( 25)"*"
2180    PRINT : PRINT
2190    GOSUB 7450
2200    INPUT "PRESS RETURN TO CONTINUE ";A$
2210    GOTO 7610
2220 C$ = "CAPRICORNUS"
2230    HOME
2240    VTAB 6: PRINT   TAB( 32)"*"
2250    VTAB 8: PRINT   TAB( 31)"*"
2260    VTAB 10: PRINT   TAB( 6)"*  ."; TAB( 13)"."; TAB( 18)"."
2270    VTAB 14: PRINT   TAB( 9)"."
2280    VTAB 15: PRINT   TAB( 13)":'"
2290    VTAB 16: PRINT   TAB( 19)"."; TAB( 26)"."
2300    VTAB 17: PRINT   TAB( 17)"*"
2310    VTAB 21: PRINT   TAB( 18)"*"
2320    PRINT : PRINT
2330    GOSUB 7450
2340    INPUT "PRESS RETURN TO CONTINUE ";A$
2350    GOTO 7610
2360 C$ = "AQUARIUS"
2370    HOME
2380    VTAB 8: PRINT   TAB( 13)"."
2390    VTAB 9: PRINT   TAB( 11)"*  *"; TAB( 19)"*"
2400    VTAB 10: PRINT   TAB( 15)"*"
2410    VTAB 12: PRINT   TAB( 28)"*"; TAB( 36)"."
2420    VTAB 14: PRINT   TAB( 9)"."
2430    VTAB 15: PRINT   TAB( 37)"."
2440    VTAB 16: PRINT   TAB( 32)"."
2450    VTAB 18: PRINT   TAB( 10)"."
2460    VTAB 20: PRINT   TAB( 8)"*"
2470    PRINT : PRINT : PRINT
2480    GOSUB 7450
2490    INPUT "PRESS RETURN TO CONTINUE ";A$
2500    GOTO 7610
2510 C$ = "ORION"
2520 S$ = "RIGEL":S2$ = "BETELGEUX":S3$ = "ORION NEBULA"
2530    HOME
2540    VTAB 4: PRINT   TAB( 8)"."; TAB( 16)".:"; TAB( 30)"."
2550    VTAB 5: PRINT   TAB( 31)"."
2560    VTAB 6: PRINT   TAB( 10)"X"; TAB( 31)"."
2570    VTAB 7: PRINT   TAB( 19)"."; TAB( 21)"*"; TAB( 31)"."
2580    VTAB 9: PRINT   TAB( 16)"."
2590    VTAB 10: PRINT   TAB( 30)". ."
2600    VTAB 13: PRINT   TAB( 19)"*"
2610    VTAB 14: PRINT   TAB( 17)"*"
2620    VTAB 15: PRINT   TAB( 15)"*"; TAB( 21)"."
2630    VTAB 17: PRINT   TAB( 16)"."
2640    VTAB 18: PRINT   TAB( 16)":"
2650    IF FL = 1 THEN   VTAB 19: PRINT TAB( 24)"";
2660    IF FL = 1 THEN   FLASH : PRINT "X": NORMAL : GOTO 2680
2670    VTAB 19: PRINT   TAB( 24)"X"
2680    VTAB 20: PRINT   TAB( 12)"*"
2690    PRINT : PRINT
2700    IF FL = 1 THEN FL = 0: GOTO 7540
```

```
┌─ CONST (continued) ─────────────────────────────────────────────┐
│                                                                  │
│  2710    GOSUB 7450                                              │
│  2720    INPUT "PRESS RETURN TO CONTINUE ";A$                   │
│  2730    GOTO 2530                                               │
│  2740 C$ = "LEPUS"                                               │
│  2750    HOME                                                    │
│  2760    VTAB 5: PRINT  TAB( 25)"."                              │
│  2770    VTAB 6: PRINT  TAB( 22)".   ."                          │
│  2780    VTAB 7: PRINT  TAB( 7)"."; TAB( 10)"."                  │
│  2790    VTAB 8: PRINT  TAB( 4)"."                               │
│  2800    VTAB 9: PRINT  TAB( 24)"*"                              │
│  2810    VTAB 11: PRINT  TAB( 16)"*"                             │
│  2820    VTAB 13: PRINT  TAB( 9)"."                              │
│  2830    VTAB 14: PRINT  TAB( 18)"*"                             │
│  2840    VTAB 15: PRINT  TAB( 12)"."                             │
│  2850    VTAB 16: PRINT  TAB( 27)"*"                             │
│  2860    PRINT : PRINT : PRINT                                   │
│  2870    GOSUB 7450                                              │
│  2880    INPUT "PRESS RETURN TO CONTINUE ";A$                   │
│  2890    GOTO 7610                                               │
│  2900 C$ = "CANIS MAJOR"                                         │
│  2910 S$ = "SIRIUS"                                              │
│  2920    HOME                                                    │
│  2930    VTAB 7: PRINT  TAB( 17)"."                              │
│  2940    IF FL = 1 THEN  VTAB 8: PRINT TAB( 21)"";               │
│  2950    IF FL = 1 THEN  FLASH : PRINT "X": NORMAL : GOTO 2970   │
│  2960    VTAB 8: PRINT  TAB( 21)"X"                              │
│  2970    VTAB 10: PRINT  TAB( 23)"."; TAB( 27)"*"                │
│  2980    VTAB 12: PRINT  TAB( 16)"*"                             │
│  2990    VTAB 13: PRINT  TAB( 12)"."; TAB( 18)"."                │
│  3000    VTAB 14: PRINT  TAB( 14)"*"                             │
│  3010    VTAB 15: PRINT  TAB( 9)"*"                              │
│  3020    VTAB 16: PRINT  TAB( 15)"*"                             │
│  3030    VTAB 18: PRINT  TAB( 15)"."; TAB( 24)"*"                │
│  3040    VTAB 20: PRINT  TAB( 20)".": PRINT : PRINT              │
│  3050    IF FL = 1 THEN FL = 0: GOTO 7540                        │
│  3060    GOSUB 7450                                              │
│  3070    INPUT "PRESS RETURN TO CONTINUE ";A$                   │
│  3080    GOTO 2920                                               │
│  3090 C$ = "CANIS MINOR"                                         │
│  3100 S$ = "PROCYON"                                             │
│  3110    HOME                                                    │
│  3120    VTAB 8: PRINT  TAB( 21)"."                              │
│  3130    VTAB 9: PRINT  TAB( 22)"*"                              │
│  3140    IF FL = 1 THEN  VTAB 12: PRINT TAB( 16)"";              │
│  3150    IF FL = 1 THEN  FLASH : PRINT "X": NORMAL : GOTO 3170   │
│  3160    VTAB 12: PRINT  TAB( 16)"X"                             │
│  3170    VTAB 15: PRINT  TAB( 6)"."                              │
│  3180    VTAB 22                                                 │
│  3190    IF FL = 1 THEN FL = 0: GOTO 7540                        │
│  3200    GOSUB 7450                                              │
│  3210    INPUT "PRESS RETURN TO CONTINUE ";A$                   │
│  3220    GOTO 3110                                               │
│  3230 C$ = "CETUS"                                               │
│  3240 S$ = "MIRA"                                                │
│  3250    HOME                                                    │
│                                                                  │
└──────────────────────────────────────────────────────────────────┘
```

─────────────── **CONST** *(continued)* ───────────────

```
3260   VTAB 2: PRINT   TAB( 5)"."; TAB( 8)"."
3270   VTAB 5: PRINT   TAB( 4)"*"
3280   VTAB 6: PRINT   TAB( 8)"*"
3290   VTAB 8: PRINT   TAB( 10)"."
3300   IF FL = 1 THEN   VTAB 9: PRINT TAB( 14)"";
3310   IF FL = 1 THEN   FLASH : PRINT "." : NORMAL : GOTO 3330
3320   VTAB 9: PRINT   TAB( 14)"."
3330   VTAB 13: PRINT   TAB( 32)"*"
3340   VTAB 14: PRINT   TAB( 25)"*"; TAB( 35)"*"
3350   VTAB 15: PRINT   TAB( 8)"."; TAB( 26)"*"
3360   VTAB 18: PRINT   TAB( 26)"*"; TAB( 39)"*"
3370   VTAB 21: PRINT   TAB( 22)".": PRINT : PRINT
3380   IF FL = 1 THEN FL = 0: GOTO 7540
3390   GOSUB 7450
3400   INPUT "PRESS RETURN TO CONTINUE ";A$
3410   GOTO 3250
3420 C$ = "CORVUS"
3430   HOME
3440   VTAB 7: PRINT   TAB( 14)"."
3450   VTAB 8: PRINT   TAB( 15)"*"
3460   VTAB 10: PRINT   TAB( 21)"*"
3470   VTAB 15: PRINT   TAB( 13)"*"; TAB( 22)"*"
3480   VTAB 17: PRINT   TAB( 22)"*"
3490   VTAB 22
3500   GOSUB 7450
3510   INPUT "PRESS RETURN TO CONTINUE ";A$
3520   GOTO 7610
3530 C$ = "CRATER"
3540   HOME
3550   VTAB 13: PRINT   TAB( 13)"*"
3560   VTAB 16: PRINT   TAB( 10)"*"
3570   VTAB 17: PRINT   TAB( 20)"*"
3580   VTAB 20: PRINT   TAB( 15)"*"
3590   PRINT : PRINT : PRINT : PRINT
3600   PRINT : PRINT
3610   GOSUB 7450
3620   INPUT "PRESS RETURN TO CONTINUE ";A$
3630   GOTO 7610
3640 C$ = "HYDRA"
3650   HOME
3660   VTAB 5: PRINT   TAB( 33)". ."
3670   VTAB 6: PRINT   TAB( 34).."
3680   VTAB 7: PRINT   TAB( 31)"*"
3690   VTAB 9: PRINT   TAB( 27)"."
3700   VTAB 10: PRINT   TAB( 22)"."; TAB( 28)"."
3710   VTAB 11: PRINT   TAB( 25)"."
3720   VTAB 14: PRINT   TAB( 11)"."
3730   VTAB 16: PRINT   TAB( 5)"."
3740   VTAB 17: PRINT   TAB( 18)". ."
3750   PRINT : PRINT
3760   VTAB 21: PRINT   TAB( 14)"."; TAB( 17)"."
3770   VTAB 23: PRINT   TAB( 14)"*"; TAB( 17)"*"
3780   PRINT : PRINT
3790   GOSUB 7450
3800   INPUT "PRESS RETURN TO CONTINUE ";A$
```

```
┌─ CONST (continued) ──────────────────────────────────────────┐
│                                                               │
│   3810   GOTO 7610                                            │
│   3820   C$ = "SERPENS"                                       │
│   3830   HOME                                                 │
│   3840   VTAB 5: PRINT  TAB( 20)"."                           │
│   3850   VTAB 6: PRINT  TAB( 18)"."                           │
│   3860   VTAB 7: PRINT  TAB( 16)"."; TAB( 19)"."              │
│   3870   VTAB 12: PRINT  TAB( 21)"*"                          │
│   3880   VTAB 14: PRINT  TAB( 18)"."                          │
│   3890   VTAB 15: PRINT  TAB( 19)"*"                          │
│   3900   VTAB 17: PRINT  TAB( 17)"*"                          │
│   3910   VTAB 21: PRINT  TAB( 18)"*"                          │
│   3920   PRINT : PRINT                                        │
│   3930   GOSUB 7450                                           │
│   3940   INPUT "PRESS RETURN TO CONTINUE ";A$                 │
│   3950   GOTO 7610                                            │
│   3960   C$ = "DELPHINUS"                                     │
│   3970   HOME                                                 │
│   3980   VTAB 8: PRINT  TAB( 17)"."                           │
│   3990   VTAB 9: PRINT  TAB( 21)"."                           │
│   4000   VTAB 10: PRINT  TAB( 18)"."                          │
│   4010   VTAB 11: PRINT  TAB( 22)".."                         │
│   4020   VTAB 13: PRINT  TAB( 23)"."                          │
│   4030   VTAB 15: PRINT  TAB( 24)"."                          │
│   4040   PRINT : PRINT                                        │
│   4050   GOSUB 7450                                           │
│   4060   INPUT "PRESS RETURN TO CONTINUE ";A$                 │
│   4070   GOTO 7610                                            │
│   4080   C$ = "AQUILA"                                        │
│   4090   S$ = "ALTAIR"                                        │
│   4100   HOME                                                 │
│   4110   VTAB 4: PRINT  TAB( 27)"."                           │
│   4120   VTAB 5: PRINT  TAB( 25)"*"                           │
│   4130   VTAB 7: PRINT  TAB( 16)"*"                           │
│   4140   IF FL = 1 THEN   VTAB 8: PRINT TAB( 15)"";           │
│   4150   IF FL = 1 THEN  FLASH : PRINT "X": NORMAL : GOTO 4170│
│   4160   VTAB 8: PRINT  TAB( 15)"X"                           │
│   4170   VTAB 9: PRINT  TAB( 20)"."                           │
│   4180   VTAB 10: PRINT  TAB( 14)"*"                          │
│   4190   VTAB 14: PRINT  TAB( 14)"."; TAB( 22)"*"             │
│   4200   VTAB 15: PRINT  TAB( 9)"*"                           │
│   4210   VTAB 19: PRINT  TAB( 27)"*"                          │
│   4220   VTAB 20: PRINT  TAB( 29)"*"                          │
│   4230   PRINT : PRINT                                        │
│   4240   IF FL = 1 THEN FL = 0: GOTO 7540                     │
│   4250   GOSUB 7450                                           │
│   4260   INPUT "PRESS RETURN TO CONTINUE ";A$                 │
│   4270   GOTO 4100                                            │
│   4280   C$ = "LYRA"                                          │
│   4290   S$ = "VEGA"                                          │
│   4300   HOME                                                 │
│   4310   VTAB 5: PRINT  TAB( 15)"."                           │
│   4320   VTAB 9: PRINT  TAB( 19)"."                           │
│   4330   VTAB 10: PRINT  TAB( 8)".";                          │
│   4340   IF FL = 1 THEN  PRINT  TAB( 21)"";                   │
│   4350   IF FL = 1 THEN  FLASH : PRINT "X": NORMAL : GOTO 4370│
│                                                               │
└───────────────────────────────────────────────────────────────┘
```

─────────────────── **CONST** (continued) ───────────────────

```
4360   PRINT   TAB( 21)"X"
4370   VTAB 11: PRINT   TAB( 7)"."; TAB( 18)"."
4380   VTAB 12: PRINT   TAB( 15)"."; TAB( 27)"."
4390   VTAB 16: PRINT   TAB( 17)"*"
4400   VTAB 17: PRINT   TAB( 14)"*"
4410   PRINT : PRINT : PRINT
4420   IF FL = 1 THEN FL = 0: GOTO 7540
4430   GOSUB 7450
4440   INPUT "PRESS RETURN TO CONTINUE ";A$
4450   GOTO 4300
4460 C$ = "CYGNUS"
4470 S$ = "DENEB"
4480   HOME
4490   VTAB 5: PRINT   TAB( 30)"."
4500   VTAB 6: PRINT   TAB( 28)"."
4510   VTAB 9: PRINT   TAB( 20)"."
4520   VTAB 10: PRINT   TAB( 20)"."
4530   VTAB 11: PRINT   TAB( 6)".";
4540   IF FL = 1 THEN   PRINT   TAB( 15)"";
4550   IF FL = 1 THEN   FLASH : PRINT "X";: NORMAL : GOTO 4570
4560   PRINT   TAB( 15)"X";
4570   PRINT   TAB( 26)"*"
4580   VTAB 12: PRINT   TAB( 10)"."
4590   VTAB 13: PRINT   TAB( 11)"."
4600   VTAB 14: PRINT   TAB( 19)"*"
4610   VTAB 15: PRINT   TAB( 7)"."
4620   VTAB 17: PRINT   TAB( 24)"."
4630   VTAB 18: PRINT   TAB( 14)"*"; TAB( 20)"."
4640   VTAB 20: PRINT   TAB( 8)"."; TAB( 18)"."
4650   VTAB 22: PRINT   TAB( 30)"*"
4660   PRINT : PRINT
4670   IF FL = 1 THEN FL = 0: GOTO 7540
4680   GOSUB 7450
4690   INPUT "PRESS RETURN TO CONTINUE ";A$
4700   GOTO 4480
4710 C$ = "HERCULES"
4720   HOME
4730   VTAB 5: PRINT   TAB( 18)"*"; TAB( 30)". ."
4740   VTAB 6: PRINT   TAB( 31)"."
4750   VTAB 7: PRINT   TAB( 28)"."; TAB( 36)"."
4760   VTAB 9: PRINT   TAB( 26)"*"
4770   VTAB 10: PRINT   TAB( 13)"*"; TAB( 20)"*"
4780   VTAB 13: PRINT   TAB( 26)"*"
4790   VTAB 14: PRINT   TAB( 11)"."; TAB( 22)"*"
4800   VTAB 15: PRINT   TAB( 9)".  ."
4810   VTAB 16: PRINT   TAB( 13)"*"
4820   VTAB 17: PRINT   TAB( 19)"*"
4830   VTAB 19: PRINT   TAB( 4)"."; TAB( 30)"*"
4840   VTAB 20: PRINT   TAB( 32)"*": PRINT : PRINT
4850   GOSUB 7450
4860   INPUT "PRESS RETURN TO CONTINUE ";A$
4870   GOTO 7610
4880 C$ = "OPHIUCHUS"
4890   HOME
4900   VTAB 5: PRINT   TAB( 17)"*"; TAB( 26)"."
```

```
┌─ CONST (continued) ─────────────────────────────────────────┐
│                                                              │
│  4910   VTAB 6: PRINT   TAB( 25)"."                          │
│  4920   VTAB 9: PRINT   TAB( 31)"."                          │
│  4930   VTAB 10: PRINT  TAB( 15)"*"                          │
│  4940   VTAB 11: PRINT  TAB( 10)". ."; TAB( 14)"*"           │
│  4950   VTAB 12: PRINT  TAB( 36)"*"                          │
│  4960   VTAB 13: PRINT  TAB( 34)"*"                          │
│  4970   VTAB 17: PRINT  TAB( 30)"*"                          │
│  4980   VTAB 18: PRINT  TAB( 14)"."                          │
│  4990   VTAB 20: PRINT  TAB( 22)"*"                          │
│  5000   PRINT : PRINT                                        │
│  5010   GOSUB 7450                                           │
│  5020   INPUT "PRESS RETURN TO CONTINUE ";A$                 │
│  5030   GOTO 7610                                            │
│  5040  C$ = "URSA MINOR                                      │
│  5050  S$ = "POLARIS"                                        │
│  5060   HOME                                                 │
│  5070   IF FL = 1 THEN  VTAB 7: PRINT TAB( 22)"";            │
│  5080   IF FL = 1 THEN  FLASH : PRINT "X": NORMAL : GOTO 5100│
│  5090   VTAB 7: PRINT   TAB( 22)"X"                          │
│  5100   VTAB 9: PRINT   TAB( 19)"."                          │
│  5110   VTAB 11: PRINT  TAB( 17)"."                          │
│  5120   VTAB 14: PRINT  TAB( 14)"."; TAB( 17)"*"             │
│  5130   VTAB 15: PRINT  TAB( 20)"."                          │
│  5140   VTAB 16: PRINT  TAB( 18)"*"                          │
│  5150   VTAB 17: PRINT  TAB( 15)"*"                          │
│  5160   PRINT : PRINT : PRINT                                │
│  5170   IF FL = 1 THEN FL = 0: GOTO 7540                     │
│  5180   GOSUB 7450                                           │
│  5190   INPUT "PRESS RETURN TO CONTINUE ";A$                 │
│  5200   GOTO 5060                                            │
│  5210  C$ = "URSA MAJOR"                                     │
│  5220  S$ = "DUBHE":S2$ = "ALCOR"                            │
│  5230   HOME                                                 │
│  5240   VTAB 8: PRINT   TAB( 35)"."                          │
│  5250   IF FL = 1 THEN  VTAB 10: PRINT TAB( 25)"";           │
│  5260   IF FL = 1 THEN  FLASH : PRINT "*";: NORMAL : PRINT  TAB( 34)"." │
│         : GOTO 5280                                          │
│  5270   VTAB 10: PRINT  TAB( 25)"*"; TAB( 34)"."             │
│  5280   VTAB 11: PRINT  TAB( 8)"*"                           │
│  5290   VTAB 12: PRINT  TAB( 12)"*"; TAB( 17)"*"             │
│  5300   VTAB 13: PRINT  TAB( 3)"*"                           │
│  5310   VTAB 14: PRINT  TAB( 25)"*"                          │
│  5320   VTAB 15: PRINT  TAB( 18)"*"                          │
│  5330   VTAB 16: PRINT  TAB( 35)"."                          │
│  5340   VTAB 17: PRINT  TAB( 18)"."                          │
│  5350   VTAB 19: PRINT  TAB( 25)"."; TAB( 33)"."             │
│  5360   VTAB 20: PRINT  TAB( 32)"."                          │
│  5370   VTAB 24: PRINT  TAB( 24)":"                          │
│  5380   PRINT : PRINT                                        │
│  5390   IF FL = 1 THEN FL = 0: GOTO 7540                     │
│  5400   GOSUB 7450                                           │
│  5410   INPUT "PRESS RETURN TO CONTINUE ";A$                 │
│  5420   GOTO 5230                                            │
│  5430  C$ = "DRACO"                                          │
│  5440   HOME                                                 │
│                                                              │
└──────────────────────────────────────────────────────────────┘
```

CONST *(continued)*

```
5450   VTAB 5: PRINT   TAB( 36)"*"
5460   VTAB 7: PRINT   TAB( 34)"*"
5470   VTAB 10: PRINT   TAB( 14)"* ."
5480   VTAB 11: PRINT   TAB( 14)"."; TAB( 35)"*"
5490   VTAB 12: PRINT   TAB( 7)"."
5500   VTAB 13: PRINT   TAB( 6)"."; TAB( 9)"*"
5510   VTAB 14: PRINT   TAB( 8)"."; TAB( 20)"*"
5520   VTAB 15: PRINT   TAB( 24)"*"
5530   VTAB 16: PRINT   TAB( 31)"*"
5540   VTAB 17: PRINT   TAB( 25)"*"
5550   VTAB 21: PRINT   TAB( 14)"."; TAB( 17)"."
5560   VTAB 23: PRINT   TAB( 14)"*"; TAB( 17)"*"
5570   PRINT
5580   GOSUB 7450
5590   INPUT "PRESS RETURN TO CONTINUE ";A$
5600   GOTO 7610
5610  C$ = "CANES VENATICI"
5620   HOME
5630   VTAB 11: PRINT   TAB( 16)"."; TAB( 27)"."
5640   VTAB 14: PRINT   TAB( 20)"*"
5650   VTAB 20
5660   GOSUB 7450
5670   INPUT "PRESS RETURN TO CONTINUE ";A$
5680   GOTO 7610
5690  C$ = "CEPHEUS"
5700   HOME
5710   VTAB 5: PRINT   TAB( 18)"*"
5720   VTAB 11: PRINT   TAB( 24)"*"
5730   VTAB 13: PRINT   TAB( 16)"*"
5740   VTAB 17: PRINT   TAB( 25)"*"
5750   VTAB 18: PRINT   TAB( 30)"*"
5760   VTAB 19: PRINT   TAB( 17)"*"
5770   PRINT : PRINT : PRINT
5780   GOSUB 7450
5790   INPUT "PRESS RETURN TO CONTINUE ";A$
5800   GOTO 7610
5810  C$ = "CASSIOPEIA"
5820   HOME
5830   VTAB 6: PRINT   TAB( 15)"."
5840   VTAB 7: PRINT   TAB( 14)"."
5850   VTAB 12: PRINT   TAB( 11)"*"
5860   VTAB 16: PRINT   TAB( 20)"*"; TAB( 30)"*"
5870   VTAB 17: PRINT   TAB( 15)"*"
5880   VTAB 18: PRINT   TAB( 22)"."
5890   VTAB 19: PRINT   TAB( 24)"*"
5900   VTAB 21: PRINT   TAB( 26)"."
5910   PRINT : PRINT : PRINT
5920   GOSUB 7450
5930   INPUT "PRESS RETURN TO CONTINUE ";A$
5940   GOTO 7610
5950  C$ = "PERSEUS"
5960  S$ = "ALGOL"
5970   HOME
5980   VTAB 5: PRINT   TAB( 23)"."
5990   VTAB 7: PRINT   TAB( 21)"*  ."
```

─ **CONST** *(continued)* ─

```
6000   VTAB 10: PRINT   TAB( 6)"."; TAB( 18)"X"; TAB( 21)"."; TAB( 25)"."
6010   VTAB 11: PRINT   TAB( 6)"."
6020   VTAB 12: PRINT   TAB( 8)"."; TAB( 13)"*"
6030   VTAB 13: PRINT   TAB( 22)"."
6040   VTAB 16: PRINT   TAB( 14)"."
6050   IF FL = 1 THEN   VTAB 17: PRINT TAB( 23)"";
6060   IF FL = 1 THEN   FLASH : PRINT "*": NORMAL : GOTO 6080
6070   VTAB 17: PRINT   TAB( 23)"*"
6080   VTAB 18: PRINT   TAB( 11)"*"
6090   VTAB 19: PRINT   TAB( 24)"."
6100   VTAB 21: PRINT   TAB( 12)"."
6110   VTAB 23: PRINT   TAB( 13)"*": PRINT
6120   IF FL = 1 THEN FL = 0: GOTO 7540
6130   GOSUB 7450
6140   INPUT "PRESS RETURN TO CONTINUE ";A$
6150   GOTO 5970
6160 C$ = "ANDROMEDA"
6170   HOME
6180   VTAB 5: PRINT   TAB( 14)"."
6190   VTAB 6: PRINT   TAB( 15)"."
6200   VTAB 7: PRINT   TAB( 33)"."
6210   VTAB 8: PRINT   TAB( 33)"."
6220   VTAB 9: PRINT   TAB( 34)"."
6230   VTAB 10: PRINT   TAB( 9)"*"
6240   VTAB 11: PRINT   TAB( 14)"."; TAB( 24)"."
6250   VTAB 12: PRINT   TAB( 23)"."
6260   VTAB 14: PRINT   TAB( 20)"*"; TAB( 27)"."
6270   VTAB 15: PRINT   TAB( 26)"."
6280   VTAB 16: PRINT   TAB( 33)"*"
6290   VTAB 17: PRINT   TAB( 26)".": PRINT : PRINT
6300   GOSUB 7450
6310   INPUT "PRESS RETURN TO CONTINUE ";A$
6320   GOTO 7610
6330 C$ = "PEGASUS"
6340   HOME
6350   VTAB 6: PRINT   TAB( 22)"."
6360   VTAB 7: PRINT   TAB( 19)"*"
6370   VTAB 9: PRINT   TAB( 21)"."
6380   VTAB 10: PRINT   TAB( 22)"."
6390   VTAB 13: PRINT   TAB( 8)"*"; TAB( 19)"*"
6400   VTAB 16: PRINT   TAB( 23)"*"
6410   VTAB 17: PRINT   TAB( 34)"*"
6420   VTAB 18: PRINT   TAB( 28)"*"
6430   VTAB 22
6440   GOSUB 7450
6450   INPUT "PRESS RETURN TO CONTINUE ";A$
6460   GOTO 7610
6470 C$ = "BOOTES"
6480 S$ = "ARCTURUS"
6490   HOME
6500   VTAB 6: PRINT   TAB( 21)". ."
6510   VTAB 8: PRINT   TAB( 22)"."
6520   VTAB 10: PRINT   TAB( 17)"*"
6530   VTAB 11: PRINT   TAB( 20)"*"
6540   VTAB 13: PRINT   TAB( 14)"*"
6550   VTAB 15: PRINT   TAB( 21)"."
6560   VTAB 17: PRINT   TAB( 19)"*"
```

─────────── **CONST** *(continued)* ───────────

```
6570    IF FL = 1 THEN  VTAB 20: PRINT TAB( 25)"";
6580    IF FL = 1 THEN  FLASH : PRINT "X": NORMAL : GOTO 6600
6590    VTAB 20: PRINT  TAB( 25)"X"
6600    VTAB 21: PRINT  TAB( 30)"*"
6610    PRINT : PRINT
6620    IF FL = 1 THEN FL = 0: GOTO 7540
6630    GOSUB 7450
6640    INPUT "PRESS RETURN TO CONTINUE ";A$
6650    GOTO 6490
6660  C$ = "CORONA"
6670  S$ = "ALPHEKKA"
6680    HOME
6690    VTAB 7: PRINT  TAB( 20)"."
6700    VTAB 8: PRINT  TAB( 9)"."
6710    VTAB 10: PRINT  TAB( 9)"."; TAB( 22)"."
6720    VTAB 12: PRINT  TAB( 24)"*"
6730    VTAB 13: PRINT  TAB( 14)"*"
6740    VTAB 14: PRINT  TAB( 16)"* *";
6750    IF FL = 1 THEN  PRINT  TAB( 21)"";
6760    IF FL = 1 THEN  FLASH : PRINT "X": NORMAL : GOTO 6780
6770    PRINT  TAB( 21)"X"
6780    VTAB 21
6790    IF FL = 1 THEN FL = 0: GOTO 7540
6800    GOSUB 7450
6810    INPUT "PRESS RETURN TO CONTINUE ";A$
6820    GOTO 6680
6830  C$ = "AURIGA"
6840  S$ = "CAPELLA"
6850    HOME
6860    VTAB 5: PRINT  TAB( 16)"."
6870    IF FL = 1 THEN  VTAB 10: PRINT TAB( 22)"";
6880    IF FL = 1 THEN  FLASH : PRINT "X": NORMAL : GOTO 6900
6890    VTAB 10: PRINT  TAB( 22)"X"
6900    VTAB 11: PRINT  TAB( 14)"*"
6910    VTAB 12: PRINT  TAB( 25)"."
6920    VTAB 13: PRINT  TAB( 24)".."
6930    VTAB 15: PRINT  TAB( 16)".."
6940    VTAB 16: PRINT  TAB( 14)"*"
6950    VTAB 18: PRINT  TAB( 26)"*"
6960    VTAB 22
6970    IF FL = 1 THEN FL = 0: GOTO 7540
6980    GOSUB 7450
6990    INPUT "PRESS RETURN TO CONTINUE ";A$
7000    GOTO 6850
7010  C$ = "ERIDANUS"
7020    HOME
7030    VTAB 5: PRINT  TAB( 25)"."
7040    VTAB 6: PRINT  TAB( 26)"."
7050    VTAB 7: PRINT  TAB( 19)"."
7060    VTAB 10: PRINT  TAB( 11)"."; TAB( 14)"."
7070    VTAB 11: PRINT  TAB( 6)"."
7080    VTAB 12: PRINT  TAB( 17)"."
7090    VTAB 13: PRINT  TAB( 24)"."; TAB( 28)"."; TAB( 34)"."
7100    VTAB 15: PRINT  TAB( 24)"."
7110    VTAB 16: PRINT  TAB( 20)"."
7120    VTAB 24: PRINT  TAB( 15)"."
7130    PRINT
```

CONST *(continued)*

```
7140  GOSUB 7450
7150  INPUT "PRESS RETURN TO CONTINUE ";A$
7160  GOTO 7610
7170  C$ = "MONOCEROS"
7180  HOME
7190  VTAB 6: PRINT  TAB( 24)"."
7200  VTAB 8: PRINT  TAB( 25)"."
7210  VTAB 10: PRINT  TAB( 22)"."
7220  VTAB 11: PRINT  TAB( 18)"."
7230  VTAB 12: PRINT  TAB( 6)"."
7240  VTAB 14: PRINT  TAB( 28)"."
7250  VTAB 15: PRINT  TAB( 25)"."
7260  VTAB 16: PRINT  TAB( 12)"."
7270  PRINT : PRINT
7280  GOSUB 7450
7290  INPUT "PRESS RETURN TO CONTINUE ";A$
7300  GOTO 7610
7310  C$ = "PISCIS AUSTRALIS"
7320  S$ = "FOMALHAUT"
7330  HOME
7340  VTAB 12: PRINT  TAB( 16)"."
7350  IF FL = 1 THEN  VTAB 14: PRINT TAB( 13)"";:
7360  IF FL = 1 THEN  FLASH : PRINT "X": NORMAL : GOTO 7380
7370  VTAB 14: PRINT  TAB( 13)"X"
7380  VTAB 15: PRINT  TAB( 18)"."; TAB( 24)"."; TAB( 28)"."
7390  VTAB 16: PRINT  TAB( 14)"."
7400  VTAB 20
7410  IF FL = 1 THEN FL = 0: GOTO 7540
7420  GOSUB 7450
7430  INPUT " PRESS RETURN TO CONTINUE";A$
7440  GOTO 7330
7450  REM  SUB QUERY
7460  INPUT "NAME THIS CONSTELLATION ";A$
7470  IF A$ = C$ THEN  FLASH : PRINT "CORRECT": NORMAL :FL = 1: RETURN
7480 A$ = ""
7490  PRINT CL$;: INPUT "TRY AGAIN ";A$
7500  IF A$ = C$ THEN  GOTO 7470
7510  PRINT "YOU ARE STILL WRONG"
7520  PRINT " THE CONSTELLATION IS ";C$
7530 FL = 1: RETURN
7540  REM  SUB STAR QUERY
7550  INPUT "NAME THIS STAR ";A$
7560  IF A$ = S$ THEN  FLASH : PRINT "CORRECT": NORMAL : GOTO 7600
7570  INPUT "WRONG; TRY AGAIN ";A$
7580  IF A$ = S$ THEN  GOTO 7560
7590  PRINT "STILL WRONG, THE STAR IS ";S$
7600  INPUT "PRESS RETURN TO CONTINUE ";A$
7610  HOME : PRINT : PRINT : PRINT : PRINT
7620  INPUT "ANOTHER TEST? Y/N ";A$
7630  IF A$ = "N" THEN  HOME : GOTO 7700
7640  IF A$ < > "Y" THEN  PRINT "INVALID RESPONSE": PRINT : GOTO 7620
7650  HOME
7660  IF FL = 1 THEN  GOTO 7680
7670 FL = 0
7680 Z = Z + 1
7690 X =  FRE (0):Y =  FRE (A$): GOTO 420
7700  END
```

```
8000   REM   SUPPLEMENTARY PROGRAM
8010   REM   FOR CONSTELLATIONS OF
8020   REM   THE SOUTHERN HEMISPHERE
8030   HOME
8040   VTAB 10: PRINT "SOUTHERN HEMISPHERE CONSTELLATIONS"
8050   PRINT : PRINT : PRINT : PRINT
8060   PRINT   TAB( 6)"BY ERIC BURGESS F.R.A.S."
8070   PRINT   TAB( 6)"ALL RIGHTS RESERVED BY"
8080   PRINT   TAB( 6)"S & T SOFTWARE SERVICE"
8090   PRINT : PRINT : PRINT
8100   INPUT "PRESS RETURN WHEN READY ";A$
8110   W = 1
8120   A =   RND (W) * 14
8130   A =   INT (A) + 1
8140   FL = 0
8150   ON A GOSUB 8160,8260,8370,8530,8670,8780,8880,9010,9200,9370,9500,
       9620,9720,9830
8160   C$ = "CHAMAELEON"
8170   HOME
8180   VTAB 11: PRINT   TAB( 21)"*"
8190   VTAB 12: PRINT   TAB( 28)"*"
8200   VTAB 14: PRINT   TAB( 9)"*"; TAB( 20)"*"; TAB( 27)"*"
8210   VTAB 15: PRINT   TAB( 8)"*"
8220   VTAB 22
8230   GOSUB 9940
8240   INPUT "PRESS RETURN TO CONTINUE ";A$
8250   GOTO 10030
8260   C$ = "MUSCA"
8270   HOME
8280   VTAB 9: PRINT   TAB( 18)"*"; TAB( 22)"*"
8290   VTAB 12: PRINT   TAB( 19)"*"
8300   VTAB 13: PRINT   TAB( 20)"*"
8310   VTAB 14: PRINT   TAB( 17)"."
8320   VTAB 15: PRINT   TAB( 13)"."
8330   VTAB 22
8340   GOSUB 9940
8350   INPUT "PRESS RETURN TO CONTINUE ";A$
8360   GOTO 10030
8370   C$ = "HYDRUS"
8380   HOME
8390   S$ = "ACHERNAR"
8400   VTAB 5: PRINT   TAB( 11)"*"
8410   VTAB 7: PRINT   TAB( 26)"."; TAB( 31)"*"
8420   VTAB 13: PRINT   TAB( 22)"."
8430   VTAB 14: PRINT   TAB( 19)"."; TAB( 24)"."
8440   VTAB 19: PRINT   TAB( 19)"*"
8450   IF FL = 1 THEN   VTAB 21: PRINT TAB( 14)"";
8460   IF FL = 1 THEN   FLASH : PRINT TAB( 14)"";
8470   IF FL = 1 THEN   FLASH : PRINT "X": NORMAL : GOTO 8490
8480   VTAB 21: PRINT   TAB( 14)"X"
8490   IF FL = 1 THEN FL = 0: GOTO 10080
8500   GOSUB 9940
8510   INPUT "PRESS RETURN TO CONTINUE ";A$
```

CONSH (continued)

```
8520    HOME : GOTO 8400
8530 C$ = "PAVO"
8540    HOME
8550    VTAB 6: PRINT  TAB( 25)"*"
8560    VTAB 8: PRINT  TAB( 18)"*"
8570    VTAB 12: PRINT  TAB( 19)"*"
8580    VTAB 13: PRINT  TAB( 25)"*"; TAB( 31)"*"; TAB( 36)"*"
8590    VTAB 14: PRINT  TAB( 5)"*"
8600    VTAB 15: PRINT  TAB( 9)"."
8610    VTAB 16: PRINT  TAB( 13)"."; TAB( 17)"*"
8620    VTAB 17: PRINT  TAB( 12)"."
8630    VTAB 20: PRINT  TAB( 28)"*"
8640    GOSUB 9940
8650    INPUT "PRESS RETURN TO CONTINUE ";A$
8660    GOTO 10030
8670 C$ = "TUCANA"
8680    HOME
8690    VTAB 9: PRINT  TAB( 11)"."; TAB( 26)"."
8700    VTAB 10: PRINT  TAB( 28)"*"
8710    VTAB 12: PRINT  TAB( 31)"*"
8720    VTAB 14: PRINT  TAB( 9)"*"
8730    VTAB 16: PRINT  TAB( 21)"*"
8740    VTAB 22
8750    GOSUB 9940
8760    INPUT "PRESS RETURN TO CONTINUE ";A$
8770    GOTO 10030
8780 C$ = "APUS"
8790    HOME
8800    VTAB 11: PRINT  TAB( 11)"."
8810    VTAB 12: PRINT  TAB( 11)"."
8820    VTAB 13: PRINT  TAB( 13)"*"; TAB( 22)"* *"
8830    VTAB 15: PRINT  TAB( 25)"*"
8840    VTAB 22
8850    GOSUB 9940
8860    INPUT "PRESS RETURN TO CONTINUE ";A$
8870    GOTO 10030
8880 C$ = "CENTAURUS"
8890    HOME
8900    VTAB 5: PRINT  TAB( 31)"*"
8910    VTAB 9: PRINT  TAB( 20)"."; TAB( 26)"*"; TAB( 31)"X"
8920    VTAB 12: PRINT  TAB( 28)"."
8930    VTAB 15: PRINT  TAB( 10)"."; TAB( 22)"*"
8940    VTAB 16: PRINT  TAB( 23)"."
8950    VTAB 17: PRINT  TAB( 8)"*"
8960    VTAB 19: PRINT  TAB( 23)"*"
8970    VTAB 21: PRINT  TAB( 20)". ."; TAB( 29)"*"
8980    GOSUB 9940
8990    INPUT "PRESS RETURN TO CONTINUE ";A$
9000    GOTO 10030
9010 C$ = "CARINA"
9020 S$ = "CANOPUS"
9030    HOME
9040    VTAB 5: PRINT  TAB( 20)"*"; TAB( 27)"*"
9050    VTAB 8: PRINT  TAB( 15)"."; TAB( 19)"*"
9060    VTAB 9: PRINT  TAB( 23)"*"; TAB( 30)"*"
```

```
                                              ──── CONSH (continued) ────

      9070   VTAB 12: PRINT   TAB( 26)"* *"
      9080   VTAB 14: PRINT   TAB( 15)"*"; TAB( 20)"**"; TAB( 24)"."
      9090   VTAB 17: PRINT   TAB( 18)"*"; TAB( 24)"*"
      9100   VTAB 19: PRINT   TAB( 12)"*"
      9110   IF FL = 1 THEN   VTAB 20: PRINT TAB( 3)"";
      9120   IF FL = 1 THEN   FLASH : PRINT TAB( 3)"";
      9130   IF FL = 1 THEN   FLASH : PRINT "X": NORMAL : GOTO 9150
      9140   VTAB 20: PRINT   TAB( 3)"X"
      9150   VTAB 21
      9160   IF FL = 1 THEN FL = 0: GOTO 10080
      9170   GOSUB 9940
      9180   INPUT "PRESS RETURN TO CONTINUE ";A$
      9190   GOTO 9030
      9200   C$ = "LUPUS"
      9210   HOME
      9220   VTAB 6: PRINT   TAB( 22)"*"
      9230   VTAB 7: PRINT   TAB( 12)"."
      9240   VTAB 8: PRINT   TAB( 14)"."
      9250   VTAB 9: PRINT   TAB( 22)"."
      9260   VTAB 10: PRINT   TAB( 15)"*"; TAB( 20)"."; TAB( 23)"."
      9270   VTAB 11: PRINT   TAB( 21)"."
      9280   VTAB 12: PRINT   TAB( 23)"*"
      9290   VTAB 13: PRINT   TAB( 18)"*"
      9300   VTAB 14: PRINT   TAB( 23)"*"; TAB( 26)"*"
      9310   VTAB 17: PRINT   TAB( 22)"."; TAB( 32)"*"
      9320   VTAB 19: PRINT   TAB( 29)"."
      9330   VTAB 22
      9340   GOSUB 9940
      9350   INPUT "PRESS RETURN TO CONTINUE ";A$
      9360   GOTO 10030
      9370   C$ = "ARA"
      9380   HOME
      9390   VTAB 6: PRINT   TAB( 26)"."
      9400   VTAB 9: PRINT   TAB( 24)"*"
      9410   VTAB 11: PRINT   TAB( 17)"*"
      9420   VTAB 13: PRINT   TAB( 23)"*"
      9430   VTAB 14: PRINT   TAB( 18)"*"; TAB( 23)"*"
      9440   VTAB 16: PRINT   TAB( 18)"."
      9450   VTAB 19: PRINT   TAB( 24)"*"; TAB( 30)"."
      9460   VTAB 22
      9470   GOSUB 9940
      9480   INPUT "PRESS RETURN TO CONTINUE ";A$
      9490   GOTO 10030
      9500   C$ = "PHOENIX"
      9510   HOME
      9520   VTAB 7: PRINT   TAB( 22)"*"
      9530   VTAB 13: PRINT   TAB( 28)"*"
      9540   VTAB 14: PRINT   TAB( 8)"*"; TAB( 16)"."; TAB( 22)"*"
      9550   VTAB 15: PRINT   TAB( 31)"."
      9560   VTAB 16: PRINT   TAB( 12)"*"; TAB( 33)"."
      9570   VTAB 18: PRINT   TAB( 12)"*"; TAB( 26)"*"
      9580   VTAB 22
      9590   GOSUB 9940
      9600   INPUT "PRESS RETURN TO CONTINUE ";A$
      9610   GOTO 10030
```

---- **CONSH** (continued) ----

```
9620 C$ = "TRIANGULUM AUSTRALIS"
9630  HOME
9640  VTAB 9: PRINT  TAB( 22)"."
9650  VTAB 10: PRINT  TAB( 14)"*"; TAB( 25)"*"
9660  VTAB 12: PRINT  TAB( 16)"."
9670  VTAB 14: PRINT  TAB( 18)"*"; TAB( 22)"."
9680  VTAB 22
9690  GOSUB 9940
9700  INPUT "PRESS RETURN TO CONTINUE ";A$
9710  GOTO 10030
9720 C$ = "GRUS"
9730  HOME
9740  VTAB 7: PRINT  TAB( 22)"."
9750  VTAB 8: PRINT  TAB( 20)"*"
9760  VTAB 12: PRINT  TAB( 12)"*"; TAB( 21)"*"
9770  VTAB 14: PRINT  TAB( 28)"."
9780  VTAB 16: PRINT  TAB( 27)"."
9790  VTAB 22
9800  GOSUB 9940
9810  INPUT "PRESS RETURN TO CONTINUE ";A$
9820  GOTO 10030
9830 C$ = "CRUX"
9840  HOME
9850  VTAB 8: PRINT  TAB( 20)"X"
9860  VTAB 11: PRINT  TAB( 17)"."
9870  VTAB 12: PRINT  TAB( 25)"*"
9880  VTAB 13: PRINT  TAB( 15)"*"
9890  VTAB 15: PRINT  TAB( 20)"*"; TAB( 25)"."
9900  VTAB 22
9910  GOSUB 9940
9920  INPUT "PRESS RETURN TO CONTINUE ";A$
9930  GOTO 10030
9940  REM  SUBROUTINE FOR QUERY
9950  INPUT "NAME THIS CONSTELLATION ";A$
9960  IF A$ = C$ THEN  FLASH : PRINT "CORRECT": NORMAL :FL = 1: RETURN
9970 A$ = ""
9980  PRINT CL$;: INPUT "TRY AGAIN ";A$
9990  IF A$ = C$ THEN  GOTO 9960
10000  PRINT "YOU ARE STILL WRONG"
10010  PRINT "THE CONSTELLATION IS ";C$
10020 FL = 1: RETURN
10030  HOME : PRINT : PRINT : PRINT
10040  INPUT "ANOTHER TEST? Y/N ";A$
10050  IF A$ = "N" THEN  HOME : GOTO 10160
10060  IF A$ < > "Y" THEN  PRINT "INVALID RESPONSE": PRINT : GOTO 10030
10070 W = W + A: GOTO 8120
10080  REM  SUB STAR QUERY
10090  INPUT "NAME THIS STAR ";A$
10100  IF A$ = S$ THEN  FLASH : PRINT "CORRECT": NORMAL : GOTO 10140
10110  INPUT "WRONG; TRY AGAIN ";A$
10120  IF A$ = S$ THEN  GOTO 10100
10130  PRINT "STILL WRONG, THE STAR IS ";S$
10140  INPUT "PRESS RETURN TO CONTINUE ";A$
10150  HOME : PRINT : PRINT : PRINT : GOTO 10030
10160  HOME
10170  END
```

Photo Credit: NASA

Anyone interested in astronomy often wants to refer to data about the planets and their satellites: How large is the Moon? How far away is it from Earth? How massive are other satellites compared with our Moon? The questions are unlimited. The following program gives a menu from which to select information about the planets and their satellites and about the larger asteroids, the Sun, and the Moon. This Apollo 15 picture shows the brightest crater on the Moon, Aristarchus.

Program 22: PLNDT

Solar System Data

Amateur astronomers often find themselves trying to locate data about a planet or a satellite for observation, for calculations, or merely for information. Sometimes the item of data you want does not appear in the books you have access to, or cannot be located easily. With this program you can always have these important items of astronomical data available for quick access—in the time it takes to load a disk into your system. The program offers several menus—the solar system, planets, satellites, rings, and astronomical constants. There are also submenus allowing quick access to information of greater detail. Typical displays of the program are shown in Figure 22.1a–d.

The data are the most current available, but the remark (REM) statements will allow you to update data as it becomes available or is refined in the future.

Not all of the small satellites of Jupiter and Saturn are included, but when you are keying in the program these could be added, if you wish, at the appropriate places. Note that in the data statements 0.0 is used where values are either extremely small or not clearly defined. These data also can be amended if you wish.

Take great care when entering the data. A single misplaced comma or period can produce wild data throughout the program.

The listing for the PLNDT program follows.

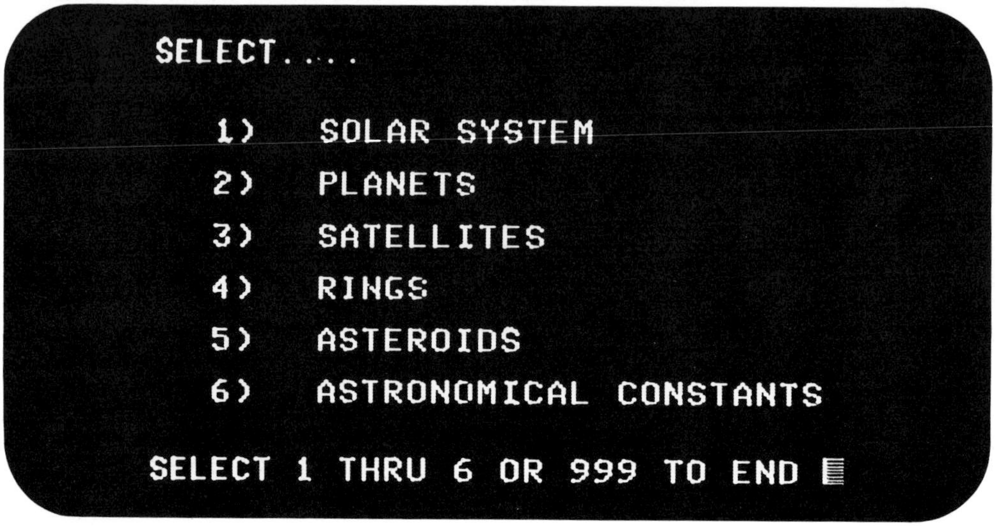

A series of menu displays generated during a search for planetary data.

Figure 22.1a

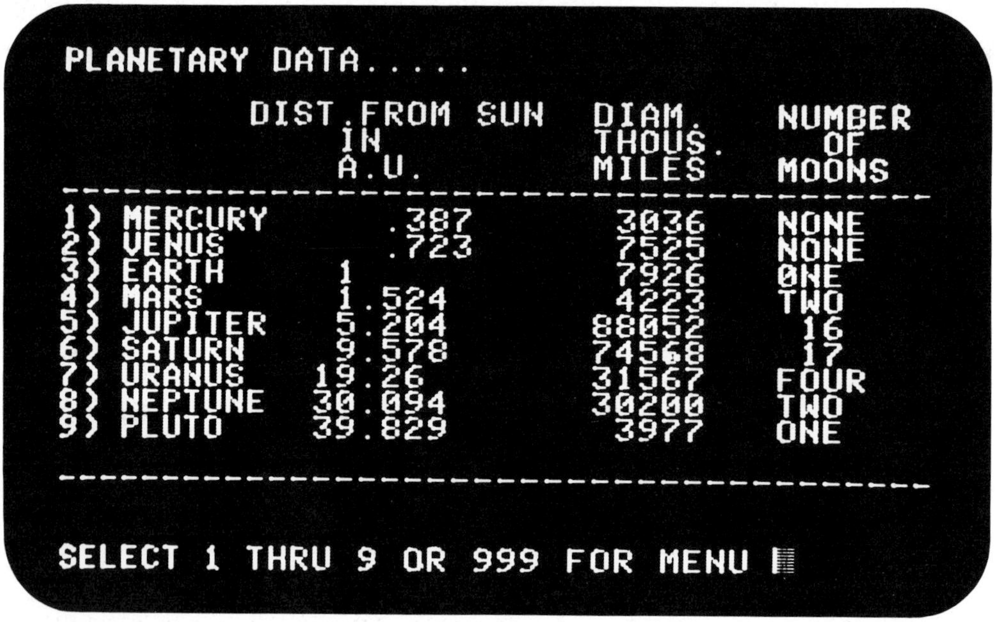

Figure 22.1b

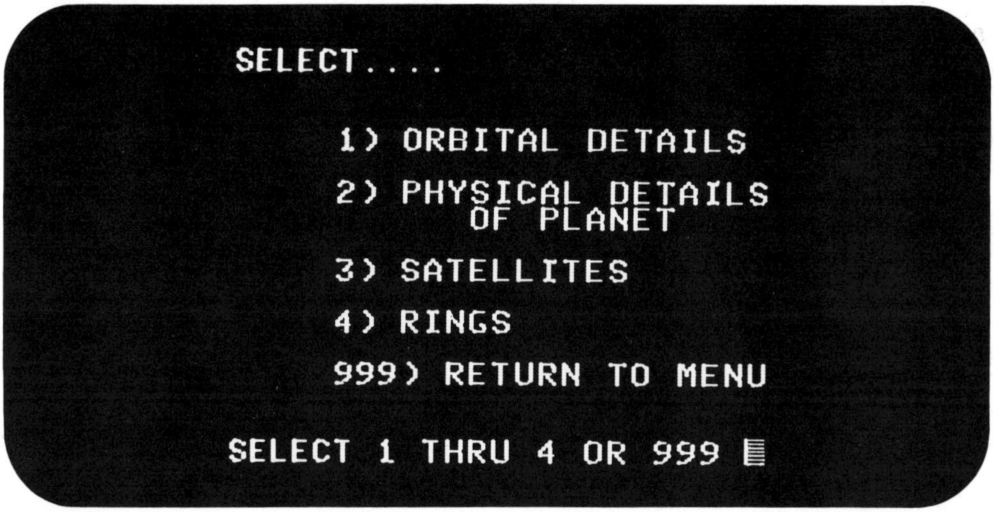

```
            SELECT....

               1) ORBITAL DETAILS

               2) PHYSICAL DETAILS
                     OF PLANET

               3) SATELLITES

               4) RINGS

               999) RETURN TO MENU


           SELECT 1 THRU 4 OR 999 ▤
```

Figure 22.1c

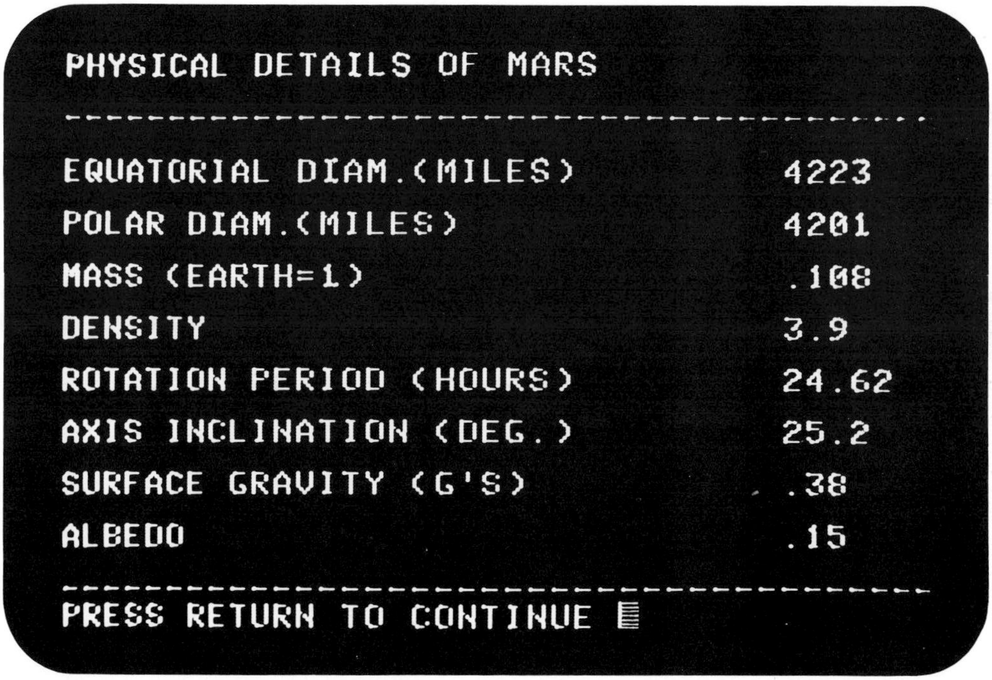

```
PHYSICAL DETAILS OF MARS
------------------------------------------
EQUATORIAL DIAM.(MILES)          4223
POLAR DIAM.(MILES)               4201
MASS (EARTH=1)                   .108
DENSITY                          3.9
ROTATION PERIOD (HOURS)          24.62
AXIS INCLINATION (DEG.)          25.2
SURFACE GRAVITY (G'S)            .38
ALBEDO                           .15
------------------------------------------
PRESS RETURN TO CONTINUE ▤
```

Figure 22.1d

```
┌─ PLNDT ─────────────────────────────────────────────────┐
│                                                          │
│  10    HOME : PRINT : PRINT                              │
│  20    DIM PD(10,16),JD(12,6),SD(18,8)                   │
│  30    DIM UD(5,6),ND(2,6),DM(2,6),DA(10,9)              │
│  40    PRINT "PLEASE WAIT; LOADING DATA"                 │
│  50    GOSUB 1680: GOSUB 1910: GOSUB 2080: GOSUB 2310    │
│  60    GOSUB 2420: GOSUB 2490: GOSUB 2560                │
│  70    HOME : PRINT : PRINT                              │
│  80    PRINT : PRINT                                     │
│  90    PRINT  TAB( 8)"--------------------"              │
│  100   PRINT  TAB( 8)"I    SOLAR SYSTEM    I"            │
│  110   PRINT  TAB( 8)"--------------------"              │
│  120   PRINT : PRINT                                     │
│  130   PRINT  TAB( 9)"AN ASTRONOMY PROGRAM"              │
│  140   PRINT  TAB( 6)"BY ERIC BURGESS F.R.A.S."          │
│  150   PRINT : PRINT                                     │
│  160   PRINT  TAB( 7)"ALL RIGHTS RESERVED BY"            │
│  170   PRINT  TAB( 7)"S & T SOFTWARE SERVICE"            │
│  180   PRINT : PRINT                                     │
│  190   PRINT "PROVIDES DATA ON SOLAR SYSTEM OBJECTS"     │
│  200   FOR K = 1500 TO 0 STEP  - 1: NEXT K               │
│  210   HOME : PRINT : PRINT                              │
│  220   PRINT "SELECT...."                                │
│  230   PRINT : PRINT                                     │
│  240   PRINT  TAB( 4)"1)   SOLAR SYSTEM"                 │
│  250   PRINT : PRINT  TAB( 4)"2)   PLANETS"              │
│  260   PRINT : PRINT  TAB( 4)"3)   SATELLITES"           │
│  270   PRINT : PRINT  TAB( 4)"4)   RINGS"                │
│  280   PRINT : PRINT  TAB( 4)"5)   ASTEROIDS"            │
│  290   PRINT : PRINT  TAB( 4)"6)   ASTRONOMICAL CONSTANTS"│
│  300   PRINT : PRINT                                     │
│  310   INPUT "SELECT 1 THRU 6 OR 999 TO END ";M          │
│  320   IF M = 999 THEN M = 7                             │
│  330   ON M GOTO 340,2710,5590,580,3510,7730,8100        │
│  340   HOME : PRINT                                      │
│  350   PRINT "SOLAR SYSTEM... ": PRINT                   │
│  360   PRINT  TAB( 20)"DIST. FROM SUN"                   │
│  370   PRINT  TAB( 24)"IN A.U."                          │
│  380   PRINT "--------------------------------------"    │
│  390   PRINT "1)    SUN"; TAB( 27)"N/A"                  │
│  400   PRINT "2)    MERCURY"; TAB( 28)PD(1,1)            │
│  410   PRINT "3)    VENUS"; TAB( 28)PD(2,1)              │
│  420   PRINT "4)    EARTH"; TAB( 26)PD(3,1)              │
│  430   PRINT "5)    MARS"; TAB( 26)PD(4,1)               │
│  440   PRINT "6)    ASTEROIDS"; TAB( 26)PD(10,1)         │
│  450   PRINT "7)    JUPITER"; TAB( 26)PD(5,1)            │
│  460   PRINT "8)    SATURN"; TAB( 26)PD(6,1)             │
│  470   PRINT "9)    URANUS"; TAB( 25)PD(7,1)             │
│  480   PRINT "10)   NEPTUNE"; TAB( 25)PD(8,1)            │
│  490   PRINT "11)   PLUTO"; TAB( 25)PD(9,1)              │
│  500   PRINT                                             │
│  510   PRINT "--------------------------------------"    │
│  520   PRINT                                             │
│  530   INPUT "SELECT 1 THRU 11 OR 999 FOR MENU ";N       │
│                                                          │
└──────────────────────────────────────────────────────────┘
```

PLNDT (continued)

```
540   IF N > 1 AND N < 6 THEN P = N - 1: GOTO 560
550   IF N > 6 THEN P = N - 2
560   IF N = 999 THEN N = 12
570   ON N GOTO 1450,2940,3050,3160,3280,3510,3840,4180,4710,5030,5260,210
580   HOME : PRINT : PRINT
590   PRINT "PLANETARY RINGS"
600   PRINT : PRINT
610   PRINT "------------------------"
620   PRINT  TAB( 4)"1)    JUPITER"
630   PRINT : PRINT  TAB( 4)"2)    SATURN"
640   PRINT : PRINT  TAB( 4)"3)    URANUS"
650   PRINT : PRINT  TAB( 4)"4)    OTHERS"
660   PRINT "------------------------"
670   PRINT : PRINT
680   INPUT "SELECT 1 THRU 4 OR 999 FOR MENU ";Z
690   IF Z = 999 THEN Z = 5
700   ON Z GOTO 710,900,1130,1320,210
710   HOME : PRINT : PRINT : PRINT
720   PRINT "JUPITER'S RING SYSTEM"
730   PRINT
740   PRINT "---------------------------------------"
750   PRINT  TAB( 8)"INNER DIAM."; TAB( 20)"WIDTH"
760   PRINT  TAB( 12)"MILES"; TAB( 20)"MILES"
770   PRINT
780   PRINT "RING 1"; TAB( 10)"29,200"; TAB( 20)"3,100"
790   PRINT
800   PRINT "RING 2"; TAB( 10)"32,435"; TAB( 20)"  500"
810   PRINT
820   PRINT "A VERY TENUOUS RING SYSTEM EXTENDS"
830   PRINT "FROM THE INNERMOST RING TO THE PLANET"
840   PRINT
850   PRINT "---------------------------------------"
860   PRINT
870   INPUT "PRESS RETURN TO CONTINUE ";A$
880   IF F = 1 THEN F = 0: GOTO 3840
890   GOTO 580
900   REM  SATURN'S RINGS
910   HOME : PRINT : PRINT : PRINT
920   PRINT "RING SYSTEM OF SATURN"
930   PRINT "---------------------------------------"
940   PRINT : PRINT  TAB( 8)"INNER DIAM."; TAB( 20)"WIDTH"
950   PRINT  TAB( 12)"MILES"; TAB( 20)"MILES"
960   PRINT
970   PRINT "D RING"; TAB( 10)" 41,600"; TAB( 20)"  4,350"
980   PRINT "C RING"; TAB( 10);" 49,580"; TAB( 20)" 11,120"
990   PRINT "B RING"; TAB( 10)" 56,850"; TAB( 20)" 15,840"
1000  PRINT "A RING"; TAB( 10)" 75,740"; TAB( 20)"  9,130"
1010  PRINT "F RING"; TAB( 10)" 87,180"; TAB( 20)"     60"
1020  PRINT "G RING"; TAB( 10)"105,630"; TAB( 20)"  6,210"
1030  PRINT "E RING"; TAB( 10)"111,845"; TAB( 20)"186,400"
1040  PRINT
1050  PRINT "DIVISIONS"
1060  PRINT "CASINI"; TAB( 10)" 72950"; TAB( 20)"  2,800"
1070  PRINT "ENCKE"; TAB( 10)" 82,790"; TAB( 20)"    200"
1080  PRINT
```

─── **PLNDT** *(continued)* ───

```
1090   PRINT "------------------------------------"
1100   INPUT "PRESS RETURN TO CONTINUE ";A$
1110   IF F = 1 THEN F = 0: GOTO 4180
1120   GOTO 580
1130   REM   URANUS' RINGS
1140   HOME : PRINT : PRINT : PRINT
1150   PRINT "URANUS' RING SYSTEM"
1160   PRINT "------------------------------------"
1170   PRINT   TAB( 8)"INNER DIAM."; TAB( 20)"WIDTH"
1180   PRINT   TAB( 12)"MILES"; TAB( 20)"MILES"
1190   PRINT
1200   PRINT "RING 6"; TAB( 10)"26,300"; TAB( 22)"5"
1210   PRINT "RING 7"; TAB( 10)"26,500"; TAB( 22)"5"
1220   PRINT "RING 1"; TAB( 10)"27,800"; TAB( 22)"6"
1230   PRINT "RING 2"; TAB( 10)"28,700"; TAB( 22)"6"
1240   PRINT "RING 3"; TAB( 10)"29,600"; TAB( 22)"6"
1250   PRINT "RING 4"; TAB( 10)"30,600"; TAB( 22)"6"
1260   PRINT "RING 5"; TAB( 10)"31,500"; TAB( 20)"100"
1270   PRINT
1280   PRINT "------------------------------------"
1290   INPUT "PRESS RETURN TO CONTINUE ";A$
1300   IF F = 1 THEN F = 0: GOTO 4710
1310   GOTO 580
1320   REM   OTHERS
1330   HOME : PRINT : PRINT : PRINT
1340   PRINT "OTHER RINGS?"
1350   PRINT "------------"
1360   PRINT : PRINT
1370   PRINT "NO PLANETS OTHER THAN JUPITER, SATURN,"
1380   PRINT "AND URANUS APPEAR TO HAVE RINGS"
1390   PRINT "BUT THERE HAVE BEEN SPECULATIONS"
1400   PRINT "THAT NEPTUNE MAY HAVE A RING SYSTEM"
1410   PRINT "------------"
1420   INPUT "PRESS RETURN TO CONTINUE ";A$
1430   IF F = 1 THEN F = 0: GOTO 5030
1440   GOTO 580
1450   REM   DATA FOR SUN
1460   HOME : PRINT : PRINT : PRINT
1470   PRINT "THE SUN - OUR NEAREST STAR"
1480   PRINT "------------------------------------"
1490   PRINT "DISTANCE (MILLION MI)"; TAB( 30)" 92.953"
1500   PRINT
1510   PRINT "DIAMETER (MILES)"; TAB( 30)"864,000"
1520   PRINT
1530   PRINT "MASS (EARTH=1)"; TAB( 30)"332,958"
1540   PRINT
1550   PRINT "ROTATION PERIOD (DAYS)"; TAB( 30)" 25.38"
1560   PRINT
1570   PRINT "AXIS INCLINATION (DEG..)"; TAB( 30)" 7.25"
1580   PRINT "   (TO ECLIPTIC)"
1590   PRINT
1600   PRINT "SUNSPOT CYCLE (YEARS)"; TAB( 30)" 11"
1610   PRINT
1620   PRINT "SPECTRAL TYPE"; TAB( 30)"   G"
1630   PRINT
```

PLNDT (continued)

```
1640    PRINT "----------------------------------------"
1650    PRINT
1660    INPUT "PRESS RETURN TO CONTINUE ";A$
1670    GOTO 210
1680    REM   DATA FOR PLANETS
1690    FOR XP = 1 TO 9: FOR YP = 1 TO 16
1700    READ PD(XP,YP)
1710    NEXT YP,XP
1720    REM   MERCURY
1730    DATA .387,3036,3036,.055,5.4,.06,.27,58.7,0,.2056,7.0044,.241,48.0994,
        77.1511,4.0923,158.2455
1740    REM   VENUS
1750    DATA .723,7525,7525,.817,5.2,.65,.85,243,178,.0068,3.9444,.6115,
        76.5038,131.2958,1.6021,226.0461
1760    REM   EARTH
1770    DATA 1,7926,7899,1,5.5,.55,1,23.92,23.4,.0167,0,1,0,102.604,.9856,
        357.2227
1780    REM   MARS
1790    DATA 1.524,4223,4201,.108,3.9,.15,.38,24.62,25.2,.0934,1.85,1.881,
        49.4066,335.699,.5240,151.1544
1800    REM   JUPITER
1810    DATA 5.204,88052,82417,318.4,1.31,.42,2.33,9.84,3.07,.0478,1.3057,
        11.86,100.2448,14.5847,.0831,132.7203
1820    REM   SATURN
1830    DATA 9.578,74568,67111,95.2,.7,.45,.92,10.22,26.74,.0555,2.4864,29.46,
        113.5115,95.4709,.0334,69.5615
1840    REM   URANUS
1850    DATA 19.260,31567,30052,15,1.31,.46,.85,10.82,97.93,.00503,.7716,84.01,
        74.0054,172.9222,.01166,54.9953
1860    REM   NEPTUNE
1870    DATA 30.094,30200,29596,17.2,1.66,.53,1.12,15.67,28.8,.0066,1.7730,
        164.8,131.5228,58.512,.00597,199.748
1880    REM   PLUTO
1890    RETURN
1900    DATA 39.829,3977,3977,.1,4.9,.14,.44,153.84,0,.2548,17.1372,248.6,
        109.9658,223.0141,.00392,346.4629
1910    REM   DATA FOR JOVIAN SATELLITES
1920    FOR JS = 1 TO 12: FOR SJ = 1 TO 6
1930    READ JD(JS,SJ)
1940    NEXT SJ,JS
1950    DATA   112.8,.0028,.455,.498,.00008,125
1960    DATA   262.2,0,.027 ,1.77,.98,2262
1970    DATA   417.2,.0003,.47,3.55,.65,1864
1980    DATA   665.5,.0015,.18,7.15,2.1,3274.6
1990    DATA   1170.6,.0075,.25,16.69,1.3,3107
2000    DATA   7137,.158,27.6,250.57,.00003,106
2010    DATA   7299,.207,24.8,259.65,0.0,50
2020    DATA   7370,.13,29,263.55,0.0,15
2030    DATA   13204,.169,147,631,0.0,12
2040    DATA   14005,.207,164,692,0.0,20
2050    DATA   14608,.378,145,739,0.0,22
2060    DATA   14707,.275,153,758,0.0,17
2070    RETURN
2080    REM   DATA FOR SATURN SATELLITES
2090    FOR SS = 1 TO 18: FOR SA = 1 TO 8
```

PLNDT *(continued)*

```
2100    READ SD(SS,SA)
2110    NEXT SA: NEXT SS
2120    DATA  85.5,0.0,0.0,.602,0.0,48,0.0,.4
2130    DATA  86.6,0.0,0.0,.613,0.0,37,0.0,.6
2140    DATA 88.0,0.0,0.0,.629,0.0,56,0.0,.6
2150    DATA 94.1,0.0,0.0,.694,0.0,75,0.0,.4
2160    DATA   94.1,0.0,0.0,.694,0.0,125,0.0,.4
2170    DATA  115.3,.02,1.5,.943,.00005,244,1.44,.6
2180    DATA  115.6,0.0,0.0,0.0,0.0,0.0,0.0,0.0
2190    DATA 147.9,.004,.0002,1.37,.001,310,1.16,.9
2200    DATA 183.1,0,.0008,1.888,.0088,659,1.21,.8
2210    DATA 183.1,0.0,0.0,1.888,0.0,18,0.0,.6
2220    DATA 183.1,0.0,0.0,1.888,0.0,17,0.0,.8
2230    DATA 234.5,.002,.0002,2.737,.0143,696,1.43,.62
2240    DATA 234.9,0.0,0.0,2.739,0.0,21,0.0,.5
2250    DATA 327.5,.001,.35,4.518,.02,951,1.33,.65
2260    DATA 759.2,.029,.333,15.945,1.87,3200,1.88,.2
2270    DATA 920.2,.104,0.0,21.27,0.0,186,0.0,.3
2280    DATA 2212.5,.028,14.72,79.58,.02,907,1.16,.5
2290    DATA 8048.9,.163,150.05,550.45,0.0,137,0.0,.06
2300    RETURN
2310    REM   DATA FOR OTHER SATELLITES
2320    REM   URANUS
2330    FOR US = 1 TO 5: FOR SU = 1 TO 6
2340    READ UD(US,SU)
2350    NEXT SU: NEXT US
2360    DATA 80.65,.01,0.0,1.414,.0004,186
2370    DATA 118.6,.0033,0.0,2.52,.018,497
2380    DATA 165.3,.011,0.0,4.144,.0073,342
2390    DATA 270.9,.0018,0.0,8.706,.0593,621
2400    DATA 362.5,.0007,0.0,13.463,.034,559
2410    RETURN
2420    REM   NEPTUNE SATELLITES
2430    FOR NS = 1 TO 2: FOR SN = 1 TO 6
2440    READ ND(NS,SN)
2450    NEXT SN,NS
2460    DATA 220.9,0,159.95,5.88,1.9,2361
2470    DATA 3459.1,.75,27.71,359.9,.006,186
2480    RETURN
2490    REM   MARS SATELLITES
2500    FOR MS = 1 TO 2: FOR SM = 1 TO 6
2510    READ MD(MS,SM)
2520    NEXT SM: NEXT MS
2530    DATA .5828,.015,1.0,.3189,0.0,13.5
2540    DATA 1457.7,.0005,2.0,1.2624,0.0,7.1
2550    RETURN
2560    REM   SUB FOR ASTEROID DATA
2570    FOR AA = 1 TO 10: FOR MA = 1 TO 9
2580    READ DA(AA,MA)
2590    NEXT MA: NEXT AA
2600    DATA 2.7675,623,.0783,10.605,4.604,.21408,80.051,73.545,88.020
2610    DATA 2.7737,378,.2326,34.794,4.6195,.21336,172.712,310.045,78.086
2620    DATA 2.3612,334,.0892,7.143,3.6284,.27164,103.442,150.133,57.113
2630    DATA 3.1355,280,.1196,3.836,5.5522,.17752,283.125,317.987,289.675
2640    DATA 2.9223,155,.1370,3.092,4.9956,.19730,150.123,226.152,167.565
```

─────── **PLNDT** *(continued)* ───────

```
2650    DATA 3.1462,230,.2276,26.343,5.5808,.17661,30.676,63.506,241.231
2660    DATA 3.4363,192,.1075,3.549,6.3704,.115472,155.472,110.074,337.052
2670    DATA 3.061,172,.0674,15.213,5.3739,.18341,89.06,341.210,153.493
2680    DATA 3.1806,200,.1744,15.908,5.6723,.17376,107.387,338.661,151.999
2690    DATA 3.0596,217,.1522,17.288,5.3517,.18417,280.481,91.777,159.192
2700    RETURN
2710    HOME : PRINT
2720    PRINT "PLANETARY DATA..... "
2730    PRINT
2740    PRINT    TAB( 10)"DIST.FROM SUN"; TAB( 25)"DIAM."; TAB( 33)"NUMBER"
2750    PRINT    TAB( 14)"IN"; TAB( 25)"THOUS."; TAB( 35)"OF"
2760    PRINT    TAB( 14)"A.U."; TAB( 25)"MILES"; TAB( 33)"MOONS"
2770    PRINT "------------------------------------"
2780    PRINT "1) MERCURY"; TAB( 16)PD(1,1); TAB( 26)PD(1,2); TAB( 33)"NONE"
2790    PRINT "2) VENUS"; TAB( 16)PD(2,1); TAB( 26)PD(2,2); TAB( 33)"NONE"
2800    PRINT "3) EARTH"; TAB( 14)PD(3,1); TAB( 26)PD(3,2); TAB( 33)"ONE"
2810    PRINT "4) MARS"; TAB( 14)PD(4,1); TAB( 26)PD(4,2); TAB( 33)"TWO"
2820    PRINT "5) JUPITER"; TAB( 14)PD(5,1); TAB( 25)PD(5,2); TAB( 33)" 16"
2830    PRINT "6) SATURN"; TAB( 14)PD(6,1); TAB( 25)PD(6,2); TAB( 33)" 17"
2840    PRINT "7) URANUS"; TAB( 13)PD(7,1); TAB( 25)PD(7,2); TAB( 33)"FOUR"
2850    PRINT "8) NEPTUNE"; TAB( 13)PD(8,1); TAB( 25)PD(8,2); TAB( 33)"TWO"
2860    PRINT "9) PLUTO"; TAB( 13)PD(9,1); TAB( 26)PD(9,2); TAB( 33)"ONE"
2870    PRINT
2880    PRINT "------------------------------------"
2890    PRINT : PRINT
2900    INPUT "SELECT 1 THRU 9 OR 999 FOR MENU ";N
2910    IF N = 999 THEN N = 10
2920 P = N
2930    ON N GOTO 2940,3050,3160,3280,3840,4180,4710,5030,5260,210
2940    REM  SUB FOR MERCURY
2950 P$ = "MERCURY"
2960    GOSUB 7930
2970    IF DS = 3 THEN  HOME : PRINT "MERCURY HAS NO SATELLITE": GOTO 3020
2980    IF DS = 4 THEN  HOME : PRINT "MERCURY HAS NO RINGS": GOTO 3020
2990    ON DS GOSUB 7220,7460,7730,210
3000    IF DS < > 5 GOTO 2960
3010    GOTO 210
3020    PRINT : PRINT
3030    INPUT "PRESS RETURN TO CONTINUE ";A$
3040    GOTO 2960
3050    REM  SUB FOR VENUS
3060 P$ = "VENUS"
3070    GOSUB 7930
3080    IF DS = 3 THEN  HOME : PRINT "VENUS HAS NO SATELLITE": GOTO 3130
3090    IF DS = 4 THEN  HOME : PRINT "VENUS HAS NO RINGS": GOTO 3130
3100    ON DS GOSUB 7220,7460,7730,210
3110    IF DS < > 5 GOTO 3070
3120    GOTO 210
3130    PRINT : PRINT
3140    INPUT "PRESS RETURN TO CONTINUE ";A$
3150    GOTO 3070
3160    REM  SUB FOR EARTH
3170 P$ = "EARTH"
3180    GOSUB 7930
3190    IF DS = 3 THEN S$ = "MOON": GOTO 3240
```

PLNDT *(continued)*

```
3200   IF DS = 4 THEN  HOME : PRINT "EARTH HAS NO RINGS": GOTO 3260
3210   ON DS GOSUB 7220,7460,7730,210
3220   IF DS < > 5 GOTO 3180
3230   GOTO 210
3240 F = 1
3250   GOTO 6030
3260   PRINT : INPUT "PRESS RETURN TO CONTINUE ";A$
3270   GOTO 3180
3280   REM   MARS DETAILS
3290   HOME : PRINT : PRINT : PRINT
3300 P$ = "MARS"
3310   GOSUB 7930
3320   IF DS = 3 THEN  GOTO 3370
3330   IF DS = 4 THEN  HOME : PRINT "MARS HAS NO RINGS": GOTO 3490
3340   ON DS GOSUB 7220,7460,7730,210
3350   IF DS < > 5 GOTO 3310
3360   GOTO 210
3370   HOME : PRINT : PRINT
3380   PRINT "SATELLITES OF MARS ARE"
3390   PRINT
3400   PRINT "1) DEIMOS"
3410   PRINT
3420   PRINT "2) PHOBOS"
3430   PRINT
3440   INPUT "SELECT 1 OR 2 ";Q
3450   IF Q = 1 THEN S$ = "DEIMOS"
3460   IF Q = 2 THEN S$ = "PHOBOS"
3470 F = 1
3480   GOTO 7020
3490   PRINT : INPUT "PRESS RETURN TO CONTINUE ";A$
3500   GOTO 3310
3510   REM    SUB FOR ASTEROIDS
3520   HOME : PRINT : PRINT
3530   PRINT "ASTEROIDS LARGER THAN"
3540   PRINT  TAB( 11)"155 MI (250 KM) DIAM."
3550   PRINT "-----------------------------------"
3560   PRINT  TAB( 5)"NAME"; TAB( 20)"DIAM. MILES    KMS.
3570   PRINT
3580   PRINT "1)  CERES"; TAB( 21)DA(1,2); TAB( 32)"1003"
3590   PRINT "2)  DAVIDA"; TAB( 21)DA(9,2); TAB( 33)"323"
3600   PRINT "3)  CYBELE"; TAB( 21)DA(7,2); TAB( 33)"309"
3610   PRINT "4)  EUPHROSYNE"; TAB( 21)DA(6,2); TAB( 33)"370"
3620   PRINT "5)  HYGIEA"; TAB( 21)DA(4,2); TAB( 33)"450"
3630   PRINT "6)  INTERAMNIA"; TAB( 21)DA(10,2); TAB( 33)"350"
3640   PRINT "7)  PALLAS"; TAB( 21)DA(2,2); TAB( 33)"608"
3650   PRINT "8)  PATENTIA"; TAB( 21)DA(8,2); TAB( 33)"276"
3660   PRINT "9)  PSYCHE"; TAB( 21)DA(5,2); TAB( 33)"250"
3670   PRINT "10) VESTA"; TAB( 21)DA(3,2); TAB( 33)"538"
3680   PRINT
3690   PRINT "-----------------------------------"
3700   INPUT "SELECT 1 THRU 10 OR 999 FOR MENU ";N
3710   IF N = 1 THEN AS$ = "CERES":AA = 1
3720   IF N = 2 THEN AS$ = "DAVIDA":AA = 9
3730   IF N = 3 THEN AS$ = "CYBELE":AA = 7
3740   IF N = 4 THEN AS$ = "EUPHROSYNE":AA = 6
```

```
3750    IF N = 5 THEN AS$ = "HYGIEA":AA = 4
3760    IF N = 6 THEN AS$ = "INTERAMNIA":AA = 10
3770    IF N = 7 THEN AS$ = "PALLAS":AA = 2
3780    IF N = 8 THEN AS$ = "PATENTIA":AA = 8
3790    IF N = 9 THEN AS$ = "PSYCHE":AA = 5
3800    IF N = 10 THEN AS$ = "VESTA":AA = 3
3810    IF N = 999 THEN  GOTO 210
3820    GOSUB 5390
3830    GOTO 3510
3840    REM  SUB FOR JUPITER
3850    P$ = "JUPITER"
3860    GOSUB 7930
3870    IF DS = 3 THEN  GOTO 3920
3880    IF DS = 4 THEN F = 1: GOTO 710
3890    ON DS GOSUB 7220,7460,7730,210
3900    IF DS <  > 5 GOTO 3860
3910    GOTO 210
3920    HOME : PRINT : PRINT : PRINT
3930    PRINT "SATELLITES OF JUPITER ARE"
3940    PRINT
3950    PRINT "1) JV": PRINT "2) IO"
3960    PRINT "3) EUROPA": PRINT "4) GANYMEDE"
3970    PRINT "5) CALLISTO": PRINT "6) JVI"
3980    PRINT "7) JVII": PRINT "8) JX"
3990    PRINT "9) JXII": PRINT "10) JXI"
4000    PRINT "11) JVIII": PRINT "12) JIX"
4010    PRINT
4020    PRINT "JUPITER ALSO HAS SEVERAL VERY SMALL": PRINT "SATELLITES"
        : PRINT
4030    INPUT "SELECT 1 THRU 12 ";Q
4040    IF Q = 1 THEN S$ = "JV"
4050    IF Q = 2 THEN S$ = "IO"
4060    IF Q = 3 THEN S$ = "EUROPA"
4070    IF Q = 4 THEN S$ = "GANYMEDE"
4080    IF Q = 5 THEN S$ = "CALLISTO"
4090    IF Q = 6 THEN S$ = "JVI"
4100    IF Q = 7 THEN S$ = "JVII"
4110    IF Q = 8 THEN S$ = "JX"
4120    IF Q = 9 THEN S$ = "JXII"
4130    IF Q = 10 THEN S$ = "JXI"
4140    IF Q = 11 THEN S$ = "JVIII"
4150    IF Q = 12 THEN S$ = "JIX"
4160    F = 1
4170    GOTO 6360
4180    REM  SUB FOR SATURN
4190    P$ = "SATURN"
4200    GOSUB 7930
4210    IF DS = 4 THEN F = 1: GOTO 900
4220    IF DS = 3 THEN  GOTO 4260
4230    ON DS GOSUB 7220,7460,7730,210
4240    IF DS <  > 5 GOTO 4200
4250    GOTO 210
4260    HOME : PRINT
4270    PRINT "SATELLITES OF SATURN"
4280    PRINT
```

PLNDT *(continued)*

```
4290   PRINT "1) 1980S28    A RING SHEPHERD"
4300   PRINT "2) 1980S27    F RING SHEPHERD"
4310   PRINT "3) 1980S26    F RING SHEPHERD"
4320   PRINT "4) 1980S3    TRAILING COORBITAL"
4330   PRINT "5) 1980S1    LEADING COORBITAL"
4340   PRINT "6) MIMAS"
4350   PRINT "7) MIMAS FOLLOWER"
4360   PRINT "8) ENCELADUS"
4370   PRINT "9) TETHYS"
4380   PRINT "10) 1980S13"
4390   PRINT "11) 1980S25"
4400   PRINT "12) DIONE"
4410   PRINT "13) 1980S6"
4420   PRINT "14) RHEA"
4430   PRINT "15) TITAN"
4440   PRINT "16) HYPERION"
4450   PRINT "17) IAPETUS"
4460   PRINT "18) PHOEBE"
4470   PRINT "SEVERAL OTHER SMALL SATELLITES HAVE"
4480   PRINT "BEEN DISCOVERED"
4490   PRINT "-----------------------------------"
4500   INPUT "SELECT 1 THRU 18 ";Q
4510   IF Q = 1 THEN S$ = "1980S28"
4520   IF Q = 2 THEN S$ = "1980S27"
4530   IF Q = 3 THEN S$ = "1980S26"
4540   IF Q = 4 THEN S$ = "1980S3"
4550   IF Q = 5 THEN S$ = "1980S1"
4560   IF Q = 6 THEN S$ = "MIMAS"
4570   IF Q = 7 THEN S$ = "MIMAS FOLLOWER"
4580   IF Q = 8 THEN S$ = "ENCELADUS"
4590   IF Q = 9 THEN S$ = "TETHYS"
4600   IF Q = 10 THEN S$ = "1980S13"
4610   IF Q = 11 THEN S$ = "1980S25"
4630   IF Q = 13 THEN S$ = "1980S6"
4640   IF Q = 14 THEN S$ = "RHEA"
4650   IF Q = 15 THEN S$ = "TITAN"
4660   IF Q = 16 THEN S$ = "HYPERION"
4670   IF Q = 17 THEN S$ = "IAPETUS"
4680   IF Q = 18 THEN S$ = "PHOEBE"
4690 F = 1
4700   GOTO 6510
4710   REM  SUB FOR URANUS
4720 P$ = "URANUS"
4730   GOSUB 7930
4740   IF DS = 3 THEN  GOTO 4790
4750   IF DS = 4 THEN F = 1: GOTO 1130
4760   ON DS GOSUB 7220,7460,7730,210
4770   IF DS < > 5 THEN  GOTO 4730
4780   GOTO 210
4790   HOME : PRINT : PRINT
4800   PRINT "SATELLITES OF URANUS"
4810   PRINT "-----------------------------------"
4820   PRINT
4830   PRINT "1) MIRANDA"
4840   PRINT
```

PLNDT *(continued)*

```
4850    PRINT "2) ARIEL"
4860    PRINT
4870    PRINT "3) UMBRIEL"
4880    PRINT
4890    PRINT "4) TITANIA"
4900    PRINT
4910    PRINT "5) OBERON"
4920    PRINT
4930    PRINT "----------------------------------------"
4940    PRINT
4950    INPUT "SELECT 1 THRU  5 ";Q
4960    IF Q = 1 THEN S$ = "MIRANDA"
4970    IF Q = 2 THEN S$ = "ARIEL"
4980    IF Q = 3 THEN S$ = "UMBRIEL"
4990    IF Q = 4 THEN S$ = "TITANIA"
5000    IF Q = 5 THEN S$ = "OBERON"
5010 F = 1
5020    GOTO 6690
5030    REM  SUB FOR NEPTUNE
5040 P$ = "NEPTUNE"
5050    GOSUB 7930
5060    IF DS = 3 THEN  GOTO 5110
5070    IF DS = 4 THEN F = 1: GOTO 1320
5080    ON DS GOSUB 7220,7460,7730,210
5090    IF DS <  > 5 THEN  GOTO 5050
5100    GOTO 210
5110    HOME : PRINT : PRINT : PRINT
5120    PRINT "SATELLITES OF NEPTUNE"
5130    PRINT "----------------------------------------"
5140    PRINT
5150    PRINT "1) TRITON"
5160    PRINT
5170    PRINT "2) NEREID"
5180    PRINT
5190    PRINT "----------------------------------------"
5200    PRINT
5210    INPUT "SELECT 1 OR 2 ";Q
5220    IF Q = 1 THEN S$ = "TRITON"
5230    IF Q = 2 THEN S$ = "NEREID"
5240 F = 4
5250    GOTO 6850
5260    REM  SUB FOR PLUTO
5270 P$ = "PLUTO"
5280    GOSUB 7930
5290    IF DS = 3 THEN 5340
5300    IF DS = 4 THEN  GOTO 5350
5310    ON DS GOSUB 7220,7460,7730,210
5320    IF DS <  > 5 THEN  GOTO 5280
5330    GOTO 210
5340 F = 1:S$ = "CHARON": GOTO 6190
5350    HOME : PRINT : PRINT
5360    PRINT "PLUTO DOES NOT HAVE RINGS"
5370    PRINT : INPUT "PRESS RETURN TO CONTINUE ";A$
5380    GOTO 5280
5390    REM  SUB FOR ASTEROID DETAILS
```

PLNDT *(continued)*

```
5400    HOME : PRINT : PRINT
5410    PRINT "DETAILS OF ASTEROID ";AS$
5420    PRINT "------------------------------------"
5430    PRINT
5440    PRINT "MEAN DISTANCE FROM SUN"; TAB( 25)DA(AA,1)
5450    PRINT  TAB( 10)"(A.U.)"
5460    PRINT "DIAMETER (MILES)"; TAB( 25)DA(AA,2)
5470    PRINT "ECCENTRICITY"; TAB( 25)DA(AA,3)
5480    PRINT "ORBIT INCLINATION (DEG.)"; TAB( 25)DA(AA,4)
5490    PRINT "PERIOD (YEARS)"; TAB( 25)DA(AA,5)
5500    PRINT "DAILY MOTION (DEG.)"; TAB( 25)DA(AA,6)
5510    PRINT "LONG. ASCENDING NODE (DEG.)"; TAB( 25)DA(AA,7)
5520    PRINT "LONG. OF PERIHELION (DEG.)"; TAB( 25)DA(AA,8)
5530    PRINT "MEAN ANOMALY (DEG.)"; TAB( 25)DA(AA,9)
5540    PRINT " (EPOCH 1982, AUG 19)"
5550    PRINT "------------------------------------"
5560    PRINT
5570    INPUT "PRESS RETURN TO CONTINUE ";A$
5580    RETURN
5590    REM  SATELLITES
5600 N = 0
5610    HOME : PRINT : PRINT
5620    PRINT "SATELLITES OF SOLAR SYSTEM"
5630    PRINT "------------------------------------"
5640    PRINT "1) ARIEL"; TAB( 20)"12) MIMAS"
5650    PRINT "2) CALLISTO"; TAB( 20)"13) MIRANDA"
5660    PRINT "3) CHARON"; TAB( 20)"14) MOON"
5670    PRINT "4) DEIMOS"; TAB( 20)"15) NEREID"
5680    PRINT "5) DIONE"; TAB( 20)"16) OBERON"
5690    PRINT "6) ENCELADUS"; TAB( 20)"17) PHOBOS"
5700    PRINT "7) EUROPA"; TAB( 20)"18) PHOEBE"
5710    PRINT "8) GANYMEDE"; TAB( 20)"19) RHEA"
5720    PRINT "9) HYPERION"; TAB( 20)"20) TETHYS"
5730    PRINT "10) IAPETUS"; TAB( 20)"21) TITAN"
5740    PRINT "11) IO"; TAB( 20)"22) TITANIA"
5750    PRINT "23) TRITON"; TAB( 20)"24) UMBRIEL"
5760    PRINT "------------------------------------"
5770    PRINT
5780    INPUT "SELECT 1 THRU 24 OR 999 FOR MENU ";N
5790    IF N = 999 THEN  GOTO 210
5800    IF N = 1 THEN S$ = "ARIEL":Q = 2: GOTO 6690
5810    IF N = 2 THEN S$ = "CALLISTO":Q = 5: GOTO 6360
5820    IF N = 3 THEN S$ = "CHARON": GOTO 6190
5830    IF N = 4 THEN S$ = "DEIMOS":Q = 2: GOTO 7020
5840    IF N = 5 THEN S$ = "DIONE":Q = 12: GOTO 6510
5850    IF N = 6 THEN S$ = "ENCELADUS":Q = 8: GOTO 6510
5860    IF N = 7 THEN S$ = "EUROPA":Q = 3: GOTO 6360
5870    IF N = 8 THEN S$ = "GANYMEDE":Q = 4: GOTO 6360
5880    IF N = 9 THEN S$ = "HYPERION":Q = 16: GOTO 6510
5890    IF N = 10 THEN S$ = "IAPETUS":Q = 17: GOTO 6510
5900    IF N = 11 THEN S$ = "IO":Q = 2: GOTO 6360
5910    IF N = 12 THEN S$ = "MIMAS":Q = 6: GOTO 6510
5920    IF N = 13 THEN S$ = "MIRANDA":Q = 1: GOTO 6690
5930    IF N = 14 THEN S$ = "MOON ": GOTO 6030
5940    IF N = 15 THEN S$ = "NEREID":Q = 2: GOTO 6850
```

PLNDT *(continued)*

```
5950    IF N = 16 THEN S$ = "OBERON":Q = 5: GOTO 6690
5960    IF N = 17 THEN S$ = "PHOBOS":Q = 1: GOTO 7020
5970    IF N = 18 THEN S$ = "PHOEBE":Q = 18: GOTO 6510
5980    IF N = 19 THEN S$ = "RHEA":Q = 14: GOTO 6510
5990    IF N = 20 THEN S$ = "TETHYS":Q = 9: GOTO 6510
6000    IF N = 21 THEN S$ = "TITAN":Q = 15: GOTO 6510
6010    IF N = 22 THEN S$ = "TITANIA":Q = 4: GOTO 6690
6020    IF N = 23 THEN S$ = "TRITON":Q = 1: GOTO 6850
6030    HOME : PRINT : PRINT
6040    PRINT S$;" IS SATELLITE OF EARTH"
6050    PRINT "------------------------------------"
6060    PRINT
6070    PRINT "MEAN DISTANCE (1000 MI)"; TAB( 30)"238.853"
6080    PRINT "DIAMETER (MILES)"; TAB( 30)"2160"
6090    PRINT "MASS (MOON=1)"; TAB( 30)"1"
6100    PRINT "SIDEREAL PERIOD (DAYS)"; TAB( 30)"27.321661"
6110    PRINT "ECCENTRICITY"; TAB( 30)".0549"
6120    PRINT "INCLINATION (DEG.)"; TAB( 30)"18.28 TO"
6130    PRINT  TAB( 30)"28.58"
6140    PRINT
6150    PRINT "------------------------------------"
6160    INPUT "PRESS RETURN TO CONTINUE ";A$
6170    IF F = 1 THEN  GOTO 3160
6180    GOTO 5590
6190    HOME : PRINT : PRINT
6200    PRINT S$;" IS A SATELLITE OF PLUTO"
6210    PRINT
6220    PRINT "------------------------------------"
6230    PRINT
6240    PRINT "MEAN DISTANCE (1000 MI)"; TAB( 30)"10.563"
6250    PRINT "DIAMETER (MILES)"; TAB( 30)"745"
6260    PRINT "SIDEREAL PERIOD (DAYS)"; TAB( 30)"6.39"
6270    PRINT
6280    PRINT "FURTHER DETAILS OF CHARON ARE NOT"
6290    PRINT "     YET AVAILABLE"
6300    PRINT
6310    PRINT "------------------------------------"
6320    PRINT
6330    INPUT "PRESS RETURN TO CONTINUE ";A$
6340    IF F = 1 THEN F = 0: GOTO 5260
6350    GOTO 5590
6360    HOME : PRINT : PRINT
6370    PRINT S$;" IS A SATELLITE OF JUPITER"
6380    PRINT
6390    PRINT "------------------------------------"
6400    PRINT "MEAN DISTANCE (1000 MI)"; TAB( 30)JD(Q,1)
6410    PRINT "DIAMETER (MILES)"; TAB( 30)JD(Q,6)
6420    PRINT "MASS (MOON=1)"; TAB( 30)JD(Q,5)
6430    PRINT "SID. PERIOD (DAYS)"; TAB( 30)JD(Q,4)
6440    PRINT "ECCENTRICITY"; TAB( 30)JD(Q,2)
6450    PRINT "INCLINATION (DEG.)" TAB( 30)JD(Q,3)
6460    PRINT "------------------------------------"
6470    PRINT
6480    INPUT "PRESS RETURN TO CONTINUE ";A$
6490    IF F = 1 THEN  GOTO 3860
```

PLNDT *(continued)*

```
6500    GOTO 5590
6510    HOME : PRINT : PRINT
6520    PRINT S$;" IS A SATELLITE OF SATURN"
6530    PRINT "------------------------------------"
6540    PRINT
6550    PRINT "MEAN DISTANCE (1000 MI)"; TAB( 30)SD(Q,1)
6560    PRINT "DIAMETER (MILES)"; TAB( 30)SD(Q,6)
6570    PRINT "MASS (MOON=1)"; TAB( 30)SD(Q,5)
6580    PRINT "DENSITY"; TAB( 30)SD(Q,7)
6590    PRINT "SIDEREAL PERIOD (DAYS)"; TAB( 30)SD(Q,4)
6600    PRINT "ALBEDO"; TAB( 30)SD(Q,8)
6610    PRINT "ECCENTRICITY"; TAB( 30)SD(Q,2)
6620    PRINT "INCLINATION (DEG.)"; TAB( 30)SD(Q,3)
6630    PRINT
6640    PRINT "------------------------------------"
6650    PRINT
6660    INPUT "PRESS RETURN TO CONTINUE ";A$
6670    IF F = 1 THEN  GOTO 4180
6680    GOTO 5590
6690    HOME : PRINT : PRINT
6700    PRINT S$;" IS A SATELLITE OF URANUS"
6710    PRINT "------------------------------------"
6720    PRINT
6730    PRINT "MEAN DISTANCE (1000 MI)"; TAB( 30)UD(Q,1)
6740    PRINT "DIAMETER (MILES)"; TAB( 30)UD(Q,6)
6750    PRINT "MASS (MOON=1)"; TAB( 30)UD(Q,5)
6760    PRINT "SIDEREAL PERIOD (DAYS)"; TAB( 30)UD(Q,4)
6770    PRINT "ECCENTRICITY"; TAB( 30)UD(Q,2)
6780    PRINT "INCLINATION (DEG.)"; TAB( 30)UD(Q,3)
6790    PRINT
6800    PRINT "------------------------------------"
6810    PRINT
6820    INPUT "PRESS RETURN TO CONTINUE ";A$
6830    IF F = 1 THEN F = 0: GOTO 4200
6840    GOTO 5590
6850    HOME : PRINT : PRINT
6860    PRINT S$;" IS A SATELLITE OF NEPTUNE"
6870    PRINT
6880    PRINT "------------------------------------"
6890    PRINT
6900    PRINT "MEAN DISTANCE (1000 MI)"; TAB( 30)ND(Q,1)
6910    PRINT "DIAMETER (MILES)"; TAB( 30)ND(Q,6)
6920    PRINT "MASS (MOON=1)"; TAB( 30)ND(Q,5)
6930    PRINT "SIDEREAL PERIOD (DAYS)"; TAB( 30)ND(Q,4)
6940    PRINT "ECCENTRICITY"; TAB( 30)ND(Q,2)
6950    PRINT "INCLINATION (DEG.)"; TAB( 30)ND(Q,3)
6960    PRINT
6970    PRINT "------------------------------------"
6980    PRINT
6990    INPUT "PRESS RETURN TO CONTINUE ";A$
7000    IF F = 4 THEN  GOTO 5030
7010    GOTO 5590
7020    HOME : PRINT : PRINT
7030    PRINT S$;" IS A SATELLITE OF MARS"
7040    PRINT
```

———— **PLNDT** *(continued)* ————

```
7050    PRINT "-------------------------------------"
7060    PRINT
7070    PRINT "MEAN DISTANCE (1000 MI)"; TAB( 30)MD(Q,1)
7080    PRINT "DIAMETER (MILES)"; TAB( 30)MD(Q,6)
7090    PRINT "MASS (MOON=1)"; TAB( 30)MD(Q,5)
7100    PRINT "SIDEREAL PERIOD (DAYS)"; TAB( 30)MD(Q,4)
7110    PRINT "ECCENTRICITY"; TAB( 30)MD(Q,2)
7120    PRINT "INCLINATION (DEG.)"; TAB( 30)MD(Q,3)
7130    PRINT
7140    PRINT "-------------------------------------"
7150    PRINT
7160    INPUT "PRESS RETURN TO CONTINUE ";A$
7170    IF F = 1 THEN F = 0: GOTO 3310
7180    IF F = 2 THEN F = 0: GOTO 3860
7190    IF F = 3 THEN F = 0: GOTO 4730
7200    IF F = 4 THEN F = 0: GOTO 5050
7210    RETURN
7220    REM  SUB TO DISPLAY ORBIT DETAILS
7230    HOME : PRINT : PRINT : PRINT
7240    PRINT "ORBITAL DETAILS FOR ";P$
7250    PRINT
7260    PRINT "-------------------------------------"
7270    PRINT
7280    PRINT "MEAN DISTANCE FROM SUN (A.U.)";
7290    PRINT  TAB( 33)PD(P,1)
7300    PRINT
7310    PRINT "ECCENTRICITY"; TAB( 33)PD(P,10)
7320    PRINT
7330    PRINT "INCLINATION (DEGREES)"; TAB( 33)PD(P,11)
7340    PRINT
7350    PRINT "SIDEREAL PERIOD (YEARS)"; TAB( 33)PD(P,12)
7360    PRINT
7370    PRINT "LONG. ASCENDING NODE (DEG.)"; TAB( 32)PD(P,13)
7380    PRINT
7390    PRINT "LONG. OF PERIHELION (DEG.)"; TAB( 32)PD(P,14): PRINT
7400    PRINT "MEAN MOTION (DEG./DAY)"; TAB( 33)PD(P,15): PRINT
7410    PRINT "MEAN ANOMALY (DEG.)"; TAB( 32)PD(P,16): PRINT  TAB( 5)
        "(EPOCH 1980,1,1)"
7420    PRINT
7430    PRINT "-------------------------------------"
7440    INPUT "PRESS RETURN TO CONTINUE ";A$
7450    RETURN
7460    REM  SUB FOR PLANET DETAILS
7470    HOME : PRINT : PRINT
7480    PRINT "PHYSICAL DETAILS OF ";P$
7490    PRINT
7500    PRINT "-------------------------------------"
7510    PRINT
7520    PRINT "EQUATORIAL DIAM.(MILES)"; TAB( 33)PD(P,2)
7530    PRINT
7540    PRINT "POLAR DIAM.(MILES)"; TAB( 33)PD(P,3)
7550    PRINT
7560    PRINT "MASS (EARTH=1)"; TAB( 33)PD(P,4)
7570    PRINT
7580    PRINT "DENSITY"; TAB( 33)PD(P,5)
```

PLNDT *(continued)*

```
7590    PRINT
7600    IF P$ = "MERCURY" OR P$ = "VENUS" THEN  GOTO 7620
7610    PRINT "ROTATION PERIOD (HOURS)"; TAB( 33)PD(P,8): GOTO 7630
7620    PRINT "ROTATION PERIOD (DAYS)"; TAB( 33)PD(P,8)
7630    PRINT
7640    PRINT "AXIS INCLINATION (DEG.)"; TAB( 33)PD(P,9)
7650    PRINT
7660    PRINT "SURFACE GRAVITY (G'S)"; TAB( 33)PD(P,7)
7670    PRINT
7680    PRINT "ALBEDO"; TAB( 33)PD(P,6)
7690    PRINT
7700    PRINT "-------------------------------------"
7710    INPUT "PRESS RETURN TO CONTINUE ";A$
7720    RETURN
7730    HOME : PRINT : PRINT : PRINT
7740    PRINT "ASTRONOMICAL CONSTANTS"
7750    PRINT "-------------------------------------"
7760    PRINT
7770    PRINT "ASTRONOMICAL UNIT"; TAB( 25)"149.6 MILL. KM"
7780    PRINT  TAB( 8)"(A.U.)"; TAB( 26)"92.953 MILL. MI"
7790    PRINT
7800    PRINT "VELOCITY OF LIGHT"; TAB( 25)"299792.5 KM/SEC"
7810    PRINT  TAB( 25)"186635.4 MI/SEC"
7820    PRINT
7830    PRINT "EARTH EQUATORIAL RADIUS"; TAB( 25)"6378.16 KM"
7840    PRINT  TAB( 25)"3963.3 MILES"
7850    PRINT
7860    PRINT "MASS EARTH/MASS MOON"; TAB( 25)"81.3"
7870    PRINT
7880    PRINT "MASS SUN/MASS EARTH"; TAB( 25)"332958"
7890    PRINT
7900    PRINT "-------------------------------------"
7910    INPUT "PRESS RETURN TO CONTINUE ";A$
7920    GOTO 210
7930    REM  SUB FOR DETAIL SELECTION
7940    HOME : PRINT : PRINT : PRINT
7950 DS = 0
7960    PRINT "SELECT...."
7970    PRINT : PRINT
7980    PRINT  TAB( 6)"1) ORBITAL DETAILS"
7990    PRINT
8000    PRINT  TAB( 6)"2) PHYSICAL DETAILS"
8010    PRINT  TAB( 12)"OF PLANET"
8020    PRINT
8030    PRINT  TAB( 6)"3) SATELLITES"
8040    PRINT : PRINT  TAB( 6)"4) RINGS"
8050    PRINT : PRINT  TAB( 6)"999) RETURN TO MENU"
8060    PRINT : PRINT
8070    INPUT "SELECT 1 THRU 4 OR 999 ";DS
8080    IF DS = 999 THEN DS = 5
8090    RETURN
8100    HOME
8110    END
```

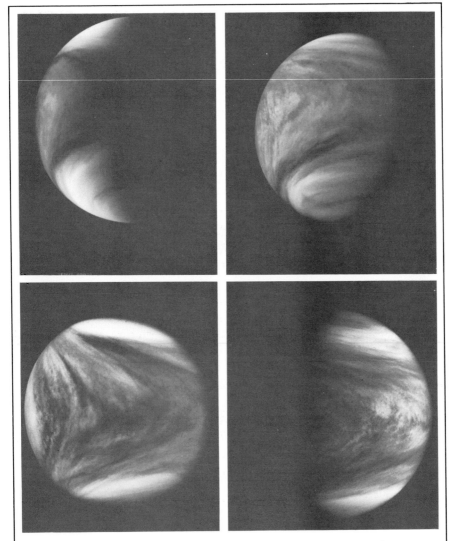

While amateurs are not able to obtain pictures of the planets equal to this series of Venus taken by the Pioneer Orbiter, with suitable telescope/camera combinations and a judicious choice of film and exposure, good planetary pictures can be obtained. This program helps by providing rapid calculations for different lens combinations, exposures, and film types. These four images of Venus show variation of the illumination of the disk, as seen in ultraviolet light, from 25 December 1978 to 24 March 1979. The features shown are typical of the ultraviolet markings of the planet's clouds revealed by spacecraft photography; in normal light the clouds seem bland and uninteresting.

Program 23: PHOTO

Photographing the Planets

Obtaining good telescopic photographs of the planets has always presented a challenge to amateur astronomers. Photography is difficult because the angular size of the planets is so small. Long effective focal lengths of the telescope/camera combination are required to produce images on the photographic film that are large enough to show any detail. But long effective focal lengths mean long exposures, and then tremors of the telescope or atmospheric turbulence spoil the image detail.

Three common methods of planetary photography are:

1. Placing the photographic film at the prime focus of the telescope's objective

2. Attaching a camera (with lens) to the eyepiece of the telescope

3. Attaching a camera (without lens) to the eyepiece of the telescope (projection photography)

This program allows you to calculate exposures for the various planets for each of these three methods of photography. You have to input the diameter (in inches) of the telescope's objective (clear aperture), the focal length (in inches) of the objective, the focal length (in mm) of the eyepiece used, the focal length (in mm) of the camera lens used, and the distance (in inches) from the eyepiece to the film plane if using projection photography. Also, you must give the name of the planet and its diameter in arc seconds at the date you are making the photograph. You can determine planetary diameters in arc seconds from distance data. You must divide the diameter at unit distance of one astronomical unit by the distance of the planet in astronomical units (given by Program 11). Diameters of the planets at unit distance are: Mercury, 6.68; Venus, 16.82; Mars, 9.36; Jupiter, 190; Saturn, 158; Uranus, 68; and Neptune, 73. Pluto is too small to be of significance. In addition, you must select an appropriate film and provide the ASA rating.

The basic equations used in the calculations are as follows:

Method 1 $EFL = FLO$

Method 2 $EFL = FLO \times FC/FE$

Method 3 $EFL = FLO \times DE/FE$

EFL is the effective focal length of the system, FLO is the focal length of the telescope objective, FE is the focal length of the eyepiece, FC is the focal length of the camera lens, and DE is the distance from the eyepiece to the film plane. The effective f/number (EFN) is the effective focal length divided by the clear aperture of the telescope's objective. The exposure is computed from the equation:

$$EX = (EFN)^2/(ASA \times N)$$

where ASA is the film speed and N is a number to compensate for the albedo (reflecting power) of each planet (see instruction lines 1010 through 1070).

The program provides a guide to the exposure you will need, the resolution to be expected, and the size of the image on the film plane. With this information you can effectively plan a photographic session in advance and thus avoid many disappointments. A typical display is shown in Figure 23.1.

The listing for the PHOTO program follows.

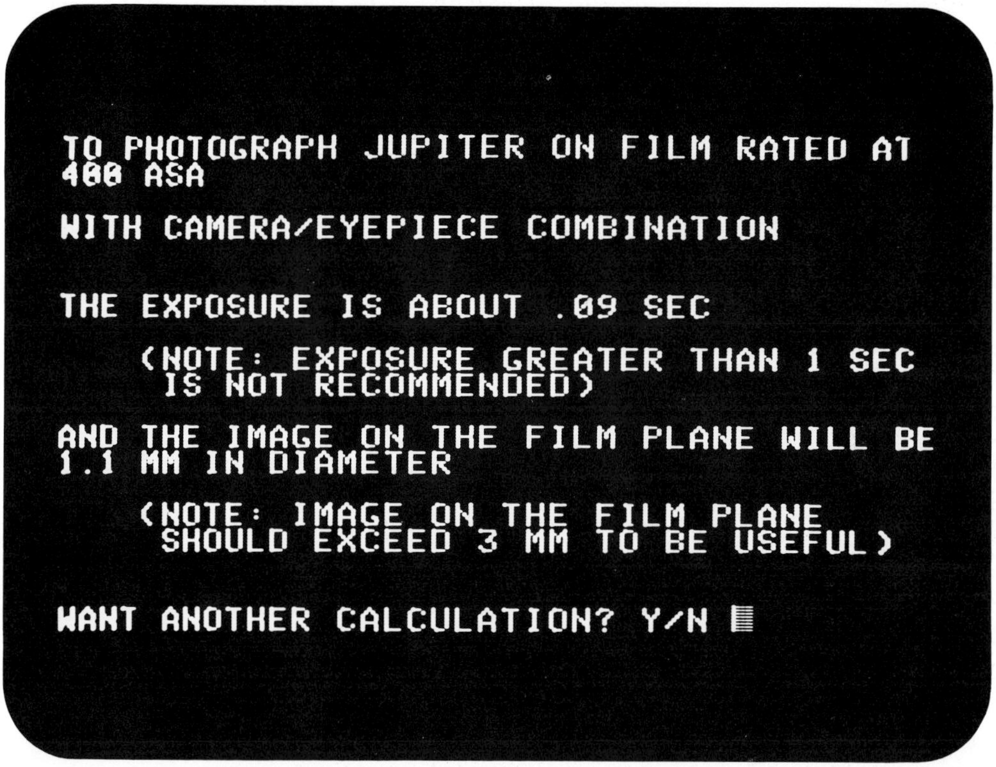

```
TO PHOTOGRAPH JUPITER ON FILM RATED AT
400 ASA

WITH CAMERA/EYEPIECE COMBINATION

THE EXPOSURE IS ABOUT .09 SEC
     (NOTE: EXPOSURE GREATER THAN 1 SEC
        IS NOT RECOMMENDED)

AND THE IMAGE ON THE FILM PLANE WILL BE
1.1 MM IN DIAMETER

     (NOTE: IMAGE ON THE FILM PLANE
        SHOULD EXCEED 3 MM TO BE USEFUL)


WANT ANOTHER CALCULATION? Y/N ▮
```

To help you obtain better planetary photographs, PHOTO develops displays such as this for three types of camera/telescope arrangements.

Figure 23.1

PHOTO

```
10   REM   PHOTO GUIDE FOR PLANETS
20   HOME : PRINT : PRINT : PRINT : PRINT
30   PRINT : PRINT
40   PRINT  TAB( 9)"-----------------"
50   PRINT  TAB( 9)"I    PHOTO     I"
60   PRINT  TAB( 9)"-----------------"
70   PRINT : PRINT
80   PRINT  TAB( 7)"AN ASTRONOMY PROGRAM"
90   PRINT : PRINT  TAB( 6)"BY ERIC BURGESS F.R.A.S."
100  PRINT : PRINT
110  PRINT  TAB( 6)"ALL RIGHTS RESERVED BY"
120  PRINT  TAB( 6)"S & T SOFTWARE SERVICE"
130  PRINT : PRINT
140  PRINT : PRINT
150  INPUT "WANT INSTRUCTIONS? Y/N ";A$
160  IF A$ = "N" THEN  GOTO 310
170  IF A$ <  > "Y" THEN  PRINT "INVALID RESPONSE"; PRINT : GOTO 150
180  HOME : PRINT
190  PRINT : PRINT : PRINT
200  PRINT "THIS PROGRAM ASKS FOR PARTICULARS OF"
210  PRINT "YOUR ASTROPHOTO INSTALLATION AND THEN"
220  PRINT "COMPUTES THE SIZE OF THE IMAGE OF ANY"
230  PRINT "SELECTED PLANET ON THE FILM PLANE"
240  PRINT : PRINT : PRINT
250  PRINT "FOR AN ASA RATING WHICH YOU INPUT"
260  PRINT "IT CALCULATES AN AVERAGE EXPOSURE"
270  PRINT "FOR THE PLANET AND THE INSTRUMENT"
280  PRINT "CONFIGURATION YOU ARE USING."
290  PRINT : PRINT : PRINT
300  INPUT "PRESS RETURN TO CONTINUE ";A$
310  HOME : PRINT : PRINT
320  PRINT "WHICH METHOD ARE YOU USING?"
330  PRINT : PRINT
340  PRINT " 1) FILM AT PRIME FOCUS OF"
350  PRINT "    TELESCOPE'S OBJECTIVE"
360  PRINT
370  PRINT " 2) FILM IN CAMERA AND CAMERA"
380  PRINT "    AT EYEPIECE OF TELESCOPE"
390  PRINT "    WITH CAMERA LENS IN USE"
400  PRINT
410  PRINT " 3) IMAGE PROJECTED FROM EYEPIECE"
420  PRINT "    OF TELESCOPE INTO CAMERA"
430  PRINT "    WITHOUT A CAMERA LENS"
440  PRINT : PRINT
450  INPUT "SELECT 1, 2, OR 3 ";A
460  HOME : PRINT : PRINT
470  PRINT "GIVE FOLLOWING DATA "
480  PRINT
490  INPUT "CLEAR APERTURE OF OBJECTIVE (IN) ";AO
500  AO = AO * 30.4
510  PRINT
520  INPUT "FOCAL LENGTH OF OBJECTIVE (IN) ";FO
530  FO = FO * 30.4
```

PHOTO *(continued)*

```
540 NF = FO / AO
550  IF A <  > 2 THEN  GOTO 580
560  PRINT
570  INPUT "FOCAL LENGTH OF CAMERA LENS (MM) ";FC
580  IF A = 1 THEN  GOTO 670
590  PRINT
600  INPUT "FOCAL LENGTH OF EYEPIECE (MM) ";FE
610  IF A = 1 THEN  GOTO 670
620  IF A = 2 THEN  GOTO 670
630  PRINT
640  INPUT "DISTANCE EYEPIECE TO FILM PLANE (IN) ";DE
650 DE = DE * 25.4
660  REM  CALC. EFFECTIVE F NUMBER
670  IF A = 1 THEN EFL = NF
680  IF A = 2 THEN EFL = NF * FC / FE: GOTO 700
690  IF A = 3 THEN EFL = NF * DE/FE
700  HOME : PRINT : PRINT
710  PRINT "YOU MUST NOW ENTER THE NAME OF"
720  PRINT "THE PLANET YOU ARE PHOTOGRAPHING"
730  PRINT "AND ITS DIAMETER IN ARCSECS"
740  PRINT
750  PRINT "(YOU CAN FIND THIS FROM ONE OF THE"
760  PRINT " PROGRAMS IN THIS SERIES)"
770  PRINT : PRINT
780  INPUT "WHAT IS PLANET'S NAME ";PL$
790  IF PL$ = "PLUTO" THEN  PRINT "NO DATA ON PLUTO; PICK ANOTHER PLANET "
     : GOTO 780
800  PRINT
810  INPUT "WHAT IS DIAMETER IN ARCSECS ";DP
820  F = EFL * AO
830  HOME : PRINT : PRINT : PRINT
840 DP = DP / 3600
850 DP =  TAN (DP / 57.29578)
860  I = F * DP
870  R = 1450 / EFL
880  R =  VAL ( LEFT$ ( STR$ (R),4))
890  PRINT : PRINT
900  PRINT "RESOLUTION FOR ";PL$;" AT"
910  PRINT "FILM PLANE WILL BE "; INT (R);" LINES/MM"
920  PRINT : PRINT
930  PRINT "YOU ARE ADVISED TO PICK A FILM WITH"
940  PRINT "AT LEAST THREE TIMES THIS RESOLUTION"
950  R =  VAL ( LEFT$ ( STR$ (R * 3),4))
960  PRINT : PRINT "NAMELY "; INT (R);" LINES/MM TO AVOID"
970  PRINT "GRAIN SPOILING THE IMAGE"
980  PRINT : PRINT : PRINT
990  INPUT "WHAT IS ASA RATING OF CHOSEN FILM? ";ASA
1000  HOME : PRINT : PRINT
1010  IF PL$ = "MERCURY" THEN N = 260
1020  IF PL$ = "VENUS" THEN N = 960
1030  IF PL$ = "MARS" THEN N = 40
1040  IF PL$ = "JUPITER" THEN N = 12
1050  IF PL$ = "SATURN" THEN N = 3.5
1060  IF PL$ = "URANUS" THEN N = .81
1070  IF PL$ = "NEPTUNE" THEN N = .38
```

PHOTO *(continued)*

```
1080 N = N * ASA
1090 EX = EFL ^ 2 / N
1100  HOME : PRINT : PRINT
1110  PRINT "TO PHOTOGRAPH ";PL$;" ON FILM RATED AT "
1120  PRINT ASA;" ASA"
1130  PRINT
1140  IF A = 1 THEN EQ$ = "FILM AT PRIME FOCUS OF OBJECTIVE"
1150  IF A = 2 THEN EQ$ = "CAMERA/EYEPIECE COMBINATION"
1160  IF A = 3 THEN EQ$ = "PROJECTION FROM EYEPIECE"
1170  PRINT "WITH ";EQ$
1180  PRINT : PRINT
1190 EX =  VAL ( LEFT$ ( STR$ (EX),3))
1200  PRINT "THE EXPOSURE IS ABOUT ";EX;" SEC"
1210  PRINT
1220  PRINT "     (NOTE: EXPOSURE GREATER THAN 1 SEC"
1230  PRINT "      IS NOT RECOMMENDED)"
1240  PRINT
1250  PRINT "AND THE IMAGE ON THE FILM PLANE WILL BE"
1260 I =  VAL ( LEFT$ ( STR$ (I),3))
1270  PRINT I;" MM IN DIAMETER"
1280  PRINT
1290  PRINT "     (NOTE: IMAGE ON THE FILM PLANE"
1300  PRINT "      SHOULD EXCEED 3 MM TO BE USEFUL)"
1310  PRINT : PRINT
1320  INPUT "WANT ANOTHER CALCULATION? Y/N ";A$
1330  IF A$ = "N" THEN  GOTO 1370
1340  IF A$ < > "Y" THEN  PRINT "INVALID RESPONSE": PRINT : GOTO 1320
1350  GOTO 310
1360  HOME
1370  END
```

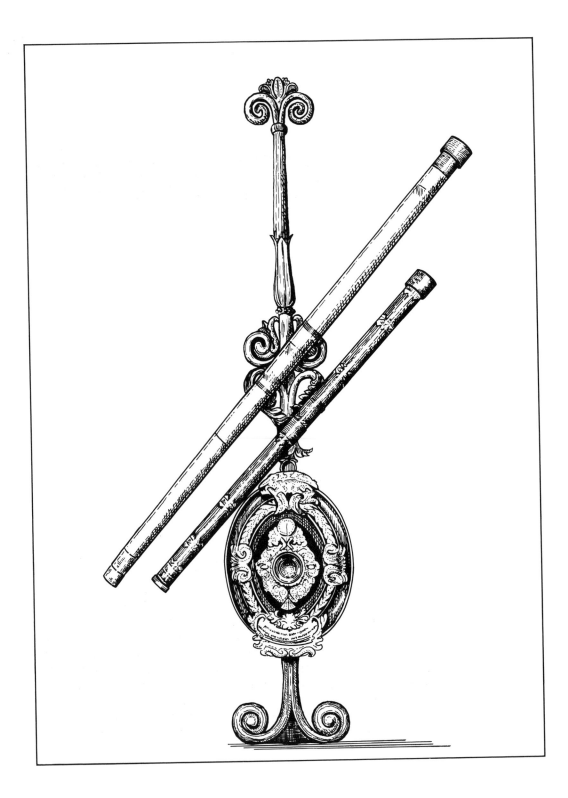

APPENDIX

MANY READERS WILL HAVE computers other than the Apple for which the programs in this book are listed. Because the programs have been written in the popular BASIC language, they are readily adaptable to other computers. Some computers have the ability to display more information on the monitor screen, others less. The more difficult programs to adapt are those that use screen graphics. In addition to knowing BASIC, you also need to be familiar with the way your computer handles graphic displays.

Three important programs are repeated in this part of the book to aid in adapting these programs to other computers. They provide the same information as the programs designed for the Apple computer, but they use different programming techniques. By studying these programs (and the displays they produce) in conjunction with those for the Apple computer, you should have no major difficulty in converting one or the other to suit your computer.

On the lonely plains of the Moon, Apollo 17 astronaut Schmitt is working at the Lunar Roving Vehicle in the Taurus-Littrow area. The peak in the center background was named Family Mountain. On the left horizon is a part of the South Massif. Amateur astronomers can still derive much pleasure from watching the sunrise and sunset shadows across these great lunar plains and their surrounding mountains.

Program 16A: SKYPLA

Alternative Skyplot Program_____

This version of the sky plotting program was written for an Exidy Sorcerer and makes full use of the graphics capabilities of that computer. It is a program complete within itself; it does not require any machine language routines as in Program 16. Hence it is more readily adaptable to other computers. The program has all the instructions needed to run it. It is also adequately supplied with remark (REM) statements in case you should wish to modify it in any way for your use or special needs. When you run the program you can request detailed instructions if you wish.

You will be asked for date and other input information, as well as whether you want to change the location parameters of time zone, latitude, and longitude. Next you are asked to select a horizon of 180 degrees centered on east, west, south, or north. As the program begins its calculations the screen displays the message COMPUTING .. PLEASE WAIT. During this period it loads arrays.

The horizon chart is generated next. For the particular horizon you have requested, the azimuth is shown below the display and the elevation at the left. Date and time information is displayed above the chart. Then the stars are plotted; this takes about five minutes. Next the program plots the Sun and any planets above that horizon at the time and date selected. Finally it plots the Moon and shows it as) before full, o when close to full, and (after full, selected in instructions 5850 through 5870. Since the projection is Mercator, constellations toward the zenith are somewhat distorted (stretched out horizontally).

Variables have been initially set for your latitude, longitude, and time zone. While running the program you can change the variables to other latitudes, longitudes, and time zones. The program should not be expected to run accurately at latitudes exceeding 85 degrees north and south.

Planets are identified by letters: m, Mercury; V, Venus; M, Mars; J, Jupiter; S, Saturn; U, Uranus; N, Neptune; and P, Pluto. Note that if planets are within one pixel of each other, the outermost planet will overprint the innermost; only the symbol for the outer planet will be displayed.

Note that the numbers in lines 2420 and 2440 through 2510 refer to the Sorcerer computer's graphic symbols, which create the horizon silhouette. By referring to Figure 16A.1 you can select suitable graphic symbols from those available on your computer. You can, of course,

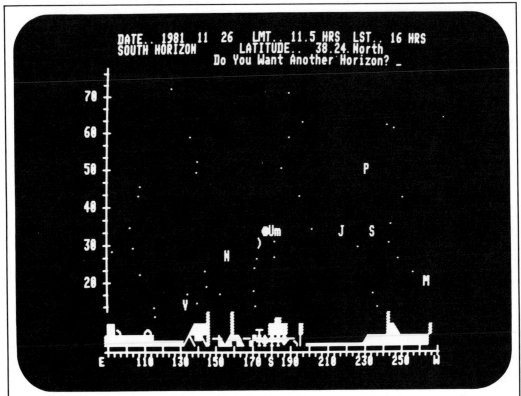

SKYPLA is a version of SKYSET/SKYPLT suitable for adaptation to other computers. Compare this figure with Figure 16.1, showing the same configuration of the planets, the Sun, and the Moon. Note the use of graphic symbols to provide a horizon silhouette.

—Figure 16A.1——

design appropriate silhouettes for your own locality and for the different horizon views.

As with other programs containing many numerical data statements, take particular care in keying in the numbers and differentiating between commas and periods.

To demonstrate the program and compare the display with that generated by the Apple (Program 16), select the display of 26 November 1981 at 11.5 hours for the south horizon (Figure 16A.1). This shows all the planets, the Sun, and the Moon above the horizon at the same time. If you then ask for the same date and time in the Southern Hemisphere (for example, for −40 degrees latitude) and request the north horizon, you will see these same planets and constellations inverted.

The listing of the SKYPLA program follows.

─ **SKYPLA** ─

```
10 CLEAR
20 DEFFNASN(X)=ATN(X/SQR(-X*X+1))
30 DEFFNACO(X)=-ATN(X/SQR(-X*X+1))+3.14159/2
40 DEFFNRAD(X)=(X)*3.14159/180
50 DEFFNDEG(X)=(X)*180/3.14159
60 PI=3.14159
70 PRINTCHR$(12)
80 PRINT:PRINT:PRINT:PRINT:PRINT:PRINT:PRINT:PRINT
90 PRINTTAB(20)"S K Y    P L O T":PRINT:PRINT
100 PRINTTAB(14)"ASTRONOMY PROGRAM FOR SORCERER":PRINT:PRINT
110 PRINT:PRINT
120 ZN=8:LA=FNRAD(38.24):L$="SEBASTOPOL,CAL.":LO=122.49
130 PRINT:PRINT
140 PRINT"Initial conditions are set for..."L$:PRINT
150 PRINT
160 PRINT"                         LATITUDE"FNDEG(LA)
170 PRINT"                         LONGITUDE"LO
180 PRINT"                         TIME ZONE"ZN:PRINT:PRINT
190 PRINT"You can change these later if you wish"
200 DF=0
210 FORJ7=2000TO1STEP-1:NEXTJ7
220 PRINTCHR$(12):PRINT:PRINT:PRINT:PRINT:PRINT:PRINT
230 :PRINTTAB(21);" SKY PLOT"
240 PRINT:PRINTTAB(25);"by"
250 PRINT:PRINTTAB(18);"ERIC BURGESS F.R.A.S.":PRINT
260 PRINT:PRINTTAB(11)"Plots the stars for a requested"
270 PRINTTAB(12)"time and date, and shows the"
280 PRINTTAB(10)"planets, Sun, and Moon above the"
290 PRINTTAB(13)"         horizon."
300 PRINT:PRINTTAB(10)"Close to full the Moon is shown as o"
310 PRINTTAB(10)"Before full it is shown as ) and"
320 PRINTTAB(12)"after full it is shown as (":PRINT:PRINT
330 PRINTTAB(13)"   All rights reserved"
340 PRINTTAB(13)"     S & T Software"
350 PRINT:PRINT
360 FORJ=2500TO1STEP-1:NEXTJ:PRINTCHR$(12):PRINT
370 GOTO1490
380 REM DAYS FROM EPOCH
390 G=365*Y+D+((M-1)*31)
400 IFM>=3THENGOTO420
410 G=G+INT((Y-1)/4)-INT(.75*INT((Y-1)/100+1)):RETURN
420 G=G-INT(2.3+M*.4)+INT(Y/4)-INT(.75*INT((Y/100)+1))
430 RETURN
440 REM CALCUL OF LST
450 SG=.065711
460 T2=SG*NS+12.064707+(((ZN+T1)/24)*SG)+T1
470 IFT2>24THENT2=T2-24:GOTO470
480 IFT2<-24THENT2=T2+24:GOTO480
490 IFT2<0THENT2=T2+24
500 T2$=STR$(T2)
510 T2=VAL(LEFT$(T2$,5))
520 RETURN
530 REM ENTER VAR.FOR CALC LST
540 GOSUB380:NS=G-722895:GOSUB440
```

SKYPLA *(continued)*

```
550 REM ENTER VAR.FOR CALC NO. OF DAYS
560 MO=M:DA=D:YR=Y
570 GOSUB380:ND=G-715875+T1/24
580 REM STORE DATA FOR CALCS
590 IFF4=1GOTO870
600 RESTORE
610 DIMPD(9,9)
620 FORYY=0TO8
630 FORXX=0TO8
640 READPD(YY,XX)
650 NEXTXX,YY
660 REM ORBITAL ELEMENTS OF PLANETS, MERCURY-PLUTO
670 DATA.071425,3.8494,.388301,1.34041,.3871
680 DATA.07974,2.73514,.122173,.836013
690 DATA.027962,3.02812,.013195,2.28638,.7233
700 DATA.00506,3.85017,.059341,1.33168
710 DATA.017202,1.74022,.032044,1.78547,1,.017,3.33926,0
720 DATA0,.009146,4.51234,.175301,5.85209,1.5237
730 DATA.141704,1.04656,.03142,.858702
740 DATA.001451,4.53364,.090478,.23911,5.2028
750 DATA.24934,1.76188,.01972,1.74533
760 DATA.000584,4.89884,.105558,1.61094,9.5385
770 DATA.534156,3.1257,.043633,1.977458
780 DATA.000205,2.46615,.088593,2.96706,19.182
790 DATA.901554,4.49084,.01396,1.28805
800 DATA.000104,3.78556,.016965,.773181,30.06
810 DATA.27054,2.33498,.031416,2.29162
820 DATA.000069,3.16948,.471239,3.91303,39.44
830 DATA9.86,5.23114,.300197,1.91812
840 FORI9=1TO9
850 READP(I9):NEXTI9
860 DATA109,86,83,77,74,83,85,78,80
870 F4=1
880 I=1
890 FORJ=0TO8
900 GOSUB1000
910 A(I)=A:D(I)=D:L(I)=L
920 I=I+1:NEXT
93U FORI=1TO9
940 IFI=3THENNEXT
950 PRINTCHR$(12):PRINT"COMPUTING...PLEASE WAIT"
960 GOSUB1110
970 Z(I)=Z:Q(I)=Q:X(I)=X:R(I)=R:V(I)=V:AL(I)=AL:AZ(I)=AZ
980 NEXT
990 RETURN
1000 REM SUB FOR A,D,L
1010 A=ND*PD(J,0)+PD(J,1)
1020 IFA>PI*2THENA=(A/(PI*2)-INT(A/(PI*2)))*PI*2
1030 IFA<0THENA=A+PI*2:GOTO1030
1040 C=PD(J,2)*SIN(A-PD(J,3))
1050 A=A+C
1060 IFA>PI*2THENA=A-PI*2
1070 IFA<0THENA=A+PI*2:GOTO1070
1080 D=PD(J,4)+PD(J,5)*SIN(A-PD(J,6))
1090 L=PD(J,7)*SIN(A-PD(J,8))
```

```
1100 RETURN
1110 REM ELEV AND AZ OF SUN AND PLANETS
1120 F=0:Z=A(3)-A(I)
1130 IFABS(Z)>PIANDZ<0THENZ=Z+(PI*2)
1140 IFABS(Z)>PIANDZ>0THENZ=Z-(PI*2)
1150 Q=SQR(D(I)^2+D(3)^2-2*D(I)*D(3)*COS(Z))
1160 P=(D(I)+D(3)+Q)/2
1170 X=2*FNACO(SQR(((P*(P-D(I)))/(D(3)*Q))))
1180 T=X*(12/PI)
1190 IFZ<0THENR=FNDEC(A(3)+PI-X)/15
1200 IFZ>0THENR=FNDEC(A(3)+PI+X)/15
1210 IFR>24THENR=R-24:GOTO1220
1220 IFR<-24THENR=R+24:GOTO1220
1230 IFR<0THENR=R+24:GOTO1230
1240 IFZ<0THENV=SIN(A(3)+PI-X)*23.44194+FNDEG(L(I))
1250 IFZ>0THENV=SIN(A(3)+PI+X)*23.44194+FNDEG(L(I))
1260 HA=T2-R
1270 IFHA<-12THENHA=HA+24
1280 IFHA>12THENHA=HA-24
1290 HA=FNRAD(HA*15):V=FNRAD(V)
1300 AL=FNASN(SIN(V)*SIN(LA)+COS(V)*COS(LA)*COS(HA))
1310 AZ=FNACO((SIN(V)-SIN(LA)*SIN(AL))/(COS(LA)*COS(AL)))
1320 IFHA>0THENAZ=PI*2-AZ
1330 AL=FNDEG(AL):AZ=FNDEG(AZ)
1340 RS=FNDEG(A(3)+PI)/15
1350 IFRS>24THENRS=RS-24:GOTO1350
1360 IFRS<-24THENRS=RS+24:GOTO1360
1370 IFRS<0THENRS=RS+24:GOTO1370
1380 VS=SIN(A(3)+PI)*23.44194
1390 HS=T2-RS
1400 IFHS<-12THENHS=HS+24
1410 IFHS>12THENHS=HS-24
1420 HS=FNRAD(HS*15):VS=FNRAD(VS)
1430 AS=FNASN(SIN(VS)*SIN(LA)+COS(VS)*COS(LA)*COS(HS))
1440 ZS=FNACO((SIN(VS)-SIN(LA)*SIN(AS))/(COS(LA)*COS(AS)))
1450 IFHS>0THENZS=PI*2-ZS
1460 AS=FNDEG(AS):ZS=FNDEG(ZS)
1470 RETURN
1480 REM BEGIN MAIN PROGRAM
1490 PRINT:PRINT"PLEASE WAIT":PRINTCHR$(12)
1500 IFDF=0THENPRINT:PRINT:PRINT:GOTO1640
1510 PRINT:INPUT"DO YOU WANT TO SEE VARIABLES";AB$
1520 IFAB$="N"THENGOTO1560
1530 PRINT:PRINT"DATE..";YR;MO;DA;"        LMT..";T1;"HRS";
1540 PRINT"      TIME ZONE...";ZN
1550 PRINT"LAT...";FNDEG(LA);:PRINT"     LONG...";LO:PRINT
1560 PRINT
1570 INPUT"DO YOU WANT TO ENTER OR CHANGE ANY VARIABLES   ";A$
1580 IFA$="N"GOTO1970
1590 PRINT:PRINT
1600 PRINT"DO YOU WANT TO CHANGE THE DATE OF: ";YR;MO;DA
1610 PRINT
1620 INPUT"ANSWER 'Y'OR 'N'";A1$
1630 IFA1$="N"THENGOTO1740
1640 PRINT:PRINT"WHAT IS THE DATE REQUIRED"
```

SKYPLA *(continued)*

```
1650 PRINT:INPUT".....THE YEAR";Y
1660 PRINT:INPUT"....THE MONTH";M
1670 PRINT:INPUT"......THE DAY";D
1680 MO=M:DA=D:YR=Y
1690 IFY>1800GOTO1740
1700 PRINT:PRINT:PRINT"IS ";Y;" THE YEAR YOU WANT?"
1710 INPUTY$
1720 IFY$="Y"GOTO1740
1730 IFY$="N"THENPRINT:GOTO1650
1740 PRINT:PRINT:GOSUB3440
1750 IFDF=0THEN1780
1760 PRINT"                 THE LMT OF: ";T1;"   ";
1770 PRINT:INPUTA1$:IFA1$="N"THENGOTO1830
1780 PRINT"NOW YOU MUST INPUT THE LMT FOR THE DISPLAY":GOSUB3320
1790 PRINT:PRINT"IF YOU WANT TO CHANGE VARIABLES"
1800 PRINT"ANSWER 'Y' WHEN ASKED, OTHERWISE ANSWER 'N'"
1810 IFDF=0THEN1830
1820 PRINT:Y$="LMT":GOSUB3320
1830 PRINT:PRINT
1840 PRINT"               THE TIME ZONE OF: ";ZN:INPUTA1$
1850 IFA1$="N"THENGOTO1870
1860 PRINT:INPUT"WHAT IS THE TIME ZONE (WEST COAST IS 8)";ZN
1870 PRINT:PRINT
1880 PRINTTAB(15)"THE LONGITUDE OF: ";LO:INPUTA1$
1890 IFA1$="N"THENGOTO1910
1900 PRINT:GOSUB3510
1910 PRINT:PRINT
1920 PRINT"               THE LATITUDE OF: ";FNDEG(LA)
1930 INPUTA1$
1940 IFA1$="N"THENGOTO1980
1950 PRINT:INPUT"WHAT IS THE LATITUDE (SEBASTOPOL;38.24)";LA
1960 LA=FNRAD(LA)
1970 :
1980 IFHR=1ANDA$="N"THENHR=0:A$="":GOTO2680
1990 IFHR=1ANDA$="Y"THENHR=0:A$="":GOTO2120
2000 PRINTCHR$(12):PRINT:PRINT:PRINT:PRINT
2010 PRINT:PRINT"WHICH HORIZON DO YOU WANT TO SEE?"
2020 PRINT
2030 PRINT:PRINTTAB(5)"1 EAST HORIZON ...  0 TO 180 DEG."
2040 PRINT:PRINTTAB(5)"2 WEST HORIZON ...180 TO 360 DEG."
2050 PRINT:PRINTTAB(5)"3 SOUTH HORIZON .. 90 TO 270 DEG."
2060 PRINT:PRINTTAB(5)"4 NORTH HORIZON .. 270 TO 90 DEG."
2070 PRINT:INPUT"SELECT    ";HZ$
2080 IF(HZ$="1"ORHZ$="2"ORHZ$="3"ORHZ$="4")THEN2100
2090 PRINT"INVALID ENTRY":GOTO2010
2100 HZ=VAL(HZ$)
2109 DS=0:PRINT:PRINT
2110 PRINT"Do You Want Stars Displayed As Well As Planets,"
2111 PRINT"Sun, and Moon (Y/N)";
2112 INPUT DS$:IF DS$="N"THENDS=1:GOTO2120
2113 IF DS$<>"Y"THENPRINT"Invalid Response":PRINT:GOTO2110
2120 REM BEGIN CALCS
2130 FL=3
2140 PRINTCHR$(12):PRINT"COMPUTING...PLEASE WAIT":PRINT
2150 IFFP=1GOTO2330
2160 DIMPO(30,64)
```

SKYPLA (continued)

```
2170 P=-3968
2180 FORYY=29TO0STEP-1
2190 FORXX=0TO63
2200 LETPO(YY,XX)=P
2210 P=P+1
2220 NEXTXX
2230 PRINTCHR$(12):PRINT"COMPUTING...PLEASE WAIT"
2240 NEXTYY
2250 DIMPP(8,8)
2260 P=192
2270 FORX8=0TO7
2280 FORY8=0TO7
2290 LETPP(Y8,X8)=P
2300 P=P+1
2310 NEXTY8,X8
2320 FP=1
2330 REM SET UP GRAPHICS
2335 IFDS=1THEN2670
2340 GOSUB5030
2350 GOTO2670
2360 PRINT
2370 REM PRINT COORDINATES
2380 PRINT:PRINT:PRINT
2390 FORI3=80TO20STEP-10
2400 PRINTI3"-"
2405 REM PRINTS SORCERER GRAPHIC FOR VERTICAL SCALE
2406 REM (See Figure 16A.1)
2410 FORJ3=1TO2
2420 PRINT"    9"
2430 NEXTJ3,I3
2440 PRINT"    ";
2445 REM NEXT LINES PRINT SORCERER GRAPHIC CHARACTERS FOR
2446 REM THE HORIZON; YOU MUST PICK GRAPHICS TO SUIT YOUR
2447 REM OWN COMPUTER THROUGH LINE 2490 (See Figure 16A.1)
2450 PRINT"            5    5      8            5"
2460 PRINT"    ";
2470 PRINT"('        115   5   ; 111 5          1 5"
2480 PRINT"    ";
2490 PRINT"1111111188888+,),5)&$$$#::::::::::1111111111 "
2495 REM NEXT TWO LINES ARE SORCERER GRAPHICS FOR HORIZONTAL
2496 REM SCALE (See Figure 16A.1)
2500 PRINT"   -;;-;;-;;-;;-;;-;;-;;-;;-;;-;;-;;-;;-;;-";
2510 PRINT";;-;;-;;9"
2520 IFHZ=2GOTO2580
2530 IFHZ=3GOTO2610
2540 IFHZ=4GOTO2640
2550 PRINT"   N    20    40    60    80 E 100   120   140 ";
2560 PRINT"  160    S"
2570 F6=1:PRINTCHR$(17):RETURN
2580 PRINT"   S    200   220   240   260 W 280   300   320 ";
2590 PRINT"  340    N"
2600 F6=1:PRINTCHR$(17):RETURN
2610 PRINT"   E    110   130   150   170 S 190   210   230 ";
2620 PRINT"  250    W"
2630 F6=1:PRINTCHR$(17):RETURN
```

SKYPLA (continued)

```
2640 PRINT"    W    290    310    330    350  N   10    30     50 ";
2650 PRINT"   70     E"
2660 F6=1:PRINTCHR$(17):RETURN
2670 GOSUB530
2680 PRINTCHR$(12)
2690 IFHZ=1THENHZ$="EAST HORIZON"
2700 IFHZ=2THENHZ$="WEST HORIZON"
2710 IFHZ=3THENHZ$="SOUTH HORIZON"
2720 IFHZ=4THENHZ$="NORTH HORIZON"
2730 GOSUB3440
2740 GOSUB2380
2750 U$="North"
2760 IFLA<0THENU$="South"
2770 PRINT"      DATE.."YR;MO;DA;"  LMT.."T1;"HRS";
2780 PRINT"  LST.."T2"HRS":PRINT"      "HZ$;
2790 PRINT"      LATITUDE.. ";ABS(FNDEG(LA));U$
2800 PRINT"      Please Wait...Computing Positions"
2810 PRINTCHR$(17)
2820 REM PRINT SKY DOME
2825 IF DS=1THEN2850
2830 REM POKE STARS
2840 GOSUB3580
2850 REM POKE SUN
2860 IFAS<0GOTO2950
2870 BS=ZS
2880 IFHZ=3AND(BS>90ANDBS<270)THENX1=BS-90:Y1=AS:GOTO2930
2890 IFHZ=4AND(BS>270ANDBS=<360)THENX1=BS-270:Y1=AS:GOTO2930
2900 IFHZ=4AND(BS>0ANDBS<85)THENX1=BS+90:Y1=AS:GOTO2930
2910 IFHZ=1ANDBS<=180THENX1=BS:Y1=AS:GOTO2930
2920 IFHZ=2ANDBS>180THENX1=BS-180:Y1=AS:GOTO2930
2925 GOTO2950
2930 Y2=28/90*Y1+2.5:X2=55/180*X1+4
2940 POKEPO(INT(Y2),INT(X2)),132
2950 REM SELECT AND POKE PLANETS
2960 FORI8=1TO9
2970 BZ(I8)=AZ(I8)
2980 IFAL(I8)<50RAL(I8)>80THENGOTO3110
2990 IFI8=3THENGOTO3110
3000 IFHZ=1ANDBZ(I8)<=180THENGOTO3060
3010 IFHZ=2ANDBZ(I8)>180THENBZ(I8)=BZ(I8)-180:GOTO3060
3020 IFHZ=3ANDBZ(I8)>90ANDBZ(I8)<270THENBZ(I8)=BZ(I8)-90:GOTO3060
3030 IFHZ=4AND(BZ(I8)>270ANDBZ(I8)=<360)THENGOTO3050
3040 IFHZ=4AND(BZ(I8)>0ANDBZ(I8)<85)THENBZ(I8)=BZ(I8)+90:GOTO3060
3045 GOTO3110
3050 BZ(I8)=BZ(I8)-270
3060 REM POKE ROUTINE
3070 X1=BZ(I8):Y1=AL(I8)
3080 IF(X1<50RX1>175)THENGOTO3110
3090 Y2=28/90*Y1+2.5:X2=55/180*X1+4
3100 POKEPO(INT(Y2),INT(X2)),P(I8)
3110 NEXTI8
3120 GOSUB5230
3130 GOTO3150
3140 REM END SEQUENCE
3150 PRINT:PRINT
3160 PRINTTAB(22)"Do You Want Another Horizon";:INPUTA$
```

SKYPLA *(continued)*

```
3170 M=MO:D=DA:Y=YR
3180 FP=1
3190 IFA$="N"THENGOTO3250
3200 HR=1:PRINTCHR$(12):PRINT:PRINT:PRINT"WHICH HORIZON"
3210 PRINT:PRINTTAB(7)"EAST... 1":PRINTTAB(7)"WEST... 2"
3220 PRINTTAB(7)"SOUTH.. 3":PRINTTAB(7)"NORTH.. 4"
3230 PRINT:INPUT"SELECT   ";HZ$:HZ=VAL(HZ$)
3240 PRINTCHR$(12):HR=1:PRINT:PRINT:PRINT:GOTO1570
3250 PRINTCHR$(12):PRINT:PRINT:PRINT
3260 INPUT"WANT ANOTHER DATE Y/N";A$
3270 IFA$="Y"THENGOTO1510
3280 IFA$<>"N"THENPRINT:PRINT"INVALID REPLY":GOTO3260
3290 PRINTCHR$(12):PRINT:PRINT"SKYPLOT SIGNING OFF"
3300 END
3310 REM TIME INPUT
3320 PRINT:PRINT"DO YOU WANT INPUT IN DEC.HRS (1)"
3330 PRINT"              OR IN HR,MI,SE (2)":PRINT
3340 INPUTPT$:PT=VAL(PT$)
3350 IFPT=1THENGOTO3380
3360 IFPT=2THENGOTO3400
3370 PRINT"INVALID REPLY":PRINT:GOTO3320
3380 PRINT:PRINT"WHAT IS THE LMT -HH.XXXX"
3390 PRINT:INPUTT1:PRINT:GOTO3430
3400 PRINT:PRINT"WHAT IS THE LMT -HR,MI,SE"
3410 PRINT:INPUTHR,MI,SE
3420 T1=HR+MI/60+SE/3600:PRINT
3430 RETURN
3440 REM PRINTING HR,MI,SE
3450 RR=INT(T1)
3460 IM=(T1-INT(T1))*60
3470 ES=(IM-INT(IM))*60
3480 ES=INT(ES)
3490 IM=INT(IM)
3500 RETURN
3510 REM CORRECT ZN FOR LONGITUDE
3520 PRINT:INPUT"WHAT IS LONGITUDE, SEBASTOPOL IS 122.49";LO
3530 LGC=(ZN*15)-LO
3540 IFLGC<0THENGOTO3560
3550 ZN=ZN+ABS(LGC/15):GOTO3570
3560 ZN=ZN+LGC/15
3570 PRINT:RETURN
3580 REM DEVELOP HA, AZ AND AL FOR STARS
3590 IFSF=1THEN4620
3600 DIMST(237,2)
3610 FORSY=0TO236:FORSX=0TO1
3620 READST(SY,SX)
3630 NEXTSX,SY
3640 REM DATA ON RA AND DEC OF STARS
3650 REM URSA MINOR
3660 DATA2,89,18,86,17,82,16,78,15,75,15.4,72,16.3,76
3670 REM CEPHEUS
3680 DATA20.8,61,21.5,70
3690 REM CASSIOPEIA
3700 DATA1.9,63,1.4,60,0.9,60,0.6,56,0.1,59
3710 REM PERSEUS
```

```
┌─ SKYPLA (continued) ─────────────────────────────────────────────┐

3720 DATA3.3,50,3.0,53,3.7,48,3.1,41,3.9,40,3.9,32
3730 REM URSA MAJOR
3740 DATA11,57,11,63,11.9,54,12.2,58,12.9,57,13.4,55,13.7,50
3750 REM DRACO
3760 DATA16,59,16.4,62,17.1,66,17.5,52,17.9,51,18.3,73,19.2,68
3770 REM CEPHEUS
3780 DATA23.8,78,21.3,62,22.1,58,22.8,67
3790 REM ANDROMEDA
3800 DATA2,42,1.1,35,.6,31
3810 REM TRIANGULUM
3820 DATA2.1,35,1.8,29,2.2,34
3830 REM PEGASUS
3840 DATA22.7,30,0.1,29,0.2,14,21.7,10,22.2,6,22.7,10,23.0,4
3850 DATA23,28
3860 REM AURIGA
3870 DATA5.2,46,5.9,45,5.9,37,4.9,33,5,41
3880 REM BOOTES
3890 DATA14.5,39,15,40,15.3,33,14.2,20,13.9,19,14.7,27,15.5,27
3900 DATA15.4,29
3910 REM CORONA
3920 DATA15.6,27
3930 REM HERCULES
3940 DATA16.7,39,16.7,31,17,31,17.2,37,17.2,25,16.5,21,16.4,19
3950 REM LYRA
3960 DATA18.7,39,18.8,33,19,32
3970 REM CYGNUS
3980 DATA20.7,45,20.3,40,19.8,45,20.8,34,19.5,28
3990 REM TAURUS
4000 DATA3.6,24,3,4,2.7,3,4.5,17,5.4,29,5.6,21
4010 DATA3.7,24
4020 DATA4.3,15,4.45,19
4030 REM ARIES
4040 DATA2.1,23,1.8,21,1.8,19
4050 REM ERIDANUS
4060 DATA3.9,-13,3.3,-20
4070 REM PISCES
4080 DATA1.5,-9,1.2,-10
4090 REM CETUS
4100 DATA.7,-18,1.1,-10,1.3,-9,2,2
4110 REM ORION
4120 DATA5.9,8,5.4,8,5.75,-2,5.6,-1,5.45,0,5.8,-10,5.6,-6
4130 DATA5.6,10,5.5,-21,5.2,-9
4140 REM CANIS MAJOR
4150 DATA6.7,-17,6.3,-18,6.9,-29,7.2,-27,7.4,-29
4160 REM CANIS MINOR
4170 DATA7.6,7,7.4,9
4180 REM GEMINI
4190 DATA7.6,32,7.7,28,7.3,22,6.7,25,6.6,16,6.4,22,6.3,22
4200 REM LEO
4210 DATA10.1,12,10.1,17,10.3,20,10.3,24,11.2,20,11.2,16
4220 DATA11.8,15,9.8,28,9.7,26
4230 REM CANCER
4240 DATA8.7,29,8.6,21
4250 REM HYDRA
4260 DATA9.5,-9,8.7,7,8.9,7,9.2,2,10.4,-17

└──────────────────────────────────────────────────────────────────┘
```

SKYPLA (continued)

```
4270 REM VIRGO
4280 DATA11.8,2,13.4,-11,13,11,12.9,3,12.7,-1,12.3,-1,13.1,-5
4290 REM CRATER
4300 DATA10.8,-16,10.9,-18,11.3,-15,11.4,-18
4310 REM CORVUS
4320 DATA12.5,-16,12.2,-17,12.5,-23,12.2,-22
4330 REM SERPENS
4340 DATA15.8,17,15.5,10,15.7,7,15.8,5,15.8,-3
4350 REM LIBRA
4360 DATA15.3,-9,14.8,-16
4370 REM OPHIUCHUS
4380 DATA17.5,12,17.2,25,17.6,5,17.7,3
4390 REM SAGITTARIUS
4400 DATA18.3,-30,18,-30,18.4,-25,18.9,-26,19,-30,19.1,-21
4410 DATA18.3,-21
4420 REM SCORPIO
4430 DATA16.5,-26,16.6,-28,16.4,-24,16,-20,15.9,-22,15.9,-26
4440 DATA18.6,-43,16.7,-34,18.5,-37,18.7,-40,16.7,-38
4450 DATA22.9,-30
4460 REM CAPRICORNUS
4470 DATA21.7,-18,21.6,-18,21.4,-22,20.8,-28,20.7,-26,20.3,-14
4480 DATA20.2,-12,22.9,-30
4490 REM DELPHINUS
4500 DATA20.5,11,20.6,15,20.7,15,20.6,16,20.8,16
4510 REM AQUARIUS
4520 DATA22.6,0,22.5,0,22.4,1,22.3,-2,22,0,21.5,-6
4530 REM AQUILA
4540 DATA19.8,9,19.7,10.5,19.9,6,19.1,13,18.95,14,20.1,-1
4550 REM SOUTH POLAR REGION
4560 DATA12.2,-59,12.1,-50,12.4,-57,12.7,-59,12.3,-63
4570 DATA14,-60,14.7,-60,14.7,-65,15.9,-63,15.1,-69,16.9,-69
4580 DATA20.3,-57,1.7,-57,2,-62,0.4,-63,6.3,-52,6.8,-51
4590 DATA8.8,-55,9.3,-55,9.2,-59,8.3,-60,9.1,-70,9.8,-65
4600 DATA3.9,-75,12.5,-69,12.6,-68
4610 SF=1
4620 REM GET AZ AND EL FOR EACH STAR AND POKE IT ON CHART
4630 FORK=0TO236
4640 SR=ST(K,0):SD=ST(K,1)
4650 HD=T2-SR
4660 IFHD<-12THENHD=HD+24
4670 IFHD>12THENHD=HD-24
4680 HA=HD*15
4690 HA=FNRAD(HA):SD=FNRAD(SD)
4700 SL=FNASN(SIN(SD)*SIN(LA)+COS(SD)*COS(LA)*COS(HA))
4710 SZ=(SIN(SD)-SIN(LA)*SIN(SL))/(COS(LA)*COS(SL))
4720 IFSZ=>1THENSZ=0:GOTO4760
4730 IFSZ<=-1THENSZ=3.14159:GOTO4760
4740 SZ=FNACO(SZ)
4750 IFHA>0THENSZ=PI*2-SZ
4760 SZ=FNDEG(SZ)
4770 IFSZ>360THENSZ=SZ-360
4780 IFSZ<0THENSZ=SZ+360
4790 SL=FNDEG(SL)
4800 IFSL>70ORSL<10THEN4990
4810 IFHZ=1THEN4850
```

SKYPLA (continued)

```
4820 IFHZ=2THEN4870
4830 IFHZ=3THEN4890
4840 IFHZ=4THEN4910
4850 IF(SZ>5ANDSZ<175)THENGOTO4940
4860 GOTO4990
4870 IF(SZ>185ANDSZ<355)THENSZ=SZ-180:GOTO4940
4880 GOTO4990
4890 IF(SZ>95ANDSZ<265)THENSZ=SZ-90:GOTO4940
4900 GOTO4990
4910 IF(SZ>275ANDSZ=<360)THENSZ=SZ-270:GOTO4940
4920 IF(SZ>0ANDSZ<85)THENSZ=SZ+90:GOTO4940
4930 GOTO4990
4940 REM POKE STAR
4950 X1=SZ:Y1=SL
4960 Y2=28/90*Y1+2.5:X2=55/180*X1+4
4970 Y3=8*(Y2-INT(Y2)):X3=8*(X2-INT(X2))
4980 POKEPO(INT(Y2),INT(X2)),PP(INT(Y3),INT(X3))
4990 NEXTK
5000 X1=0:Y1=0:Y2=0:X2=0:X=0
5010 RETURN
5020 REM SET 8X8 GRID
5030 IFFS=1THENRETURN
5040 LP=-512:XP=0:AP=7:GP=128
5050 FORIP=1TO8:FORJP=1TO9
5060 IFJP=9THENAP=AP+1
5070 FORKP=1TOAP
5080 POKELP,XP
5090 LP=LP+1
5100 NEXTKP
5110 IFJP=9GOTO5200
5120 POKELP,GP
5130 LP=LP+1
5140 IFFP<>1GOTO5170
5150 PRINTCHR$(12):PRINT"COMPUTING...PLEASE WAIT"
5160 NEXTJP
5170 IFJP=9GOTO5190
5180 AP=AP-1:FP=1
5190 NEXTJP
5200 FP=0:AP=7:XP=0:GP=GP*.5
5210 NEXTIP
5220 FS=1:RETURN
5230 REM POKE MOON ON CHART
5240 ND=ND-.5
5250 LP=255.7433
5260 LZ=311.1687:LE=178.699:LM=LZ+360*ND/27.32158
5270 MD=LM:GOSUB5890:LM=MD
5280 REM CORRECT FOR ELLIPTICAL ORBIT
5300 PG=.111404*ND+LP
5310 MD=PG:GOSUB5890
5320 PG=MD
5330 PG=LM-PG
5340 DR=6.2886*(SIN(FNRAD(PG)))
5350 LM=LM+DR
5360 RQ=LM:RM=LM/15
5370 IFRM>24THENRM=RM-24:GOTO5370
5380 IFRM<0THENRM=RM+24
```

SKYPLA *(continued)*

```
5390 AL=LE-ND*.052954
5400 ND=ND+.5
5410 MD=AL:GOSUB5890
5420 AL=MD
5430 AL=RQ-AL
5440 IFAL<360THENAL=AL+360
5450 IFAL>360THENAL=AL-360
5460 HE=5.1333*SIN(AL*P1/180)
5470 DM=HE+23.1444*SIN(RQ*PI/180)
5480 HD=T2-RM
5490 IFHD<-12THENHD=HD+24
5500 IFHD>12THENHD=HD-24
5510 IF(HD>120RHD<-12)THENGOTO5940
5520 HA=FNRAD(HD*15):DM=FNRAD(DM)
5530 ML=FNASN(SIN(DM)*SIN(LA)+COS(DM)*COS(LA)*COS(HA))
5540 MZ=FNACO((SIN(DM)-SIN(LA)*SIN(ML))/(COS(LA)*COS(ML)))
5560 IFHA>0THENMZ=PI*2-MZ
5570 MZ=FNDEG(MZ)
5580 ML=FNDEG(ML)
5600 IFML>800RML<5THEN5800
5610 MX=MZ
5620 IFHZ=1THEN5660
5630 IFHZ=2THEN5680
5640 IFHZ=3THEN5700
5650 IFHZ=4THEN5720
5660 IF(MX>10ANDMX<170)THENGOTO5750
5670 GOTO5800
5680 IF(MX>190ANDMX<350)THENMX=MX-180:GOTO5750
5690 GOTO5800
5700 IF(MX>100ANDMX<260)THENMX=MX-90:GOTO5750
5710 GOTO5800
5720 IF(MX>275ANDMX=<360)THENMX=MX-270:GOTO5750
5730 IF(MX>0ANDMX<85)THENMX=MX+90:GOTO5750
5740 GOTO5800
5750 REM POKE MOON
5760 X1=MX:Y1=ML
5770 Y2=28/90*Y1+2.5:X2=55/180*X1+4
5780 GOSUB5820
5790 POKEPO(INT(Y2),INT(X2)),MS
5800 X1=0:X2=0:Y1=0:Y2=0
5810 RETURN
5820 PM=RS+12-RQ/15
5830 IFPM>24THENPM=PM-24
5840 IFPM>12THENPM=PM-24
5850 IF(PM>=-2ANDPM<=2)THENMS=111:RETURN
5855 IF LA<0ANDPM<-2THENMS=41:RETURN
5860 IFPM<-2THENMS=40:RETURN
5865 IF LA<0ANDPM>2THENMS=40:RETURN
5870 IFPM>2THENMS=41
5880 RETURN
5890 IFMD<-3600THENMD=MD+3600:GOTO5890
5900 IFMD<-360THENMD=MD+360:GOTO5900
5910 IFMD<0THENMD=MD+360
5920 IFMD>3600THENMD=MD-3600:GOTO5920
5930 IFMD>360THENMD=MD-360:GOTO5930
5940 RETURN
```

Photo Credit: NASA/Jet Propulsion Laboratory

By the end of this century man may begin to set up working bases on the planets, possibly beginning with a base on the Moon and continuing with a base on Mars, as shown in this artist's concept. In this picture solar sail spacecraft are bringing people and materials into orbit around Mars, from which descent takes place by parachute. The landed modules are then established as parts of the base. They would gradually be interconnected like the modules of bases in Antarctica or those of the oil cities on Alaska's North Slope.

Program 17A: PLNTA

Alternative Planet Finder Program_____

This program is a more advanced version of Program 17, PLNTF. It is designed for use with computers having 64 by 30 screen formats. Because of the larger screen format, the program can display more detail on the screen and a wider range of stars on the star charts.

When you select a date and a planet, the Sun, or the Moon, this program calculates where the object is located among the "fixed" stars of the celestial sphere. It selects a suitable star chart, displays its name, identifies constellations, shows the object among the stars, and names bright stars. It identifies and displays nearby constellations over 4 hours of right ascension and 60 degrees of declination, and it provides a grid of right ascension and declination. Because of the limitations of resolution of the monitor screen, the selected object's accurate right ascension and declination are displayed before the object is shown on the star chart.

The program then offers several options. You can elect to have other planets, the Sun, and the Moon displayed on the chart if they are located in the same chart region on the date requested, and you can elect to have the grid lines deleted and again either have a single planet or all planets displayed.

When you have obtained the information, the program asks you if you want another planet on that same date. If you answer 'Y' it will then select an appropriate chart to display the new selection and offer again the ability to chart other planets in that new star chart, with or without grid lines.

Other alternatives available are to ask for another date for the same planet, or another date and another planet.

You can also determine, when selecting the date, whether you require a series of plots. If you do not, insert 1's for the interval in days and the number of plots required. If you require a series, you must input the number of plots and the time interval (in days) between each plot. If your series of plots runs off the screen, the monitor will display the name of the planet and the message OFF CHART for every plot that is off the chart. In this series mode you also have the choices of adding all other planets and the Sun in the same star chart region to show their movements during the period. Figures 17A.1 through 17A.5 illustrate the capabilities of this version of the planet finding program.

```
ENTER THE DATE

THE YEAR? 1982

THE MONTH? 2

THE DAY? 28

TIME ZONE? 8

TIME HRS? 22


Select Intervals and How Many
          Enter 1's if You Only Need One Plot

NOTE: Position of Moon shown for first interval only.

WHAT IS TIME INTERVAL (DAYS)? 1

HOW MANY INTERVALS? 1_
```

On this first display you are asked to enter the date and the interval in days if you want more than one plot. Next you enter the number of plots required.

— Figure 17A.1 —

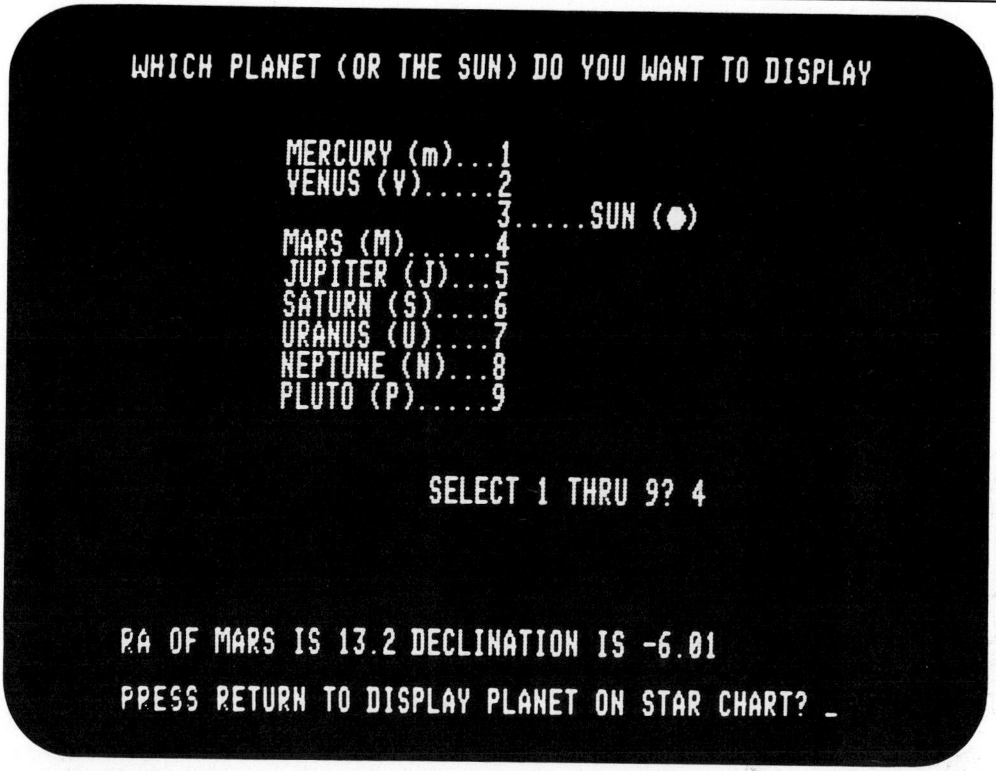

WHICH PLANET (OR THE SUN) DO YOU WANT TO DISPLAY

```
        MERCURY (m)...1
        VENUS (V).....2
                    3.....SUN (●)
        MARS (M)......4
        JUPITER (J)...5
        SATURN (S)....6
        URANUS (U)....7
        NEPTUNE (N)...8
        PLUTO (P).....9


            SELECT 1 THRU 9? 4



RA OF MARS IS 13.2 DECLINATION IS -6.01

PRESS RETURN TO DISPLAY PLANET ON STAR CHART? _
```

The program then displays the names of the planets and asks you to select one. After a short delay it provides the right ascension and declination of the selected planet.

Figure 17A.2

The program will need modification to suit each computer's display format. The parts of the program requiring modification are those that display the constellations and insert (POKE) the planets, the Sun, and the Moon in the displayed star chart.

For example, a display that has a format of only 40 full characters across the screen (PET, Atari) will have to use 10 characters for 1 hour of right ascension instead of the 15 used in this program. A computer that has only 16 character lines (some TRS-80 models) will have to reduce the vertical size of the star charts and eliminate some of the constellations; that is, it must accept star charts that are plus or minus 7 degrees of declination about the mean ecliptic position in the chart area.

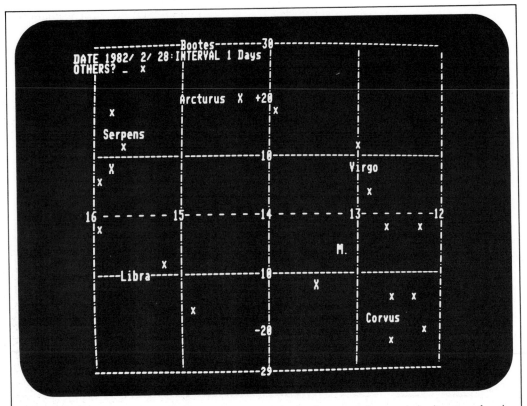

A new display is then generated showing a grid of right ascension and declination for the region of the star sphere in which the planet is located. The stars are plotted on the grid and the constellations are named. Finally, the planet is plotted at its correct position among the stars.

Figure 17A.3

The POKEs for the planets are developed in the routine 2480 through 2590, which provides *PDE,* the planet POKE number. Instruction 2870 sets up S, which is the POKE constant that can be varied to suit different machines. S is the number to POKE that will place a character on the extreme bottom right corner of your computer's display. The POKE numbers for planets assume a 64-character-wide screen. For a 40-character-wide screen, instruction 2590 (which develops the screen coordinate for declination) must be changed to

$$PDE = DE \times 40$$

The right ascension coordinate (instruction 2730) also must be changed to

$$RA = RA \times 10$$

The POKE numbers for right ascension will also have to be changed for a 40-character-wide display. Instructions 3810, 4880, and 5640 must all be changed to FOR J=1 TO 41 STEP10; instruction 4540 must be changed to FOR J=1 TO 21 STEP10; instruction 4590 must be changed to FOR J=31 TO 41 STEP10; and instruction 5250 must be changed to FOR J=11 TO 41 STEP10.

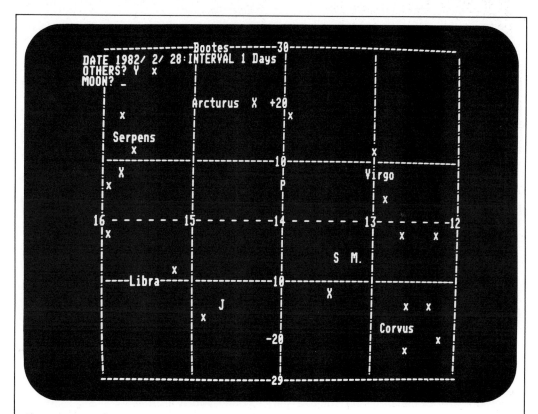

If you answer the question 'OTHERS?' with 'Y' all other planets in the same star chart region are also plotted, as shown. You can ask for the Moon to be displayed as well. If you wish you can also have the star chart displayed without the grid lines.

Figure 17A.4

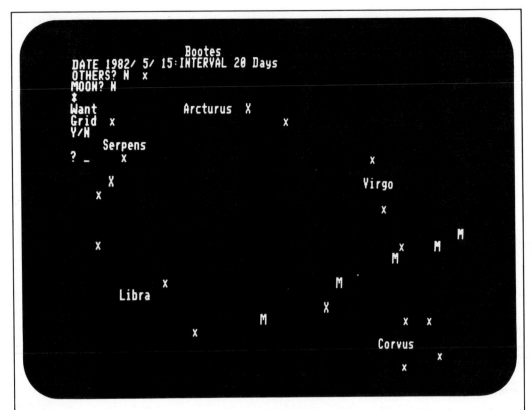

As with Program 17, you can obtain a series of plots showing the motion of the planet relative to the stars over a period of time. Mars is plotted in this display.

Figure 17A.5

Modifications are somewhat more complicated for a monitor with only 15 text lines, since the full star charts cannot be displayed. A Table of Right Ascensions and Declinations for suitable stars to build your star charts from is provided for such a computer (see pages 266–267). This will affect all the routines to plot charts, namely, instructions 3770 through 6000, the routines to plot the grid lines (2840 through 3010), and the figures of declination on the grid (3030 through 3100). Sufficient remark statements have been included in these routines for you to select those routines that need to be modified to suit your computer's display capabilities.

For a computer with a display monitor showing only 16 lines, it is suggested that the charts be made to show the following declination ranges:

Chart 1: Declination −1 to 14
Chart 2: Declination 12 to 28
Chart 3: Declination −1 to 14
Chart 4: Declination 1 to −14
Chart 5: Declination −12 to −28
Chart 6: Declination 1 to −14

This will eliminate all stars in the Table that have declinations outside these ranges. The effect will be to display only the zodiacal constellations, as in the Apple version given in Program 17. Additional resolution can be obtained by providing one chart for each zodiacal constellation instead of the six used in this version.

The easiest way to make the changes is to draw a screen map of your computer monitor and put the POKE numbers on it for the four corners of your screen. Then make six tracings of the screen, dividing it into four equal parts vertically and (for a 16-line-deep display) two parts horizontally. By placing a tracing of a star chart generated from the table of right ascensions and declinations over the screen sketch you can work out the POKE numbers to be deducted from the value you have selected for S. Then you can rewrite the instructions for the six charts of the zodiac. Also, you must check the ASCII equivalents on your computer for the characters to be POKED. For the Sun, for example, you can use a zero in place of the graphic shown in the program. Refer also to Program 17.

The basic calculations of planetary positions, as well as the subroutines and flags used to produce the different types of displays and select the chart zones, will operate in connection with any new display formats you wish to develop. The program has been sprinkled with REM statements to help you modify it to suit your system.

The listing of the PLNTA program follows.

Table of Right Ascensions and Declinations for Constellations Displayed by PLNTA Program

Chart 1: RA 0 to RA 4

Pegasus	0.1,	29	0.2,	14		
Pleiades	3.6,	24	(five close stars)			
Aries	2.1,	23	1.8,	21	1.8,	19
Taurus	3.0,	4	2.7,	3		
Eridanus	3.9,	−13	3.3,	−20		
Pisces	1.5,	−9	1.2,	−10		
Cetus	0.7,	−18				

Chart 2: RA 4 to RA 8

Orion	5.9,	8	5.4,	8	5.7,	−2	5.6,	−1
	5.5,	0	5.8,	−10	5.6,	−6	5.2,	−9
Sirius	6.7,	−17	6.3,	−18				
Taurus	4.5,	17	5.4,	29	5.6,	21		
Procyon	7.6,	7	7.4,	9				
Gemini (dropped slightly to be on chart)								
	7.5,	30	7.7,	28	7.3,	22	6.7,	25
	6.6,	16	6.4,	22	6.3,	22		

Chart 3: RA 8 to RA 12

Leo	10.1,	12	10.1,	17	10.3,	20	10.3,	24
	11.2,	20	11.2,	16	11.8,	15	9.8,	28
	9.7,	26						
Cancer	8.7,	29	8.6,	21				
Hydra	9.5,	−9	8.7,	7	8.9,	7	9.2,	2
	10.4,	−17						
Virgo	11.8,	2						
Crater	10.8,	−16	10.9,	−18	11.3,	−15	11.4,	−18

Table of Right Ascensions and Declinations for Constellations Displayed by PLNTA Program

Chart 4: RA 12 to RA 16

Virgo	13.4,	−11	13.0,	11	12.9,	3	12.7,	−1
	12.3,	−1	13.1,	−5				
Corvus	12.5,	−16	12.2,	−17	12.5,	−23	12.2,	−22
Bootes	14.2,	20	13.9,	19	14.7,	27	15.5,	27
	15.4,	29						
Serpens	15.8,	17	15.5,	10	15.7,	7	15.8,	5
	15.8,	−3						
Libra	15.3,	−9	14.8,	−16				

Chart 5: RA 16 to RA 20

Aquila	19.7,	10	19.8,	9	19.9,	8	19.0,	13
	18.9,	14						
Ophiuchus	17.5,	12	17.2,	25	17.6,	5	17.7,	3
Sagittarius	18.3,	−30	18.0,	−30	18.4,	−25	18.9,	−26
	19.0,	−30	19.1,	−21	18.3,	−21		
Scorpio	16.5,	−26	16.6,	−28	16.4,	−24	16.0,	−20
	15.9,	−22	15.9,	−26				

Chart 6: RA 20 to RA 24

Fomalhaut	22.9,	−30						
Capricornus	21.7,	−18	21.6,	−18	21.4,	−22	20.8,	−28
	20.7,	−26	20.3,	−14	20.2,	−12		
Delphinus	20.5,	11	20.6,	15	20.7,	15	20.6,	16
	20.8,	16						
Pegasus	21.7,	10	22.2,	6	22.7,	10	23.0,	4
	23.0,	28						
Aquarius	22.6,	0	22.5,	0	22.4,	1	22.3,	−2
	22.0,	0	21.5,	−6				

```
┌── PLNTA ─────────────────────────────────────────────

      10 CLEAR
      20 CLEAR(150)
      30 DIMPD(9,9)
      40 DEFFNACO(X)=-ATN(X/SQR(-X*X+1))+1.5707963
      50 DEFFNRAD(X)=.01745328*(X)
      60 DEFFNDEG(X)=57.29578*(X)
      70 PRINTCHR$(12):PRINT:PRINT:PRINT:PRINT:PRINT
      80 PRINTTAB(14)"ASTRONOMY PROGRAM FOR SORCERER":PRINT:PRINT
      90 PRINTTAB(14)"     ----------------"
     100 PRINTTAB(14)"*-*-* | PLANET FINDER | *-*-*"
     110 PRINTTAB(14)"     ----------------":PRINT
     120 PRINTTAB(17)"By ERIC BURGESS F.R.A.S."
     130 PRINT:PRINT
     140 PRINTTAB(18)"All rights reserved by"
     150 PRINT
     160 PRINTTAB(18)"S & T Software Service"
     165 PRINTTAB(10)"13361 Frati Lane Sebastopol CA 95472"
     170 PRINT:PRINT:PRINT
     180 PRINTTAB(22)"(Version 4.82)"
     190 PRINT:PRINT:PRINT:PRINT:PRINT
     200 INPUT"DO YOU WANT INSTRUCTIONS (Y/N)";A$
     210 IFA$="N"THEN630
     220 IFA$<>"Y"THENPRINT:PRINT"INVALID REPLY":GOTO190
     230 PRINTCHR$(12)
     240 PRINTTAB(12)"THIS PROGRAM PLACES A PLANET (OR THE SUN)"
     245 PRINTTAB(25)"AND THE MOON"
     250 PRINTTAB(16)"   AMONG THE CONSTELLATIONS"
     260 PRINT
     270 PRINTTAB(18)"FOR THE DATE WHICH YOU INPUT":PRINT
     280 PRINTTAB(17)"AND A SERIES OF TIME INTERVALS":PRINT
     290 PRINTTAB(11)"The program selects a zone of the celestial"
     300 PRINTTAB(11)"sphere 4 hours of right ascension wide and"
     310 PRINTTAB(11)"from 30 deg. north of the celestial equator"
     320 PRINTTAB(11)"to 29 deg. south declination. It plots the"
     330 PRINTTAB(11)"selected planet's position relative to the"
     340 PRINTTAB(11)"stars of the constellations, naming these and"
     350 PRINTTAB(11)"the brightest stars. If you answer the"
     360 PRINTTAB(11)"question 'OTHERS' with 'Y', other planets"
     370 PRINTTAB(11)"in the same chart region are plotted also."
     380 PRINTTAB(11)"The Sun is plotted if it's in the same region."
     385 PRINT
     390 PRINTTAB(11)"A grid of right ascension and declination"
     400 PRINTTAB(11)"is printed, but you can delete this grid"
     410 PRINTTAB(11)"and later bring it back again if you wish."
     420 PRINTTAB(11)"If you have the grid plotted and you answer"
     430 PRINTTAB(11)"the 'Do You Want Grid' question with 'Y'"
     440 PRINTTAB(11)"you will end the display and can then"
     450 PRINTTAB(11)"pick another planet or another date."
     460 PRINT:PRINT
     470 INPUT"WHEN READY TO CONTINUE PRESS RETURN KEY";A$
     480 PRINTCHR$(12):PRINT:PRINT:PRINT:PRINT:PRINT:PRINT
     490 PRINTTAB(11)"Right ascension is within 1 degree on chart,"
     500 PRINTTAB(11)"declination is within 2 degrees. But"
     510 PRINTTAB(11);"accurate right ascension and declination"
```

PLNTA *(continued)*

```
520 PRINTTAB(11)"are given for the selected planet before"
530 PRINTTAB(11)"the chart is displayed."
540 PRINT:PRINTTAB(11);"If in a series of plots the selected"
550 PRINTTAB(11)"planet moves off the chart, this is indicated."
555 PRINT
560 PRINTTAB(11)"The date printed on the chart is the first"
570 PRINTTAB(11);"date in a series of plotted positions."
575 PRINT
576 PRINTTAB(11)"If you answer the question MOON? with 'Y'"
577 PRINTTAB(11)"the position of the Moon for the date"
578 PRINTTAB(11)"requested is shown; ) before full moon,"
579 PRINTTAB(11)"o when close to full, and ( afterward.":PRINT
580 PRINTTAB(11)"NOTE: If two planets are close together and"
590 PRINTTAB(11)"'OTHERS' are requested, the outer planet"
600 PRINTTAB(11)"will overprint the inner planet."
610 PRINT:PRINT:PRINT
620 INPUT"PRESS RETURN TO CONTINUE";A$
630 PRINTCHR$(12):PRINT:PRINT:PRINT:PRINT
640 PRINT"ENTER THE DATE":PRINT
650 INPUT"THE YEAR";YD$:Y=VAL(YD$)
660 IFY=0THENPRINT"INVALID ENTRY":PRINT:GOTO650
670 IFY>1800GOTO730
680 PRINT"IS ";Y;" THE CORRECT YEAR?"
690 INPUTY$
700 IFY$="Y"THEN730
710 IFY$<>"N"THENPRINT"INVALID ENTRY":PRINT:GOTO680
720 IFY$="N"THENPRINT:GOTO650
730 PRINT:INPUT"THE MONTH";MD$:M=VAL(MD$)
740 IFM=0ORM>12THENPRINT"INVALID ENTRY":PRINT:GOTO730
750 PRINT:INPUT"THE DAY";DD$:D=VAL(DD$):DD=D
751 PRINT:INPUT"TIME ZONE";TZ
752 PRINT:INPUT"TIME HRS";TI
753 TI=TZ+TI:IFTI>24THEN TI=TI-24:D=D+1
760 REM STORES INITIAL DATE IN D2,M2,Y2
770 IFF9=1THEN790
780 T2=TI:D2=D:Y2=Y:M2=M
790 IFD=0ORD>31THENPRINT"INVALID ENTRY":PRINT:GOTO750
800 IFM=2ANDD>29THENPRINT"INVALID ENTRY":PRINT:GOTO750
810 PRINT:PRINT:PRINT
820 PRINT "Select Intervals and How Many"
830 PRINT "        Enter 1's if You Only Need One Plot"
835 PRINT:PRINT"NOTE: Moon's Position shown for first interval";
836 PRINT" only."
840 PRINT:PRINT
850 INPUT"WHAT IS TIME INTERVAL (DAYS)";TI$:PRINT
860 TI=VAL(TI$):IFTI=0THENPRINT"INVALID REPLY":GOTO850
870 INPUT"HOW MANY INTERVALS";IN$:PRINT
880 IN=VAL(IN$):IFIN=0THENPRINT"INVALID REPLY":GOTO870
890 REM SETS INTERVAL COUNT AT 1
900 NC=1
910 REM CALCULATING DAYS FROM 1960,1,1 EPOCH TO DATE
920 PRINTCHR$(12):PRINT"PLEASE WAIT"
930 REM FROM EPOCH 1960,1,1
940 IFM>=3GOTO990
950 REM CALCS IF MONTH IS JAN OR FEB
```

PLNTA *(continued)*

```
960  DG=365*Y+D
970  DG=DG+((M-1)*31)+INT((Y-1)/4)-INT((.75)*INT((Y-1)/100+1))
980  GOTO1020
990  REM CALS FOR MAR THRU DEC
1000 DG=365*Y+D+((M-1)*31)-INT(M*.4+2.3)
1010 DG=DG+INT(Y/4)-INT((.75)*INT((Y/100)+1))
1020 NI=DG-715875
1025 NM=NI-.5
1030 REM JUMPS PLANETARY INPUTS IF NEW INTERVAL
1040 IFF9=1THEN1690
1050 REM INPUT OF PLANETARY DATA,ORBITAL PARAMETERS, ETC.
1060 REM JUMPS PLANETARY INPUTS IF NEW DATE ETC
1070 IFFL=1GOTO1440
1080 IFF8=1GOTO1460
1090 IFF6=1GOTO1460
1100 RESTORE
1110 FORYP=0TO8:FORXP=0TO8
1120 READPD(YP,XP)
1130 NEXTXP,YP
1140 REM MERCURY
1150 DATA.071422,3.8484,.388301,1.34041,.3871,.07974,2.73514
1160 DATA.122173,.836013
1170 REM VENUS
1180 DATA.027962,3.02812,.013195,2.28638,.7233,.00506,3.85017
1190 DATA.059341,1.33168
1200 REM EARTH FOR SUN
1210 DATA.017202,1.74022,.032044,1.78547,1,.017,3.33926
1220 DATA0,0
1230 REM MARS
1240 DATA.009146,4.51234,.175301,5.85209,1.5237,.141704
1250 DATA1.04656,.03142,.858702
1260 REM JUPITER
1270 DATA.001450,4.53364,.090478,.23911,5.2028,.249374
1280 DATA1.76188,.01972,1.74533
1290 REM SATURN
1300 DATA.000584,4.89884,.105558,1.61094,9.5385,.534156
1310 DATA3.1257,.043633,1.977458
1320 REM URANUS
1330 DATA.000205,2.46615,.088593,2.96706,19.182,.901554
1340 DATA4.49084,.01396,1.28805
1350 REM NEPTUNE
1360 DATA.000104,3.78556,.016965,.773181,30.06,.270540
1370 DATA2.33498,.031416,2.29162
1380 REM PLUTO
1390 DATA.000069,3.16948,.471239,3.91303,39.44,9.86
1400 DATA5.23114,.300197,1.91812
1410 FORI9=1TO9:READP$(I9):NEXTI9
1420 DATAMERCURY,VENUS,SUN,MARS
1430 DATAJUPITER,SATURN,URANUS,NEPTUNE,PLUTO
1440 F=0
1450 IF F9=2THEN1470
1460 FL=0
1470 REM CALCULATE DATA FOR PLANETS
1480 IFF9=2THEN1690
1490 :
```

─── **PLNTA** *(continued)* ───

```
1500 IFF8=1THENF8=0:GOTO1690
1510 PRINTCHR$(12)
1520 PRINT:PRINT:PRINT
1530 PRINT"WHICH PLANET (OR THE SUN) DO YOU WANT TO DISPLAY"
1540 PRINT:PRINT
1550 PRINTTAB(10)"MERCURY (m)...1"
1560 PRINTTAB(10)"VENUS (V).....2"
1570 PRINTTAB(10)"            3.....SUN ()"
1580 PRINTTAB(10)"MARS (M)......4"
1590 PRINTTAB(10)"JUPITER (J)...5"
1600 PRINTTAB(10)"SATURN (S)....6"
1610 PRINTTAB(10)"URANUS (U)....7"
1620 PRINTTAB(10)"NEPTUNE (N)...8"
1630 PRINTTAB(10)"PLUTO (P).....9"
1640 PRINT
1650 PRINT:PRINTTAB(20):INPUT"SELECT 1 THRU 9";PS$
1660 PS=VAL(PS$)
1670 IFPS=0ORPS>9THENPRINT"INVALID SELECTION":GOTO1650
1680 REM STORES PLANET SELECTED IN P2
1690 P2=PS
1700 I=1
1710 FORJ=0TO8:GOSUB1880
1720 A(I)=A:D(I)=DS:L(I)=L
1730 I=I+1:NEXT
1740 FORI=1TO9
1750 REM SKIP EARTH
1760 IFI=3THENNEXT
1770 GOSUB2030
1780 Q(I)=Q:X(I)=X:R(I)=R:V(I)=V
1790 NEXT
1800 FORI=1TO9:A(I)=FNDEG(A(I))
1810 IFI=3THENNEXT
1820 NEXT
1830 I=PS
1840 R(3)=(A(3)-180)/15
1850 IFR(3)<0THENR(3)=R(3)+24
1860 V(3)=(SIN(FNRAD(A(3)-180)))*23.44194
1870 GOTO2240
1880 REM CALCULATE A,DS AND L
1890 REM EXPLAIN
1900 REM CALC HELIOCENTRIC LONGITUDE A
1910 A=NI*PD(J,0)+PD(J,1)
1920 IFA>6.28318THENA=((A/6.28318)-INT(A/6.28318))*6.28318
1930 IFA<0THENA=A+6.28318:GOTO1930
1940 C=PD(J,2)*SIN(A-PD(J,3))
1950 A=A+C
1960 IFA>6.28318THENA=A-6.28318
1970 IFA<0THENA=A+6.28318:GOTO1970
1980 REM CALC DIST OF PLANET FROM SUN DS
1990 DS=PD(J,4)+PD(J,5)*SIN(A-PD(J,6))
2000 REM CALC DISTANCE OF PLANET FROM ECLIPTIC L
2010 L=PD(J,7)*SIN(A-PD(J,8))
2020 RETURN
2030 REM CALCULATE Z,Q,X,T,R,V
2040 REM CALC ANGULAR DIST OF PLANET FROM SUN Z
```

── **PLNTA** *(continued)* ──────────

```
2050 Z=A(3)-A(I)
2060 IFABS(Z)>3.14159ANDZ<0THENZ=Z+6.28318
2070 IFABS(Z)>3.14159ANDZ>0THENZ=Z-6.28318
2080 REM CALC DIST OF PLANET FROM EARTH Q
2090 Q=SQR(D(I)^2+D(3)^2-2*D(I)*D(3)*COS(Z))
2100 REM CALC ANGULAR DIST OF PLANET FROM SUN X
2110 PP=(D(I)+D(3)+Q)/2
2120 X=2*FNACO(SQR(((PP*(PP-D(I)))/(D(3)*Q))))
2130 REM CALC RIGHT ASCENSION R
2140 IFZ<0THENR=FNDEG(A(3)+3.14159-X)/15
2150 IFZ>0THENR=FNDEG(A(3)+3.14159+X)/15
2160 IFR>24THENR=R-24:GOTO2160
2170 IFR<-24THENR=R+24:GOTO2170
2180 IFR<0THENR=R+24:GOTO2180
2190 REM CALCULATE DECLINATION V
2200 IFZ<0THENV=SIN(A(3)+3.14159-X)*23.44194+FNDEG(L(I))
2210 IFZ>0THENV=SIN(A(3)+3.14159+X)*23.44194+FNDEG(L(I))
2220 X=FNDEG(X)
2230 RETURN
2240 RA=R(PS)
2250 DE=V(PS)
2260 REM JUMPS PRINTING RA AND DEC IF NEW INTERVAL
2270 IFF9=1THEN2360
2280 IFF9=2THEN2360
2290 PRINT:PRINT:PRINT:PRINT
2300 RA$=STR$(RA):DE$=STR$(DE)
2310 RA$=LEFT$(RA$,5):DE$=LEFT$(DE$,5)
2320 PRINT"RA OF ";P$(PS);" IS";RA$;" DECLINATION IS ";DE$
2330 PRINT
2340 INPUT"PRESS RETURN TO DISPLAY PLANET ON STAR CHART";A$
2350 REM STORES RA FOR SELECTED PLANET IN R2
2360 R3=RA:DC=DE
2370 GOSUB2380:GOTO2480
2380 IFPS=1THENP=109:GOTO2470
2390 IFPS=2THENP=86:GOTO2470
2400 IFPS=3THENP=132:GOTO2470
2410 IFPS=4THENP=77:GOTO2470
2420 IFPS=5THENP=74:GOTO2470
2430 IFPS=6THENP=83:GOTO2470
2440 IFPS=7THENP=85:GOTO2470
2450 IFPS=8THENP=78:GOTO2470
2460 IFPS=9THENP=80:GOTO2470
2470 RETURN
2480 REM DEVELOP POKE FOR RA AND DEC
2490 IFF9=10RF5=1GOTO2510
2500 PN=P
2510 REM AND SELECT CHART FOR DISPLAY BY SETTING CH
2520 GOSUB2560
2530 IFF9=1THEN6380
2540 IF(F7=1ANDF5=1)THENRETURN
2550 GOTO2800
2560 REM POKE FOR DEC
2570 DE=(DE+29)/2
2580 DE=INT(DE)
2590 PDE=DE*64
```

PLNTA (continued)

```
2600 REM JUMPS CHANNEL SELECTION IF POKING OTHER PLANETS
2610 IFF9=1ORF5=1THEN2680
2620 IFRA>20ANDRA<23.99999THENRA=RA-20:CH=6:GOTO2710
2630 IFRA>16ANDRA<19.99999THENRA=RA-16:CH=5:GOTO2710
2640 IFRA>12ANDRA<15.99999THENRA=RA-12:CH=4:GOTO2710
2650 IFRA>8ANDRA<11.99999THENRA=RA-8:CH=3:GOTO2710
2660 IFRA>4.00001ANDRA<7.99999THENRA=RA-4:CH=2:GOTO2710
2670 CH=1
2680 IFF9=1THENGOSUB6530
2690 IFF5=1THENGOSUB6530
2700 IFF7=1THENRETURN
2710 IFF9=1THENGOSUB6470
2720 IFF5=1THENGOSUB6470
2730 RA=RA*15
2740 IF(RA-INT(RA))>.49THENRA=INT(RA)+1:GOTO2760
2750 RA=INT(RA)
2760 PL=RA+PDE
2770 IFF5=1THEN2790
2780 PX=PL
2790 RETURN
2800 REM IN EACH CHART POKE PL,P
2810 REM WHERE P IS NAME OF PLANET
2820 REM SUCH AS m,V,M,J,S
2830 PRINTCHR$(12)
2840 REM POKE HORIZONTALS
2850 IFFK=1THENFK=0:GOTO2870
2860 IFF9=2THEN3360
2870 S=2049
2880 HZ=S+1917
2890 FORJ=1TO60:POKE(-HZ+J),45:NEXTJ
2900 HZ=HZ-640
2910 IFHZ>2108GOTO2890
2920 FORJ=1TO61:POKE(-S-61+J),45:NEXTJ
2930 REM POKE VERTICALS
2940 FORK=1TO28
2950 FORJ=1TO61STEP15
2960 POKE(-S-1917+K*64+J),33
2970 NEXTJ
2980 NEXTK
2990 REM POKE EQUATOR
3000 FORJ=1TO61STEP2:POKE(-S-957+J),45
3010 NEXTJ
3020 REM POKE COORDINATES
3030 REM SET DE TO HIGHEST DECLINATION ON TOP OF CHART
3040 POKE-S-1887,51:POKE-S-1886,48
3050 POKE-S-1568,43:POKE-S-1567,50:POKE-S-1566,48
3060 POKE-S-1247,49:POKE-S-1246,48
3070 POKE-S-608,45:POKE-S-607,49:POKE-S-606,48
3080 POKE-S-288,45:POKE-S-287,50:POKE-S-286,48
3090 POKE-S-31,50:POKE-S-30,57
3100 GOTO3360
3110 PRINTCHR$(12):PRINT:PRINT:PRINT:PRINT:PRINT
3120 PRINT"DO YOU WANT ANOTHER PLANET, SAME DATE  'Y' OR 'N'"
3130 PRINT:PRINT:INPUT"...........";A$
3140 IFA$="N"THEN3190
```

PLNTA *(continued)*

```
3150 IFA$<>"Y"THENPRINT"INVALID REPLY":PRINT:GOT03120
3160 REM RESET Y,M,D TO ORIGINAL SELECTION
3170 Y=Y2:M=M2:D=D2
3180 PRINTCHR$(12):PRINT:PRINT:PRINT:GOT0920
3190 PRINTCHR$(12)
3200 PRINT:PRINT:PRINT:PRINT:PRINT
3210 PRINT"DO YOU WANT ANOTHER DATE FOR SAME PLANET 'Y' OR 'N'"
3220 PRINT:PRINT:INPUT"..........";A$
3230 IFA$="N"THENPRINTCHR$(12):PRINT:PRINT:GOT03270
3240 IFA$<>"Y"THENPRINT"INVALID REPLY":PRINT:GOT03210
3250 F8=1:GOT0630
3260 PRINT
3270 PRINT:PRINT"DO YOU WANT ANOTHER DATE AND ANOTHER PLANET"
3280 PRINT
3290 INPUT"...............";A$
3300 IFA$="N"THENPRINTCHR$(12):GOT03350
3310 IFA$<>"Y"THENPRINT"INVALID REPLY":PRINT:GOT03270
3320 PRINTCHR$(12)
3330 FL=0:F1=0:F2=0:F3=0:F4=0:F5=0:F6=0:F7=0:F9=0:F8=0
3340 FX=0:GOT0630
3350 END
3360 REM SELECT CHART REGION CH
3370 ONCHGOSUB3770,4200,4510,4850,5210,5610
3380 REM POKE OTHER PLANETS
3390 YT$=STR$(Y):MT$=STR$(M):DT$=STR$(DD)
3400 PT$="/"
3410 DA$=YT$+PT$+MT$+PT$+DT$
3420 PRINT"DATE";DA$;:PRINT":INTERVAL";TI;"Days"
3430 INPUT"OTHERS";A$
3440 IFA$="N"THENF6=1:GOT03491
3450 IFA$<>"Y"GOT03430
3460 IFA$="Y"THENF5=1
3470 F6=0
3480 REM SUB TO POKE OTHER PLANETS
3490 GOSUB6010
3491 INPUT"MOON";A$
3492 IFA$="Y"THENGOSUB7000
3500 REM CHECKS IF MORE CYCLES NEEDED
3510 IFNC<INGOT03540
3520 IFNC=>INGOT03570
3530 REM SUB TO INCREASE DATE BY INTERVAL AND REPEAT PROGRAM
3540 GOSUB6280
3550 IFF9=1THEN3510
3560 REM RESETS F9 WHEN ALL CYCLES COMPLETED
3570 :
3580 REM RESETS NC,D,M,ANDY TO INITIAL VALUES
3590 NC=1:D=D2:M=M2:Y=Y2
3600 PS=P2:RA=R3:DE=DC
3610 P=PN:PL=PX
3620 FORK=1TO500:NEXTK
3630 FORJ=1TO5:PRINT"*":FORK=1TO100:NEXT
3640 PRINT"Want"
3650 PRINT"Grid"
3660 PRINT"Y/N"
3670 PRINT:INPUTDG$
```

─ **PLNTA** *(continued)* ─

```
3680  IF(DG$="N"ANDF9=1)THENFL=1:F9=2:PS=P2:P=PN:GOTO910
3690  IFF3=1ANDDG$="N"THENF3=0:GOTO3110
3700  IFFL=1ANDDG$="N"THENF3=1
3710  IFDG$="N"THENPRINTCHR$(12)
3720  IFDG$="N"THENFL=1:PS=P2:P=PN:GOTO3370
3725  IF(DG$="Y"ANDF9=1ANDFL=0)THENF9=0:GOTO3760
3730  IF(DG$="Y"ANDF9=1)THENFL=0:FK=1:PS=P2:P=PN:F9=2:GOTO910
3740  IFFL=1ANDDG$="Y"THENFL=0:F2=1:GOTO2830
3750  IF DG$<>"Y"THEN3660
3760  GOTO3110
3770  REM CHART 1 RA AND STARS
3780  IFFL=1THENGOTO3870
3790  RA=52:GOSUB3810
3800  GOTO3870
3810  FORJ=1TO61STEP15
3820  POKE(-S-957+J),RA
3830  RA=RA-1
3840  NEXTJ
3850  RETURN
3860  REM PEGASUS
3870  POKE-S-1793,88:POKE-S-1153,88
3880  POKE-S-1544,80:POKE-S-1543,101:POKE-S-1542,103
3890  POKE-S-1541,97:POKE-S-1540,115:POKE-S-1539,117
3900  POKE-S-1538,115
3910  REM PLEIADES
3920  POKE-S-1723,46:POKE-S-1722,46:POKE-S-1719,46
3930  POKE-S-1785,46:POKE-S-1656,46
3940  POKE-S-1655,80:POKE-S-1654,108:POKE-S-1653,101
3950  POKE-S-1652,105:POKE-S-1651,97:POKE-S-1650,100
3960  POKE-S-1649,101:POKE-S-1648,115
3970  REM ARIES
3980  POKE-S-1634,88:POKE-S-1564,88:POKE-S-1499,120
3990  POKE-S-1697,65:POKE-S-1696,114:POKE-S-1695,105
4000  POKE-S-1694,101:POKE-S-1693,115
4010  POKE-S-1133,120:POKE-S-1067,120
4020  POKE-S-306,120
4030  REM TAURUS
4040  POKE-S-1207,84:POKE-S-1206,97:POKE-S-1205,117
4050  POKE-S-1204,114:POKE-S-1203,117:POKE-S-1202,115
4060  POKE-S-571,120
4070  REM ERIDANUS
4080  POKE-S-438,69:POKE-S-437,114:POKE-S-436,105
4090  POKE-S-435,100:POKE-S-434,97:POKE-S-433,110
4100  POKE-S-432,117:POKE-S-431,115
4110  REM PISCES
4120  POKE-S-786,80:POKE-S-785,105:POKE-S-784,115
4130  POKE-S-783,99:POKE-S-782,101:POKE-S-781,115
4140  POKE-S-331,88:POKE-S-594,120:POKE-S-662,120
4150  REM CETUS
4160  POKE-S-407,67:POKE-S-406,101:POKE-S-405,116
4170  POKE-S-404,117:POKE-S-403,115
4180  POKE-S-PL,P
4190  RETURN
4200  REM CHART2 RA AND STARS
4210  IFFL=1GOTO4250
```

─── **PLNTA** *(continued)* ───

```
4220 RA=56
4230 GOSUB3810
4240 REM ORION
4250 POKE-S-537,79:POKE-S-536,114:POKE-S-535,105
4260 POKE-S-534,111:POKE-S-533,110
4270 POKE-S-919,88:POKE-S-856,88:POKE-S-793,88
4280 POKE-S-1117,88:POKE-S-1174,88
4290 POKE-S-602,88:POKE-S-659,88
4300 POKE-S-727,42
4310 POKE-S-1241,120
4320 REM SIRIUS
4330 POKE-S-236,115:POKE-S-237,117:POKE-S-238,105
4340 POKE-S-239,114:POKE-S-240,105:POKE-S-241,83
4350 POKE-S-170,88:POKE-S-163,120
4360 REM TAURUS
4370 POKE-S-1416,88:POKE-S-1813,88:POKE-S-1560,88
4380 POKE-S-1802,84:POKE-S-1801,97:POKE-S-1800,117
4390 POKE-S-1799,114:POKE-S-1798,117:POKE-S-1797,115
4400 REM PROCYON
4410 POKE-S-1267,120:POKE-S-1142,88
4420 POKE-S-1080,80:POKE-S-1079,114:POKE-S-1078,111
4430 POKE-S-1077,99:POKE-S-1076,121:POKE-S-1075,111
4440 POKE-S-1074,110:POKE-S-1912,88
4450 POKE-S-1785,88:POKE-S-1699,88:POKE-S-1512,88
4460 POKE-S-1650,120
4470 POKE-S-1842,71:POKE-S-1841,101:POKE-S-1840,109
4480 POKE-S-1839,105:POKE-S-1838,110:POKE-S-1837,105
4490 POKE-S-PL,P
4500 RETURN
4510 REM CHART 3 RA AND STARS
4520 IFFL=1GOTO4630
4530 RA=50:RT=49
4540 FORJ=1TO031STEP15
4550 POKE(-S-958+J),RT:POKE(-S-957+J),RA
4560 RA=RA-1
4570 NEXTJ
4580 RA=57
4590 FORJ=46TO061STEP15
4600 POKE(-S-957+J),RA
4610 RA=RA-1
4620 NEXTJ
4630 REM LEO
4640 POKEPL,P
4650 POKE-S-1522,76:POKE-S-1521,101:POKE-S-1520,111
4660 POKE-S-1465,120:POKE-S-1585,120:POKE-S-1457,120
4670 POKE-S-1628,120:POKE-S-1694,120:POKE-S-1635,46
4680 POKE-S-1571,120:POKE-S-1504,46:POKE-S-1313,88
4690 REM CANCER
4700 POKE-S-1680,67:POKE-S-1679,97:POKE-S-1678,110
4710 POKE-S-1677,99:POKE-S-1676,101:POKE-S-1675,114
4720 POKE-S-1548,120:POKE-S-1421,120
4730 REM HYDRA
4740 POKE-S-795,72:POKE-S-794,121:POKE-S-793,100
4750 POKE-S-792,114:POKE-S-791,97:POKE-S-365,120
4760 POKE-S-1164,120:POKE-S-1166,120:POKE-S-979,120
```

PLNTA *(continued)*

```
4770 POKE-S-599,120:POKE-S-545,120
4780 REM CRATER
4790 POKE-S-496,67:POKE-S-495,114:POKE-S-494,97
4800 POKE-S-493,116:POKE-S-492,101:POKE-S-491,114
4810 POKE-S-499,120:POKE-S-427,120:POKE-S-373,120
4820 POKE-S-365,120
4830 POKE-S-PL,P
4840 RETURN
4850 REM CHART 4 FOR RA AND STARS
4860 IFFL=1GOTO4910
4870 RA=54:RT=49
4880 FORJ=1TO61STEP15
4890 POKE(-S-958+J),RT:NEXTJ
4900 GOSUB3810
4910 REM BOOTES
4920 POKE-S-1901,66:POKE-S-1900,111:POKE-S-1899,111
4930 POKE-S-1898,116:POKE-S-1897,101:POKE-S-1896,115
4940 POKE-S-1780,120:POKE-S-1842,120:POKE-S-1835,120
4950 POKE-S-1571,88:POKE-S-1501,120
4960 POKE-S-1581,65:POKE-S-1580,114:POKE-S-1579,99
4970 POKE-S-1578,116:POKE-S-1577,117:POKE-S-1576,114
4980 POKE-S-1575,117:POKE-S-1574,115
4990 REM SERPENS
5000 POKE-S-1402,83:POKE-S-1401,101:POKE-S-1400,114
5010 POKE-S-1399,112:POKE-S-1398,101:POKE-S-1397,110
5020 POKE-S-1396,115
5030 POKE-S-1529,120:POKE-S-1335,120:POKE-S-1209,88
5040 POKE-S-1147,120:POKE-S-891,120
5050 REM LIBRA
5060 POKE-S-631,76:POKE-S-630,105:POKE-S-629,98
5070 POKE-S-628,114:POKE-S-627,97
5080 POKE-S-688,120:POKE-S-427,120
5090 REM VIRGO
5100 POKE-S-1168,86:POKE-S-1167,105:POKE-S-1166,114
5110 POKE-S-1165,103:POKE-S-1164,111
5120 POKE-S-1295,120:POKE-S-1037,120:POKE-S-842,120
5130 POKE-S-836,120:POKE-S-721,46:POKE-S-534,88
5140 REM CORVUS
5150 POKE-S-333,67:POKE-S-332,111:POKE-S-331,114
5160 POKE-S-330,118:POKE-S-329,117:POKE-S-328,115
5170 POKE-S-457,120:POKE-S-453,120:POKE-S-201,120
5180 POKE-S-259,120
5190 POKE-S-PL,P
5200 RETURN
5210 REM CHART 5 FOR RA AND STARS
5220 IFFL=1GOTO5290
5230 RA=57
5240 POKE-S-957,50:POKE-S-956,48
5250 FORJ=16TO61STEP15
5260 POKE(-S-958+J),49:POKE(-S-957+J),RA
5270 RA=RA-1
5280 NEXTJ
5290 REM AQUILA
5300 POKE-S-1402,65:POKE-S-1401,113:POKE-S-1400,117
5310 POKE-S-1399,105:POKE-S-1398,108:POKE-S-1397,97
```

PLNTA (continued)

```
5320 POKE-S-1454,120:POKE-S-1391,120
5330 POKE-S-1334,120:POKE-S-1273,88:POKE-S-1210,120
5340 POKE-S-1271,65:POKE-S-1270,108:POKE-S-1269,116
5350 POKE-S-1268,97:POKE-S-1267,105:POKE-S-1266,114
5360 REM OPHIUCUS
5370 POKE-S-1501,79:POKE-S-1500,112:POKE-S-1499,104
5380 POKE-S-1498,105:POKE-S-1497,117:POKE-S-1496,99
5390 POKE-S-1495,117:POKE-S-1494,115
5400 POKE-S-1749,120:POKE-S-1305,120:POKE-S-986,120
5410 POKE-S-924,120
5420 REM SAGITTARIUS
5430 POKE-S-120,83:POKE-S-119,97:POKE-S-118,103
5440 POKE-S-117,105:POKE-S-116,116:POKE-S-115,116
5450 POKE-S-114,97:POKE-S-113,114:POKE-S-112,105
5460 POKE-S-111,117:POKE-S-110,115
5470 POKE-S-303,120:POKE-S-172,120:POKE-S-169,120
5480 POKE-S-46,120:POKE-S-228,120:POKE-S-100,120
5490 POKE-S-31,120
5500 REM SCORPIO
5510 POKE-S-330,83:POKE-S-329,99:POKE-S-328,111
5520 POKE-S-327,114:POKE-S-326,112:POKE-S-325,105
5530 POKE-S-324,111
5540 POKE-S-322,120:POKE-S-193,88:POKE-S-129,88
5550 POKE-S-134,120:POKE-S-72,88:POKE-S-10,120
5560 POKE-S-81,65:POKE-S-80,110:POKE-S-79,116
5570 POKE-S-78,97:POKE-S-77,114:POKE-S-76,101
5580 POKE-S-75,115
5590 POKE-S-PL,P
5600 RETURN
5610 REM CHART 6 FOR RA AND STARS
5620 IFFL=1GOTO5680
5630 RA=52:RT=50
5640 FORJ=1TO61STEP15
5650 POKE(-S-958+J),RT
5660 NEXTJ
5670 GOSUB3810
5680 REM PEGASUS
5690 POKE-S-1721,80:POKE-S-1720,101:POKE-S-1719,103
5700 POKE-S-1718,97:POKE-S-1717,115:POKE-S-1716,117
5710 POKE-S-1715,115
5720 POKE-S-1839,120:POKE-S-1455,120:POKE-S-1897,120
5730 POKE-S-1322,120:POKE-S-1185,120:POKE-S-1245,120
5740 REM DELPHINUS
5750 POKE-S-1552,68:POKE-S-1551,101:POKE-S-1550,108
5760 POKE-S-1549,112:POKE-S-1548,104:POKE-S-1547,105
5770 POKE-S-1546,110:POKE-S-1545,117:POKE-S-1544,115
5780 POKE-S-1488,46:POKE-S-1486,46:POKE-S-1423,46
5790 POKE-S-1421,46:POKE-S-1293,46
5800 REM  FOMALHAUT
5810 POKE-S-180,70:POKE-S-179,111:POKE-S-178,109
5820 POKE-S-177,97:POKE-S-176,108:POKE-S-175,104
5830 POKE-S-174,97:POKE-S-173,117:POKE-S-172,116
5840 POKE-S-111,88
5850 REM AQUARIUS
5860 POKE-S-997,65:POKE-S-996,113:POKE-S-995,117
```

─── **PLNTA** (continued) ───

```
5870 POKE-S-994,97:POKE-S-993,114:POKE-S-992,105
5880 POKE-S-991,117:POKE-S-990,115
5890 POKE-S-938,120:POKE-S-936,120:POKE-S-999,120
5900 POKE-S-870,120:POKE-S-929,120:POKE-S-729,120
5910 REM CAPRICORNUS
5920 POKE-S-157,67:POKE-S-156,97:POKE-S-155,112
5930 POKE-S-154,114:POKE-S-153,105:POKE-S-152,99
5940 POKE-S-151,111:POKE-S-150,114:POKE-S-149,110
5950 POKE-S-148,117:POKE-S-147,115
5960 POKE-S-412,120:POKE-S-410,120:POKE-S-214,120
5970 POKE-S-77,120:POKE-S-139,120:POKE-S-454,120
5980 POKE-S-517,120
5990 POKE-S-PL,P
6000 RETURN
6010 REM SUBROUTINE TO POKE NEARBY PLANETS
6020 PN=P
6025 OP=1
6030 FORI=1TO9
6035 IF I=P2THEN NEXTI
6040 PS=I:F5=1
6050 REM JUMPS R RANGE IF BEEN THROUGH THIS TABLE ONCE
6060 IFF4=1THEN6130
6070 IFCH=1THENR2=0:GOTO6130
6080 IFCH=2THENR2=4:GOTO6130
6090 IFCH=3THENR2=8:GOTO6130
6100 IFCH=4THENR2=12:GOTO6130
6110 IFCH=5THENR2=16:GOTO6130
6120 IFCH=6THENR2=20
6130 IFR(I)<R2ORR(I)>(R2+3.9999)THEN6220
6135 OP=0
6140 REM GET PLANET SYMBOL
6150 GOSUB2380
6160 RA=R(I):DE=V(I)
6170 REM GET POKE FOR PLANET
6180 GOSUB2560
6190 IFF7=1THENF7=0:GOTO6210
6200 POKE-S-PL,P
6205 OP=0
6210 F4=1
6220 NEXTI
6225 IFOP=1THENPRINT"NONE":OP=0
6226 OP=0
6230 PS=P2
6240 P=PN
6250 RA=R3:DE=DC
6260 F4=0:F5=0:F=1:RETURN
6270 REM INCREASES COUNTER BY ONE
6280 NC=NC+1
6290 PS=P2
6300 F9=1
6310 D=D+TI
6320 REM INCREMENTS DATE
6330 IFD>30THEND=D-30:M=M+1:GOTO6330
6340 IFM>12THENM=M-12:Y=Y+1
6350 REM GOES THROUGH PROGRAM AGAIN FOR NEW DATE
```

PLNTA *(continued)*

```
6360 GOTO930
6370 REM POKE THE PLANET'S NEW POSITION
6380 IFF7=1THENF7=0:GOTO6400
6390 POKE-S-PL,P
6400 REM POKES THE OTHER PLANETS ON NEW DATE
6410 IFF6=1THEN6430
6420 F9=0:GOSUB6010
6430 REM GOES BACK AND REPEATS IF STILL MORE INTERVALS
6440 F9=1
6450 RETURN
6470 IFRA>20ANDRA<23.9999THENRA=RA-20:RETURN
6480 IFRA>16ANDRA<19.9999THENRA=RA-16:RETURN
6490 IFRA>12ANDRA<15.9999THENRA=RA-12:RETURN
6500 IFRA>8ANDRA<11.9999THENRA=RA-8:RETURN
6510 IFRA>4ANDRA<7.9999THENRA=RA-4:RETURN
6520 RETURN
6530 ONCHGOTO6540,6560,6580,6600,6620,6640
6540 IFRA>3.9999GOTO6680
6550 RETURN
6560 IFRA<4ORRA>7.9999GOTO6680
6570 RETURN
6580 IFRA<8ORRA>11.9999GOTO6680
6590 RETURN
6600 IFRA<120RRA>15.9999GOTO6680
6610 RETURN
6620 IFRA<160RRA>19.9999GOTO6680
6630 RETURN
6640 IFRA<200RRA>23.9999GOTO6680
6650 RETURN
6660 IFF7=1THEN6700
6670 RETURN
6680 F7=1:GOTO6690
6690 PRINTP$(PS)
6700 PRINT"OFF"
6710 PRINT"CHART"
6720 RETURN
7000 REM TO POKE MOON ON CHART
7005 REM LONG OF MOON
7010 LZ=311.1687
7015 LE=178.699
7016 LP=255.7433:PG=.111404*NM+LP
7017 IFPG<-360THENPG=PG+360:GOTO7017
7018 IFPG<0THENPG=PG+360
7019 IFPG>360THENPG=PG-360:GOTO7019
7020 LMD=LZ+360*NM/27.32158
7021 PG=LMD-PG
7022 DR=6.2886*SIN(FNRAD(PG))
7023 LMD=LMD+DR
7025 IFLMD<-3600THENLMD=LMD+3600:GOTO7025
7026 IFLMD<-360THEN LMD=LMD+360:GOTO7026
7030 IFLMD<0THENLMD=LMD+360:GOTO7030
7035 IFLMD>3600THENLMD=LMD-3600:GOTO7035
7040 IFLMD>360THENLMD=LMD-360:GOTO7040
7050 RM=LMD/15
7055 RQ=RM
```

```
7060 IF RM>24THENRM=RM-24:GOTO7060
7070 IFRM<0THENRM=RM+24
7080 AL=LE-NM*.052954
7085 IFAL<-3600THENAL=AL+3600:GOTO7085
7086 IFAL<-360THENAL=AL+360:GOTO7086
7090 IF AL<0THENAL=AL+360:GOTO7090
7095 IFAL>3600THENAL=AL-3600:GOTO7095
7100 IF AL>360THENAL=AL-360:GOTO7100
7106 AL=LMD-AL
7107 IFAL<0THENAL=AL+360
7108 IFAL>360THENAL=AL-360
7110 HE=5.1333*SIN(AL*3.14159/180)
7120 DM=HE+23.1444*SIN(LMD*3.14159/180)
7130 DM=(DM+29)/2:DM=INT(DM):PDE=DM*64
7220 IFCH=2THENR2=4:GOTO7270
7230 IFCH=3THENR2=8:GOTO7270
7240 IFCH=4THENR2=12:GOTO7270
7250 IFCH=5THENR2=16:GOTO7270
7260 IFCH=6THENR2=20
7270 IFRM<R2ORRM>(R2+3.9999)THEN 7290
7271 GOSUB7310:RM=RM*15
7272 IF(RM-INT(RM))>.49THENRM=INT(RM)+1:GOTO7274
7273 RM=INT(RM)
7274 PM=RM+PDE
7275 GOSUB7500
7280 POKE-S-PM,MS
7285 RETURN
7290 PRINT"OFF":PRINT"CHART":RETURN
7310 IFRM>20ANDRM<23.9999THENRM=RM-20:RETURN
7320 IFRM>16ANDRM<19.9999THENRM=RM-16:RETURN
7330 IFRM>12ANDRM<15.9999THENRM=RM-12:RETURN
7340 IFRM>8ANDRM<11.9999THENRM=RM-8:RETURN
7350 IFRM>4ANDRM<7.9999THENRM=RM-4:RETURN
7360 RETURN
7500 RZ=R(3)
7510 DZ=RZ-RQ
7515 IFDZ<0THENDZ=DZ+24
7520 IFDZ>11ANDDZ<13THENMS=111:RETURN
7540 IFDZ>13THENMS=41:RETURN
7550 MS=40
7560 RETURN
```

Galileo

Photo Credit: NASA/Jet Propulsion Laboratory

A man-made meteor, the probe of Project Galileo is planned to penetrate into the deep atmosphere of Jupiter in the mid-1980s. This ambitious mission will explore thoroughly the environment of the giant planet of our solar system and obtain closer looks at its larger satellites over a period of several years.

Program 20A:
SSTARA

Alternative Meteor Shower Program_____

This alternative program for depicting meteor showers was written for an Exidy Sorcerer computer and makes use of POKE statements to randomly display meteors originating from the radiant before identifying the radiant.

As with Program 20, this alternative asks you to select any month and then provides the details of any annual meteor shower expected during that month. If you use PHASE with this program you will be able to find out whether or not observations of the meteor shower will be affected by a bright Moon (between full and last quarter). The program graphically displays first the meteors and then the radiant for the shower among the stars of the constellations for any selected month. It provides details of each meteor shower and names constellations. A typical display of this program is shown in Figure 20A.1.

For other computers, PRINT @ statements can be used to position the stars instead of the PRINT and TAB instructions used for the Sorcerer. The POKE statements also will have to be changed to suit your screen memory area in displaying the meteors coming from the radiant, as well as the radiant itself. In this program version, MS is the POKE location of the middle of the screen.

The listing of the SSTARA program follows.

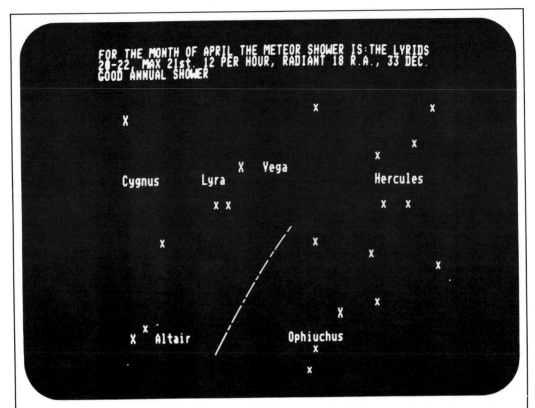

FOR THE MONTH OF APRIL THE METEOR SHOWER IS:THE LYRIDS
20-22, MAX 21st. 12 PER HOUR, RADIANT 18 R.A., 33 DEC.
GOOD ANNUAL SHOWER

In this alternative version of Program 20, meteors flash across the screen from the radiant before identifying the radiant.

Figure 20A.1

─ **SSTARA** ─

```
10 PRINTCHR$(12)
20 PRINT:PRINT:PRINT
30 PRINTTAB(11)"------------------------"
40 PRINTTAB(11)"|      M E T E O R S      |"
50 PRINTTAB(11)"------------------------"
60 PRINT:PRINT:PRINT
70 PRINTTAB(15)"An Astronomy Program"
80 PRINT:PRINTTAB(14)"By ERIC BURGESS F.R.A.S."
90 PRINT:PRINT:PRINT:
100 PRINTTAB(14)"All Rights Reserved by"
110 PRINTTAB(14)"S & T Software Service"
120 PRINT:PRINT:PRINT
130 PRINTTAB(8)"This program provides particulars of"
140 PRINTTAB(8)"Annual Meteor Showers for Any Month"
150 FOR K=1TO2000:NEXTK
160 PRINTCHR$(12):PRINT:PRINT:PRINT:PRINT
170 INPUT"SELECT MONTH (1 THROUGH 12)";M
180 PRINT
190 IFM<1ORM>12THENPRINT:PRINT"Invalid Request":PRINT:GOTO180
200 PRINTCHR$(12):PRINT
210 M2=M
220 IFM2=1THENM$="JANUARY":GOTO340
230 IFM2=2THENM$="FEBRUARY":GOTO510
240 IFM2=3THENM$="MARCH":GOTO550
250 IFM2=4THENM$="APRIL":GOTO570
260 IFM2=5THENM$="MAY":GOTO810
270 IFM2=6THENM$="JUNE":GOTO1030
280 IFM2=7THENM$="JULY":GOTO1250
290 IFM2=8THENM$="AUGUST":GOTO1470
300 IFM2=9THENM$="SEPTEMBER":GOTO1450
310 IFM2=10THENM$="OCTOBER":GOTO1650
320 IFM2=11THENM$="NOVEMBER":GOTO1870
330 IFM2=12THENM$="DECEMBER":GOTO2500
340 PRINT"FOR THE MONTH OF ";M$;" THE METEOR SHOWER IS:";
350 PRINT"THE QUANDRANTIDS"
355 PRINT"MAX ON 6th JAN. 40 PER HOUR,   RADIANT 15.5 R.A., 50 DEC
357 PRINT"GOOD ANNUAL SHOWER":PRINT
360 PRINTTAB(2)"."
370 PRINTTAB(47)"."
380 PRINTTAB(10)".";TAB(18)"Draco"
390 PRINT:PRINTTAB(24)"."
400 PRINTTAB(16)"x":PRINTTAB(54)"Ursa":PRINTTAB(53)"Major"
410 PRINTTAB(63)"X":PRINT:PRINT:PRINT:PRINT
420 PRINTTAB(31)"x"
430 PRINTTAB(38)"X":PRINT
440 PRINTTAB(27)"x";TAB(40)"Bootes"
450 PRINTTAB(42)"."
460 PRINTTAB(25)".";TAB(41)".":PRINTTAB(14)"Corona"
470 PRINTTAB(24)"x";TAB(36)"X":PRINT
480 PRINTTAB(42)"X   Arcturus"
490 GOSUB3650
500 GOTO3600
510 PRINT"DURING THE MONTH OF ";M$;" THERE IS NO MAJOR"
520 PRINT:PRINT"          ANNUAL METEOR SHOWER"
```

SSTARA *(continued)*

```
530 PRINT:PRINT:PRINT:PRINT:PRINT:PRINT:PRINT
540 GOTO3600
550 PRINT"DURING THE MONTH OF ";M$;" THERE IS NO MAJOR"
560 GOTO520
570 PRINT"FOR THE MONTH OF ";M$;" THE METEOR SHOWER IS:";
580 PRINT"THE LYRIDS"
590 PRINT"20-22, MAX 21st. 12 PER HOUR, RADIANT 18 R.A., 33 DEC.
600 PRINT"GOOD ANNUAL SHOWER"
610 PRINT:PRINT
620 PRINTTAB(35)"x";TAB(54)"x"
630 PRINTTAB(4)"X":PRINT
640 PRINTTAB(51)"x"
650 PRINTTAB(45)"x"
660 PRINTTAB(23)"X    Vega"
670 PRINTTAB(4)"Cygnus";TAB(17)"Lyra";TAB(45)"Hercules"
680 PRINT:PRINTTAB(19)"x x";TAB(46)"x";TAB(50)"x"
690 PRINT:PRINT
700 PRINTTAB(10)"x";TAB(35)"x"
710 PRINTTAB(44)"x"
720 PRINTTAB(55)"x"
730 PRINTTAB(57)".":PRINT
740 PRINTTAB(45)"x"
750 PRINTTAB(39)"X"
760 PRINTTAB(7)"x"
770 PRINTTAB(5)"X    Altair";TAB(31)"Ophiuchus"
780 PRINTTAB(35)"x":PRINTTAB(4)".":PRINTTAB(34)"x"
790 GOSUB3650
800 GOTO3600
810 PRINT"FOR THE MONTH OF ";M$;" THE METEOR SHOWER IS:";
820 PRINT"THE ETA AQUARIDS"
830 PRINT"4-13, MAX ON 6th. RADIANT AT 22.54 R.A., 0 DEC."
840 PRINT"WEAK DISPLAY"
850 PRINT:PRINT:PRINTTAB(9)"Pegasus"
860 PRINTTAB(45)"Delphinus"
870 PRINTTAB(54)".";TAB(56)"."
880 PRINTTAB(3)"X";TAB(21)"X";TAB(55)".";TAB(57)".":PRINT
890 PRINTTAB(27)".";TAB(42)"x";TAB(48)"..";TAB(60)"."
900 PRINTTAB(36)".";TAB(47)".":PRINT:PRINT
910 PRINT:PRINTTAB(29)"x";TAB(36)"x"
920 PRINTTAB(27)"x";TAB(32)"."
930 PRINTTAB(42)"Aquarius":PRINT
940 PRINTTAB(45)"X":PRINT:PRINT
950 PRINTTAB(56)".":PRINT
960 PRINTTAB(24)".";TAB(62)"x"
970 PRINTTAB(23)".";TAB(41)"x";TAB(43)".";TAB(47)".";
980 PRINTTAB(51)"."
990 PRINT:PRINTTAB(43)"Capricornus"
1000 PRINT
1010 GOSUB3650
1020 GOTO3600
1030 PRINT:PRINT:PRINT
1040 PRINT"FOR THE MONTH OF JUNE THERE ARE TWO SHOWERS;"
1050 PRINT:PRINT:PRINT
1060 PRINT:PRINTTAB(5)"THE LYRIDS"
```

SSTARA *(continued)*

```
1070 PRINT:PRINT"JUNE 10-21, MAX ON 15th, BLUISH METEORS";
1080 PRINT" AVERAGING 8 PER HOUR"
1090 PRINT:PRINT:PRINTTAB(8)"AND"
1100 PRINT:PRINT:PRINTTAB(5)"THE OPHIUCIDS"
1110 PRINT:PRINT"JUNE 17-26, MAX ON 19th";
1120 PRINT" AVERAGING 6 PER HOUR"
1130 PRINT:PRINT:PRINT
1140 PRINT:PRINT:PRINT"PLEASE SELECT LYRIDS (1)"
1150 PRINT"          OR OPHIUCIDS (2)"
1160 PRINT:INPUTA$
1170 JS=VAL(A$):IFJS<1ORJS>2THENPRINT"INVALID SELECTION":GOTO1140
1180 IFJS=1THENGOSUB3070:GOTO1200
1190 IFJS=2THENGOSUB3340
1200 INPUT"DO YOU WANT TO SEE THE OTHER SHOWER Y/N";A$
1210 PRINTCHR$(12)
1220 IFA$="Y"ANDJS=1THENGOSUB3340:GOTO1240
1230 IFA$="Y"ANDJS=2THENGOSUB3070
1240 GOTO3600
1250 PRINT"FOR THE MONTH OF ";M$;" THE METEOR SHOWER IS";
1260 PRINT" THE DELTA AQUARIDS"
1270 PRINT"JUL.25-AUG 4, MAX JUL.28th. R.A. 22.8 DEC. -16"
1280 PRINT"DIFFUSE SHOWER"
1290 PRINT
1300 PRINTTAB(25)".";TAB(29)".";TAB(42)"x"
1310 PRINT:PRINTTAB(34)". .";TAB(42)"x"
1320 PRINTTAB(39)".":PRINT
1330 PRINTTAB(52)"x":PRINTTAB(29)"."
1340 PRINTTAB(9)".";TAB(63)"."
1350 PRINTTAB(44)"Aquarius"
1360 PRINTTAB(30)"."
1370 PRINTTAB(28)"x";TAB(45)"x";TAB(49)".";TAB(57)"."
1380 PRINT" x"
1390 PRINTTAB(51)".":PRINTTAB(53)".":PRINT
1400 PRINTTAB(62)".":PRINT:PRINTTAB(31)"X":PRINT
1410 PRINTTAB(23)"Fomalhaut"
1420 PRINTTAB(43)"x":PRINT:PRINT:PRINTTAB(5)"x"
1430 GOSUB3650
1440 GOTO3600
1450 PRINT"DURING THE MONTH OF ";M$;" THERE IS NO MAJOR"
1460 GOTO520
1470 PRINT"FOR THE MONTH OF ";M$;" THE METEOR SHOWER IS";
1480 PRINT" THE PERSEIDS"
1490 PRINT"4-16, MAX 12th, 50 PER HOUR, 3.0 R.A., 58 DEC."
1500 PRINT"A MAJOR SPECTACULAR SHOWER"
1510 PRINT:PRINT:PRINT:PRINT:PRINTTAB(18)".";TAB(48)"."
1520 PRINTTAB(48)".":PRINTTAB(47)"Cassiopeia"
1530 PRINT:PRINTTAB(18)".";TAB(52)"X"
1540 PRINT:PRINTTAB(2)"x";TAB(60)"X"
1550 PRINT:PRINTTAB(32)".";TAB(38)"Perseus"
1560 PRINTTAB(29)"x":PRINTTAB(31)"."
1570 PRINTTAB(25)"X";TAB(28)"."
1580 PRINTTAB(21)"x":PRINT:PRINT
1590 PRINTTAB(48)"X":PRINTTAB(17)"X";TAB(29)"X"
1600 PRINTTAB(29)".";TAB(45)"Andromeda"
```

─── **SSTARA** *(continued)* ─────────────────

```
1610 PRINT:PRINTTAB(47)"x";TAB(63)"X"
1620 PRINT
1630 GOSUB3650
1640 GOTO3600
1650 PRINT"FOR THE MONTH OF ";M$;" THE METEOR SHOWER IS";
1660 PRINT" THE ORIONIDS"
1670 PRINT"16-26, MAX 21st. 30 PER HOUR, 6.4 R.A., 15 DEC."
1680 PRINT"SLOW METEORS WITH PERSISTENT TRAILS"
1690 PRINTTAB(41)"X     Auriga"
1700 PRINTTAB(15)"X    Gemini";TAB(57)"X"
1710 PRINT:PRINTTAB(12)"X";TAB(51)"x"
1720 PRINTTAB(28)".":PRINT
1730 PRINTTAB(18)".";TAB(35)"x";TAB(38)"."
1740 PRINTTAB(23)".";TAB(34)".";TAB(46)"x"
1750 PRINT:PRINTTAB(20)".";TAB(30)"x"
1760 PRINTTAB(61)"X":PRINTTAB(28)"."
1770 PRINT:PRINTTAB(5)".";TAB(20)"x";TAB(47)"."
1780 PRINTTAB(40)"X";TAB(50)"X"
1790 PRINTTAB(14)"X"
1800 PRINTTAB(10)"Procyon";TAB(47)"Orion"
1810 PRINT:PRINTTAB(47)"X"
1820 PRINTTAB(45)"X":PRINTTAB(43)"X";TAB(50)"."
1830 PRINTTAB(46)":";PRINTTAB(14)".";TAB(50)"X"
1840 PRINTTAB(42)"x"
1850 GOSUB3650
1860 GOTO3600
1870 PRINT:PRINT:PRINT
1880 PRINT"FOR THE MONTH OF ";M$;" THERE ARE TWO SHOWERS;"
1890 PRINT:PRINT:PRINT
1900 PRINT:PRINTTAB(5)"THE TAURIDS"
1910 PRINT:PRINT"OCT 20 - NOV 30, MAX ON 8 NOV. SLOW METEORS"
1920 PRINT"WITH SCATTERED FIREBALLS, AVERAGE 12 PER HOUR"
1930 PRINT:PRINT:PRINTTAB(8)"AND"
1940 PRINT:PRINT:PRINTTAB(5)"THE LEONIDS"
1950 PRINT:PRINT"NOV. 15-19, MAX ON 17th,"
1960 PRINT"AVERAGING 10 PER HOUR"
1970 PRINT:PRINT
1980 PRINT:PRINT:PRINT"PLEASE SELECT TAURIDS (1)"
1990 PRINT"          OR LEONIDS (2)"
2000 PRINT:INPUTA$
2010 JS=VAL(A$)
2020 IFJS<1ORJS>2THENPRINT"INVALID SELECTION":GOTO1980
2030 IFJS=1THENGOSUB2100:GOTO2050
2040 IFJS=2THENGOSUB2330
2050 INPUT"DO YOU WANT TO SEE THE OTHER SHOWER Y/N";A$
2060 PRINTCHR$(12)
2070 IFA$="Y"ANDJS=1THENGOSUB2330:GOTO2090
2080 IFA$="Y"ANDJS=2THENGOSUB2100
2090 GOTO3600
2100 REM NOV TAURIDS
2110 PRINTCHR$(12)
2120 PRINT"FIRST SHOWER FOR MONTH OF ";M$;" IS";
2130 PRINT" THE TAURIDS"
2140 PRINT"FROM OCT 20 - NOV 30, MAX 8th, 12 PER HOUR WITH"
2150 PRINT"TWO RADIANTS 3.73 R.A., 14 AND 22 DEC.
```

SSTARA *(continued)*

```
2160 PRINT"SCATTERED FIREBALLS, SLOW"
2170 PRINT:PRINTTAB(58)"X"
2180 PRINTTAB(10)". .":PRINTTAB(30)".";TAB(42)"x"
2190 PRINTTAB(42)"."
2200 PRINTTAB(29)".":PRINTTAB(56)"x"
2210 PRINTTAB(14)".";TAB(29)"x .";TAB(60)"."
2220 PRINTTAB(7)"x":PRINT:PRINT
2230 PRINTTAB(31)".::";TAB(58)"x"
2240 PRINTTAB(2)"x";TAB(12)"Taurus";TAB(21)".";
2250 PRINTTAB(52)"Aries":PRINTTAB(60)"."
2260 PRINTTAB(18)"X";TAB(22)"."
2270 PRINTTAB(20)". ."
2280 PRINTTAB(27)".":PRINTTAB(2)".";TAB(35)"."
2290 PRINTTAB(5)"x";TAB(36)"."
2300 PRINTTAB(42)"x";TAB(57)".":PRINT:PRINTTAB(2)"x"
2310 PRINT:PRINT:PRINT
2320 GOSUB3650:TF=1:GOSUB3650:RETURN
2330 REM NOV LEONIDS
2340 PRINTCHR$(12)
2350 PRINT"SECOND SHOWER FOR MONTH OF ";M$;" IS";
2360 PRINT" THE LEONIDS"
2370 PRINT"DEC. 15-19, MAX ON 17th., 10 PER HOUR"
2380 PRINT"R.A. 10.13, DEC 22, SWIFT METEORS"
2390 PRINTTAB(32)".":PRINTTAB(30)".":PRINT:PRINT:PRINT
2400 PRINTTAB(15)".";TAB(22)".":PRINTTAB(19)"."
2410 PRINT:PRINTTAB(36)".":PRINTTAB(30)".";TAB(38)"x"
2420 PRINT:PRINTTAB(29)"x":PRINTTAB(15)"x"
2430 PRINTTAB(32)".":PRINTTAB(6)"X";TAB(15)"X";TAB(20)"Leo"
2440 PRINTTAB(32)"X    Regulus"
2450 PRINTTAB(12)".";TAB(40)"."
2460 PRINTTAB(7)".";TAB(52)"x . ."
2470 PRINTTAB(57)".":PRINTTAB(6)".":PRINT
2480 PRINT:PRINT:PRINT
2490 GOSUB3650:RETURN
2500 PRINT:PRINT:PRINT
2510 PRINT"FOR THE MONTH OF ";M$;" THERE ARE TWO SHOWERS;"
2520 PRINT:PRINT:PRINT
2530 PRINT:PRINTTAB(5)"THE GEMINIDS"
2540 PRINT:PRINT"9-13, MAX ON 13th. SWIFT BRIGHT METEORS"
2550 PRINT"WITH SOME FIREBALLS, AVERAGE 60 PER HOUR"
2560 PRINT:PRINT:PRINTTAB(8)"AND"
2570 PRINT:PRINTTAB(5)"THE URSIDS"
2580 PRINT:PRINT"17-24, MAX ON 22nd. WEAK DISPLAY"
2590 PRINT"AVERAGE 5 PER HOUR"
2600 PRINT:PRINT
2610 PRINT:PRINT:PRINT"PLEASE SELECT GEMINIDS (1)"
2620 PRINT"          OR URSIDS (2)"
2630 PRINT:INPUTA$: JS=VAL(A$)
2640 IFJS=1THENGOSUB2710:GOTO2660
2650 IFJS=2THENGOSUB2880
2660 INPUT"DO YOU WANT TO SEE THE OTHER SHOWER Y/N";A$
2670 PRINTCHR$(12)
2680 IFA$="Y"ANDJS=1THENGOSUB2880:GOTO2700
2690 IFA$="Y"ANDJS=2THENGOSUB2710
2700 GOTO3600
```

SSTARA *(continued)*

```
2710 REM DEC GEMINIDS
2720 PRINTCHR$(12)
2730 PRINT"FIRST SHOWER FOR MONTH OF ";M$;" IS";
2740 PRINT" THE GEMINIDS"
2750 PRINT"FROM DEC 9-24, MAX ON 13th, 60 PER HOUR, SWIFT"
2760 PRINT"BRIGHT METEORS, 7.46 R.A., 32 DEC."
2770 PRINT:PRINT:PRINTTAB(9)".";TAB(45)"Auriga"
2780 PRINTTAB(7)".":PRINTTAB(55)"x":PRINT
2790 PRINT:PRINTTAB(58)".":PRINTTAB(55)"."
2800 PRINT:PRINTTAB(34)"x":PRINTTAB(41)"."
2810 PRINTTAB(12)".":PRINTTAB(30)"x";TAB(45)"."
2820 PRINTTAB(30)".":PRINTTAB(13)".";TAB(36)"x";TAB(52)"x   x"
2830 PRINTTAB(40)".":PRINTTAB(11)".";TAB(37)"Gemini"
2840 PRINTTAB(48)"x":PRINT
2850 PRINTTAB(47)".":PRINTTAB(61)":"
2860 PRINTTAB(35)"x";TAB(58)"X":PRINTTAB(31)"X   Procyon"
2870 GOSUB3650:RETURN
2880 REM DEC URSIDS
2890 PRINTCHR$(12)
2900 PRINT"SECOND SHOWER FOR MONTH OF ";M$;" IS";
2910 PRINT" THE URSIDS"
2920 PRINT"FROM DEC 17-24, MAX ON 22nd., 5 PER HOUR"
2930 PRINT"FAINT MEDIUM SPEED METEORS, 14.46 R.A., 78 DEC."
2940 PRINT:PRINT:PRINTTAB(22)"Polaris";TAB(32)"x"
2950 PRINT:PRINTTAB(28)".":PRINT:PRINT:PRINT
2960 PRINTTAB(24)".":PRINT
2970 PRINTTAB(58)".":PRINT:PRINT
2980 PRINTTAB(12)"Ursa";TAB(26)".";TAB(51)"."
2990 PRINTTAB(11)"Minor"
3000 PRINTTAB(22)".";TAB(28)"x":PRINT
3010 PRINTTAB(26)"x";TAB(42)"Draco"
3020 PRINT:PRINTTAB(2)"x":PRINT:PRINT
3030 PRINTTAB(37)".":PRINTTAB(10)"x":PRINT
3040 PRINTTAB(14)"x";TAB(22)"x"
3050 GOSUB3650
3060 RETURN
3070 REM JUNE LYRIDS
3080 PRINTCHR$(12)
3090 PRINT"FIRST SHOWER FOR MONTH OF ";M$;" IS";
3100 PRINT" THE LYRIDS"
3110 PRINT"FROM 10-20, MAX 15th. 8 PER HOUR";
3120 PRINT" RADIANT 18.3 R.A., 35 DEC."
3130 PRINT"BLUISH METEORS"
3140 PRINT:PRINTTAB(43)"Draco"
3150 PRINTTAB(44)"X"
3160 PRINTTAB(37)"X":PRINT:PRINT
3170 PRINTTAB(62)"x"
3180 PRINTTAB(8)"X";TAB(43)"x":PRINT
3190 PRINTTAB(2)"X";TAB(57)"x"
3200 PRINTTAB(18)"Vega";TAB(25)"X";
3210 PRINTTAB(50)"x":PRINT:PRINT
3220 PRINTTAB(21)"x   x";TAB(52)"x        x"
3230 PRINTTAB(4)"Cygnus";TAB(49)"Hercules"
3240 PRINTTAB(13)"X"
3250 PRINTTAB(51)"x":PRINT
```

─── **SSTARA** *(continued)* ───

```
3260 PRINTTAB(62)"x":PRINT:PRINT
3270 PRINTTAB(24)"."
3280 PRINTTAB(22)"x";TAB(50)"x"
3290 PRINTTAB(45)"x"
3300 PRINTTAB(11)"x";TAB(43)"Ophiuchus"
3310 PRINTTAB(9)"X";TAB(12)"Altair"
3320 GOSUB3650
3330 RETURN
3340 REM JUNE OPHIUCIDS
3350 PRINTCHR$(12)
3360 PRINT"SECOND SHOWER FOR MONTH OF ";M$;" IS";
3370 PRINT" THE OPHIUCIDS"
3380 PRINT"FROM 17-26, MAX 19th, 6 PER HOUR";
3390 PRINT" RADIANT 17.3 R.A. -20 DEC."
3400 PRINTTAB(25)"x";TAB(45)"."
3410 PRINTTAB(20)"Ophiuchus"
3420 PRINTTAB(4)".";TAB(50)"x";
3430 PRINTTAB(56)".":PRINTTAB(5)"."
3440 PRINTTAB(7)".";TAB(45)"."
3450 PRINTTAB(11)".";TAB(22)".";TAB(42)"x";TAB(60)"x"
3460 PRINT:PRINT
3470 PRINTTAB(27)".";TAB(33)"x":PRINT
3480 PRINT:PRINTTAB(2)"x";TAB(50)"x":PRINT
3490 PRINTTAB(14)"x";TAB(51)"x"
3500 PRINTTAB(5)"x";TAB(46)"x"
3510 PRINTTAB(43)"X";TAB(51)"x"
3520 PRINTTAB(3)"x";TAB(15)"x";TAB(20)"x";TAB(42)"x"
3530 PRINT:PRINTTAB(5)"Sagittarius"
3540 PRINTTAB(14)"x";TAB(38)"x"
3550 PRINTTAB(17)"x";TAB(24)".";TAB(27)"xx"
3560 PRINTTAB(24)"x";TAB(38);TAB(43)"Scorpio"
3570 PRINT:PRINTTAB(37)"."
3580 GOSUB3650
3590 RETURN
3600 INPUT"Do you want another Month Y/N";A$
3610 IFA$="Y"THEN160
3620 IF A$<>"N"THEN PRINT"Invalid Response":PRINT:GOTO3600
3630 PRINT CHR$(12)
3640 END
3650 FORH=1TO1000:NEXT
3660 REM PRINT METEORS
3670 N=0
3680 R=-1
3690 IF M$="NOVEMBER"ANDTF=1THENMS=-2722:TF=0:GOTO3710
3700 MS=-2978
3710 IFN=20THEN4290
3720 R=R-1
3730 GOSUB3770
3740 FOR G=1TOC*100:NEXT
3750 N=N+1
3760 GOTO3710
3770 REM PICK A METEOR
3780 A=RND(R)
3790 C=INT(7*RND(A)+1)
3800 B=INT(9*RND(A)+1)
```

```
┌─ SSTARA (continued) ──────────────────────────────────────────┐
│                                                                │
│  3810 ONBGOTO3980,3820,3930,4090,4040,4140,4240,3870           │
│  3820 FORJ=5TO14                                               │
│  3830 POKE MS+65*J,172:POKE MS+65*(J+1),172                    │
│  3840 POKE MS+65*J,32:POKE MS+65*(J+1),32                      │
│  3850 NEXT:GOTO4040                                            │
│  3860 RETURN                                                   │
│  3870 FORJ=3TO10                                               │
│  3880 :                                                        │
│  3890 POKE MS-65*J,172:POKE MS-65*(J+1),172                    │
│  3900 POKE MS-65*J,32:POKE MS-65*(J+1),32                      │
│  3910 NEXT:IFC=10RC=5THEN4090                                  │
│  3920 RETURN                                                   │
│  3930 FORJ=3TO9                                                │
│  3940 POKE MS-J,140:POKE MS-J+1,140                            │
│  3950 POKE MS-J,32:POKE MS-J+1,32                              │
│  3960 NEXT:IFC=10RC=6THEN4190                                  │
│  3970 RETURN                                                   │
│  3980 FORJ=4TO30 STEP 4                                        │
│  3990 POKE MS+J,45:POKEMS+J+1,45:POKEMS+J+2,45                 │
│  4000 POKE MS+J,32:POKE MS+J+1,32:POKE MS+J+2,32               │
│  4010 NEXT:IFC=6THEN3870                                       │
│  4020 GOTO4190                                                 │
│  4030 RETURN                                                   │
│  4040 FORJ=6TO30 STEP3                                         │
│  4050 POKE MS-J,45:POKE MS-J+1,45:POKE MS-J+2,45               │
│  4060 POKE MS-J,32:POKE MS-J+1,32:POKE MS-J+2,32               │
│  4070 NEXT:IFC>3THEN4140                                       │
│  4080 RETURN                                                   │
│  4090 FORJ=5TO10                                               │
│  4100 POKE MS-64*J,124:POKE MS-64*(J+1),124                    │
│  4110 POKE MS-64*J,32:POKE MS-64*(J+1),32                      │
│  4120 NEXT:IFC>2THEN4190                                       │
│  4130 RETURN                                                   │
│  4140 FORJ=4TO12                                               │
│  4150 POKE MS+64*J,124:POKE MS+64*(J+1),124                    │
│  4160 POKE MS+64*J,32:POKE MS+64*(J+1),32                      │
│  4170 NEXT:IFC=2THEN3870                                       │
│  4180 RETURN                                                   │
│  4190 FORJ=3TO(4+C)                                            │
│  4200 POKE MS-63*J,171:POKE MS-63*(J+1),171                    │
│  4210 POKE MS-63*J,32:POKE MS-63*(J+1),32                      │
│  4220 NEXT:IFC=4THEN3980                                       │
│  4230 RETURN                                                   │
│  4240 FORJ=3TO(15-C)                                           │
│  4250 POKE MS+63*J,171:POKE MS+63*(J+1),171                    │
│  4260 POKE MS+63*J,32:POKE MS+63*(J+1),32                      │
│  4270 NEXT:IFC=2THEN4140                                       │
│  4280 RETURN                                                   │
│  4290 POKE MS,111                                              │
│  4300 IF M$="NOVEMBER"ANDTF=1THENGOTO3650                      │
│  4310 RETURN                                                   │
│                                                                │
└────────────────────────────────────────────────────────────────┘
```

Observer's Guide to the Programs

In preparing for an evening of observations you might select the following programs:

1. Run TIMES to find the sidereal time at the beginning and end of your planned observations.

2. Use RISES to tell you the times of sunset, sunrise, moonrise, and moonset.

3. PHASE will tell you the phase of the Moon so you can find out if a bright Moon will interfere with your observations.

4. If the Moon is full you can find out if there will be an eclipse by using ECLIP.

5. Use SKYSET/SKYPLT to show you the sky at the time you plan to begin observing. You can see which planets will be in the sky when you plan to observe, as well as the constellations that can be seen.

6. If a planet you are interested in is not above the horizon, you can use RISES to tell you when it will rise. PLNTF will tell you where in the zodiac it is located.

7. Use PRISE to find out if Mercury or Venus is visible as an evening star, and how much time after sunset will be available for observation.

8. If either Mercury or Venus is visible the evening you plan to observe, use MERVE to provide details of distance, phase, and angular diameter.

9. If neither planet is visible you can use MVENC to find out when the next suitable elongation will take place.

10. If you are interested in Mars and you have found out that it is visible, you can ascertain its distance and angular diameter by using MARSP. This program will also tell you when Mars will next be close to Earth at opposition.

11. If you plan to observe Jupiter, you can use JSATS to identify its large satellites.

12. SSTAR will tell you if there is a major meteor shower on the night of your observations.

13. When resetting the polar axis of an equatorial mount, you can find the direction and elevation of the pole by observing transits and elongations of Polaris. PSTAR will tell you the times to do this.

14. Searching for a faint nebula or other celestial object? Look up its right ascension and declination in your star atlas and then convert to the current epoch using EPOCH. By using TIMES to get sidereal time, you can then set the hour circle on your telescope and point to the correct right ascension and declination to bring the faint object within the field of view of a finder telescope.

15. If you want to use PHOTO to photograph one of the planets, you can find the angular diameter of Mercury and Venus from MERVE and that of Mars from MARSP. Angular diameters of Jupiter, Saturn, and Uranus are found by dividing 190, 158, and 68, respectively, by the distance of the planet in astronomical units, obtained from RADEC.

16. If you are planning to observe a solar eclipse, you can find out which planets and bright stars will be visible during the eclipse by running SKYSET/SKYPLT for the time and date of totality, changing the latitude and longitude to your observing site. Making a photograph of your monitor screen will provide a guide for you to identify objects surrounding the Sun.

17. Making a trip into another hemisphere? Learn the constellations in advance by using CONST and CONSH. Use SKYSET/SKYPLT to show you what the familiar constellations will look like from the opposite hemisphere, and in which direction to look for the planets.

Bibliography

The following books are recommended for further information about astronomical calculations. Some of them provide equations that can be used to obtain much greater precision of almanac data than that provided by the programs in this book. While the programs given here are sufficient for most practical purposes, more specialized routines are sometimes needed for predictions of stellar occultations by the Moon, close conjunctions of planets, or rigorous observations of celestial objects relative to star positions. If you need such precision in computations, many of the programs provided in this book can be upgraded within the limits of your computer's capability in handling numbers.

The American Ephemeris and Nautical Almanac. U.S. Government Printing Office: Washington D.C., annually to 1980.

The Astronomical Ephemeris. H. M. Nautical Almanac Office: Royal Greenwich Observatory, H. M. Stationery Office: London, annually to 1980. (British edition of the above.)

The Astronomical Almanac. U.S. Government Printing Office: Washington, D.C., and H. M. Stationery Office: London, annually from 1981. (Combines *The American Ephemeris* and *The Astronomical Ephemeris.*)

Explanatory Supplement to the Nautical Almanac. H. M. Stationery Office: London, 1977.

Sky Catalogue 2000. Sky Publishing Corporation: Cambridge, Massachusetts, 1981.

Astrodynamics. R. M. L. Baker and M. W. Makemson. Academic Press: New York, 1960.

Mathematical Astronomy with a Pocket Calculator. Aubrey Jones. Wiley: New York, 1978.

Positional Astronomy. D. McNally. Muller: London, 1974.

Practical Astronomy With Your Calculator, Second Edition. P. Duffett-Smith. Cambridge University Press: London, 1981.

Index

Algorithm, O'Beirne, 15
Alternative meteor program, 283
Alternative planet finder program, 259
Alternative programs, 243
Alternative skyplotting program, 245
Angle reduction routine, 171
Angular diameter routine, 103
Angular distance routine, 80
Anomalistic month, 65
Aphelion, 59, 83
Apogee, 51
Apollo, 91
Apollo-Soyuz, 32, 36
Apparitions, of Venus and Mercury, 93
Aries, First Point of, 19
Azimuth and elevation routine, 136, 253

Babylonians, 3
Building in space, 42

Calculations, planetary, 113, 121, 248, 271–272
Calendar, 51, 65
   Babylonian, 3
   ecclesiastical, 15
   Egyptian, 3–4
   Gregorian, 4–5
   Jewish, 5
   Mayan, 91
   Moslem, 5
Callisto, 163
Camera/telescope combinations, 235
Cassiopeia, display of, 193
Celestial coordinates, 20
Celestial equator, 19
Celestial sphere, 75

Chart
   graphics routine, 273
   horizon, 128, 245
Constellations, 191, 295
   POKING routine for, 140, 275
   query routine for, 208, 212
   recognition of, 192
   selection routine for, 195
   in Southern Hemisphere, 194, 209
   zodiacal, 85, 146
Conversion routines, 176
Conversions
   astronomical, 175
   coordinate, 77
Coordinates, celestial, 20
Crescent moon, 66

Data
   for asteroids, 222
   for Jupiter satellites, 221
   for Mars satellites, 222
   for Neptune satellite, 222
   planetary, 79, 112, 120, 137, 148, 221, 248, 270
   for Saturn satellites, 221
   shape table, 139
   star, 138, 253
   for Uranus satellites, 222
Day
   Gregorian, 29, 33
   Julian, 29, 33, 61
   mean solar, 19
Days from epoch routines, 24, 39, 58, 88, 98, 104, 112, 135, 148, 168, 247
Declination, 19, 37, 51–52, 75, 109
Declination routine, 80, 113, 122, 136, 150, 249, 257, 272
Display
   Apple horizon plot, 129–131

Display (*continued*)
  constellations, 193
  conversion menu, 175
  date conversions, 30, 33
  Easter day, 16
  holidays, 6
  horizon chart, 129–131, 246
  Jupiter satellites, 164–165
  lunar eclipse, 60
  lunar phases, 68
  Mars data, 84
  Mars/Earth, routine for, 89
  Mercury and Venus, 94–95, 110
  Mercury and Venus, routine for, 99, 105
  meteors, 182–183, 284
  Moon's position, 52
  planet finder, 144–145, 260–264
  planetary photography, 237
  planetary positions, 76
  Polaris data, 44
  precession conversions, 38
  rise, transit, set times, 110, 118
  solar system data, 216–217
  sort routine, 170
  Sun and planets, routine for, 105
  time conversions, 21
Distance routines, 80, 87, 96–97, 104, 169
Draconic month, 65

Earth, 18
  shadow of, 59
Easter, 4, 6, 15
  routine to determine, 12
Eclipse, 59, 294
  cycle of, 59
  date routine for, 63
  duration of, 59
  magnitude of, 60
  magnitude routine for, 62
  sets of, 59
Effective focal length, 236
  routine for, 239
Elongation, 91–92
  of Polaris, 43
Elongation routine, 99, 106
Epoch, 37
Equator, celestial, 19
Equinox, 15, 19–20, 51, 75
Europa, 163
Evening star, 91

Florida coastline, 14

Galileo, 163
Ganymede, 163
Graphic routine for star plots, 132, 140, 251
Gregorian calendar, 4–5
Gregorian days routine, 26, 78

Harvest moon, 66
Heliocentric longitude routine, 80
Holidays, 5
  routine for, 11
Horizon chart, 245
Horizon plots, of planets, 127
Horizon silhouette, 245
Hour, sidereal, 51, 75
Hour angle, 19

Io, 163

Jupiter, 142, 163, 282, 295
  routine for motion of, 168
  satellites of, 162

Kalendae, 3

Latitude, 43, 192
Leap year routine, 87
Longitude, 192
Lunar cycle, 65
Lunation, 59

Mars, 82, 116, 258, 295
  aphelion of, 83
  distance of, 84
  next opposition of, 84
  orbit of, 83
  perihelion of, 83
  position routine for, 88
  routine for angular diameter of, 87
Measuring time, 1
Menu of conversions, 175
Menu of solar system data, 215
Mercator projection, 128, 245
Mercury, 91, 108, 294
  position routine for, 104
  rising and setting of, 109
Meridian, 19, 51, 75
Meteor graphic routine, 291
Meteor shower, 295
Meteoroids, 181

Meteors, 180, 283
Milky Way, 190
Modifications, program, 146, 261, 265, 283
Month, definitions of, 65
Month and year ends, 88, 106, 168
Moon, 49–51, 58, 64, 215, 244, 294
  cycles of, 3
  distance of, 51
  POKING routine for, 160, 257, 280
  position of, 52
  position routine for, 56, 136
  rise and set routine for, 124
Morning star, 91
Motions, planetary, 73

New moon, routine to determine, 70
Node, 51, 65
North point, 43

O'Beirne algorithm, 15
Observations
  program use in, 294–295
  of Venus and Mercury, 93, 109
Occultation, 166
Opposition, 83, 85
Options, planet finding, 259
Orbit, of Mars, 83
Orrery, 127

Passover, 15
Perigee, 51, 59
Perihelion, 83
Periods, of Venus and Mercury, 93
Perturbations, 51, 53
Phase
  of Moon, 66
  of Venus and Mercury, 93
Phase routine, 70, 101, 106
Photography, planetary, 235, 295
Pioneer Venus, 180
Planets
  distance routine for, 80
  finding, 143, 259
  inferior, 91
  orbital data for, 120–121, 248
  photography of, 235
  plot routine for, 89
  POKING codes for, 151
  POKING routines for, 279
  position routines for, 79, 121, 150, 248

Planets (*continued*)
  rising, transit, setting of, 117
  series plot of, 144, 264
  symbols for, 128, 245
Planetarium, 127
Polaris, 43, 191–192, 295
Pole star, 37, 43
Pope Gregory, 4, 15
Position of planets, 75
  routines to determine, 79, 121, 150, 248
Position of Venus and Mercury
  routine, 98
Precession, 19, 37
Precision of calculations, upgrading, 296
Program CHAIN, use of, 130
Programs
  modifications to, 146, 261, 265, 283
  use of in observations, 294–295
Project Galileo, 282

Radiant, 283
Right ascension, 19, 37, 51, 75, 109
Right ascension routine, 80, 113, 122, 136, 150, 249, 257, 272
  adjusting for right ascension passing zero, 114
Rising, Moon, 117
Rising, planets, 109, 117
Roman month, 3

Saros, 59
Satellites, of Jupiter, 162, 221
Saturn, 74, 126, 221
Scalinger, Julius, 29
Setting, Moon, 117
Setting, planets, 109, 117
Sidereal hour, 51, 75
Sidereal month, 65
Sidereal time routine, 114, 123, 135
Sirius, 3
Solar system, 215
Solar year, 3
Sort routine, 93, 99
Space shuttle, 2
Space telescope, 28
Sphere, celestial, 20
Star chart selection routine, 151, 153, 159, 273
Star charts, 153
Star coordinates, 129

Star names, 192
Star patterns, 191
Star POKING routine, 140, 255
Star query routine, 208, 212
Star right ascension and declination
    routine, 139, 253
Star right ascensions and declinations,
    266–267
Star, "shooting," 181
Star sphere, 19
Sun, 174
Synodic month, 65

Time, 22
    daylight saving, 22
    for Mars observation, 83
    measurement of, 1
    rising and setting routine, 123
    sidereal, 19, 22, 117
    star, 19
    universal, 22
    zone, 22

Transit, 43, 166

Umbral eclipse, 59
Upgrading precision, 296

Variables, location of, 128
Venus, 90–91, 234, 294–295
    position routine for, 104
    rising of, 109
Vernal equinox, 15, 19, 51, 75
Viewing Mercury and Venus routine,
    107
Voyager, 74, 142, 163

Wandering stars, 73
Week, day of, 8

Year, solar, 3–4

Zodiac, 143, 191

The SYBEX Library

YOUR FIRST COMPUTER
by Rodnay Zaks 264 pp., 150 illustr., Ref. C200A
The most popular introduction to small computers and their peripherals: what they do and how to buy one.

DON'T (or How to Care for Your Computer)
by Rodnay Zaks 222 pp., 100 illust., Ref. C400
The correct way to handle and care for all elements of a computer system, including what to do when something doesn't work.

INTERNATIONAL MICROCOMPUTER DICTIONARY
140 pp., Ref. X2
All the definitions and acronyms of microcomputer jargon defined in a handy pocket-size edition. Includes translations of the most popular terms into ten languages.

FROM CHIPS TO SYSTEMS:
AN INTRODUCTION TO MICROPROCESSORS
by Rodnay Zaks 558 pp., 400 illustr. Ref. C201A
A simple and comprehensive introduction to microprocessors from both a hardware and software standpoint: what they are, how they operate, how to assemble them into a complete system.

INTRODUCTION TO WORD PROCESSING
by Hal Glatzer 216 pp., 140 illustr., Ref. W101
Explains in plain language what a word processor can do, how it improves productivity, how to use a word processor and how to buy one wisely.

INTRODUCTION TO WORDSTAR™
by Arthur Naiman 208 pp., 30 illustr., Ref. W105
Makes it easy to learn how to use WordStar, a powerful word processing program for personal computers.

VISICALC® APPLICATIONS
by Stanley R. Trost   200 pp., Ref. V104
Presents accounting and management planning applications—from financial statements to master budgets; from pricing models to investment strategies.

EXECUTIVE PLANNING WITH BASIC
by X. T. Bui   192 pp., 19 illust., Ref. B380
An important collection of business management decision models in BASIC, including Inventory Management (EOQ), Critical Path Analysis and PERT, Financial Ratio Analysis, Portfolio Management, and much more.

BASIC FOR BUSINESS
by Douglas Hergert   250 pp., 15 illustr., Ref. B390
A logically organized, no-nonsense introduction to BASIC programming for business applications. Includes many fully-explained accounting programs, and shows you how to write them.

FIFTY BASIC EXERCISES
by J. P. Lamoitier   236 pp., 90 illustr., Ref. B250
Teaches BASIC by actual practice, using graduated exercises drawn from everyday applications. All programs written in Microsoft BASIC.

BASIC EXERCISES FOR THE APPLE
by J. P. Lamoitier   230 pp., 90 illustr., Ref. B500
This book is an Apple version of *Fifty BASIC Exercises.*

BASIC EXERCISES FOR THE IBM PERSONAL COMPUTER
by J. P. Lamoitier   232 pp.. 90 illustr., Ref. B510
This book is an IBM version of *Fifty BASIC Exercises.*

INSIDE BASIC GAMES
by Richard Mateosian   352 pp., 120 illustr., Ref. B245
Teaches interactive BASIC programming through games. Games are written in Microsoft BASIC and can run on the TRS-80, Apple II and PET/CBM.

THE PASCAL HANDBOOK
by Jacques Tiberghien   492 pp., 270 illustr., Ref. P320
A dictionary of the Pascal language, defining every reserved word, operator, procedure and function found in all major versions of Pascal.

INTRODUCTION TO PASCAL (Including UCSD Pascal)
by Rodnay Zaks   422 pp., 130 illustr. Ref. P310
A step-by-step introduction for anyone wanting to learn the Pascal language. Describes UCSD and Standard Pascals. No technical background is assumed.

APPLE PASCAL GAMES
by Douglas Hergert and Joseph T. Kalash   376 pp., 40 illustr., Ref. P360
A collection of the most popular computer games in Pascal, challenging the reader not only to play but to investigate how games are implemented on the computer.

CELESTIAL BASIC: Astronomy on Your Computer
by Eric Burgess   228 pp., 65 illustr., Ref. B330
A collection of BASIC programs that rapidly complete the chores of typical astronomical computations. It's like having a planetarium in your own home! Displays apparent movement of stars, planets and meteor showers.

PASCAL PROGRAMS FOR SCIENTISTS AND ENGINEERS
by Alan R. Miller   378 pp., 120 illustr., Ref. P340
A comprehensive collection of frequently used algorithms for scientific and technical applications, programmed in Pascal. Includes such programs as curve-fitting, integrals and statistical techniques.

BASIC PROGRAMS FOR SCIENTISTS AND ENGINEERS
by Alan R. Miller   326 pp., 120 illustr., Ref. B240
This second book in the "Programs for Scientists and Engineers" series provides a library of problem-solving programs while developing proficiency in BASIC.

FORTRAN PROGRAMS FOR SCIENTISTS AND ENGINEERS
by Alan R. Miller   320 pp., 120 illustr., Ref. F440
Third in the "Programs for Scientists and Engineers" series. Specific scientific and engineering application programs written in FORTRAN.

PROGRAMMING THE 6809
by Rodnay Zaks and William Labiak   520 pp., 150 illustr., Ref. C209
This book explains how to program the 6809 in assembly language. No prior programming knowledge required.

PROGRAMMING THE 6502
by Rodnay Zaks   388 pp., 160 illustr., Ref. C202
Assembly language programming for the 6502, from basic concepts to advanced data structures.

6502 APPLICATIONS BOOK
by Rodnay Zaks   286 pp., 200 illustr., Ref. D302
Real-life application techniques: the input/output book for the 6502.

ADVANCED 6502 PROGRAMMING
by Rodnay Zaks   292 pp., 140 illustr., Ref. G402A
Third in the 6502 series. Teaches more advanced programming techniques, using games as a framework for learning.

PROGRAMMING THE Z80
by Rodnay Zaks   626 pp., 200 illustr., Ref. C280
A complete course in programming the Z80 microprocessor and a thorough introduction to assembly language.

PROGRAMMING THE Z8000
by Richard Mateosian   300 pp., 124 illustr., Ref. C281
How to program the Z8000 16-bit microprocessor. Includes a description of the architecture and function of the Z8000 and its family of support chips.

THE CP/M® HANDBOOK (with MP/M™)
by Rodnay Zaks   324 pp., 100 illustr., Ref. C300
An indispensable reference and guide to CP/M—the most widely-used operating system for small computers.

MASTERING CP/M®
by Alan R. Miller   320 pp., Ref. C302
For advanced CP/M users or systems programmers who want maximum use of the CP/M operating system . . . takes up where our *CP/M Handbook* leaves off.

INTRODUCTION TO THE UCSD p-SYSTEM™
by Charles W. Grant and Jon Butah   250 pp., 10 illustr., Ref. P370
A simple, clear introduction to the UCSD Pascal Operating System; for beginners through experienced programmers.

A MICROPROGRAMMED APL IMPLEMENTATION
by Rodnay Zaks   350 pp., Ref. Z10
An expert-level text presenting the complete conceptual analysis and design of an APL interpreter, and actual listing of the microcode.

THE APPLE CONNECTION
by James W. Coffron   228 pp., 120 illustr., Ref. C405
Teaches elementary interfacing and BASIC programming of the Apple for connection to external devices and household appliances.

MICROPROCESSOR INTERFACING TECHNIQUES
by Rodnay Zaks and Austin Lesea   458 pp., 400 illust., Ref. C207
Complete hardware and software interconnect techniques, including D to A conversion, peripherals, standard buses and troubleshooting.

SELF STUDY COURSES

Recorded live at seminars given by recognized professionals in the microprocessor field.

INTRODUCTORY SHORT COURSES:
Each includes two cassettes plus special coordinated workbook (2½ hours).

S10—INTRODUCTION TO PERSONAL AND BUSINESS COMPUTING
A comprehensive introduction to small computer systems for those planning to use or buy one, including peripherals and pitfalls.

S1—INTRODUCTION TO MICROPROCESSORS
How microprocessors work, including basic concepts, applications, advantages and disadvantages.

S2—PROGRAMMING MICROPROCESSORS
The companion to S1. How to program any standard microprocessor, and how it operates internally. Requires a basic understanding of microprocessors.

S3—DESIGNING A MICROPROCESSOR SYSTEM
Learn how to interconnect a complete system, wire by wire. Techniques discussed are applicable to all standard microprocessors.

INTRODUCTORY COMPREHENSIVE COURSES:
Each includes a 300-500 page seminar book and seven or eight C90 cassettes.

SB3—MICROPROCESSORS
This seminar teaches all aspects of microprocessors: from the operation of an MPU to the complete interconnect of a system. The basic hardware course (12 hours).

SB2—MICROPROCESSOR PROGRAMMING
The basic software course: step by step through all the important aspects of micro-computer programming (10 hours).

ADVANCED COURSES:
Each includes a 300-500 page workbook and three or four C90 cassettes.

SB3—SEVERE ENVIRONMENT/MILITARY MICROPROCESSOR SYSTEMS
Complete discussion of constraints, techniques and systems for severe environmental applications, including Hughes, Raytheon, Actron and other militarized systems (6 hours).

SB5—BIT-SLICE
Learn how to build a complete system with bit slices. Also examines innovative applications of bit slice techniques (6 hours).

SB6—INDUSTRIAL MICROPROCESSOR SYSTEMS
Seminar examines actual industrial hardware and software techniques, components, programs and cost (4½ hours).

SB7—MICROPROCESSOR INTERFACING
Explains how to assemble, interface and interconnect a system (6 hours).

SOFTWARE

BAS 65™ CROSS-ASSEMBLER IN BASIC
8″ diskette, Ref. BAS 65
A complete assembler for the 6502, written in standard Microsoft BASIC under CP/M®.

8080 SIMULATORS
Turns any 6502 into an 8080. Two versions are available for APPLE II.

APPLE II cassette, Ref. S6580-APL(T)
APPLE II diskette, Ref. S6580-APL(D)

FOR A COMPLETE CATALOG
OF OUR PUBLICATIONS

U.S.A.
2344 Sixth Street
Berkeley,
California 94710
Tel: (415) 848-8233
Telex: 336311

SYBEX-EUROPE
4 Place Félix-Eboué
75583 Paris Cedex 12
France
Tel: 1/347-30-20
Telex: 211801

SYBEX-VERLAG
Heyestr. 22
4000 Düsseldorf 12
West Germany
Tel: (0211) 287066
Telex: 08 588 163